THE COXETER LEGACY
Reflections and Projections

H. S. M. (Donald) Coxeter, 1907–2003

THE COXETER LEGACY
Reflections and Projections

Chandler Davis
Erich W. Ellers
Editors

AMERICAN MATHEMATICAL SOCIETY
Providence, Rhode Island

The Fields Institute
for Research in Mathematical Sciences

2000 *Mathematics Subject Classification.* Primary 01A99, 14M25, 20E42, 20F55, 22E46, 51A20, 51M20, 52B15, 52B70, 52C23.

The photo on the back cover is the "front face" of a 4-dimensional polytope with 120 dodecahedral faces—a mobile constructed by Marc Pelletier and hanging in the atrium of the Fields Institute—as seen from below. The Fields Institute, February 2002. Photo reproduced from http://www.math.toronto.edu/~drorbn/Gallery/Symmetry/index.html and used with permission of Marc Pelletier.

Library of Congress Cataloging-in-Publication Data

The Coxeter legacy : reflections and projections / Chandler Davis, Erich W. Ellers, editors.
    p. cm.
  Includes bibliographical references and index.
  ISBN 0-8218-3722-2 (alk. paper)
  1. Coxeter groups. 2. Group theory. 3. Algebraic geometry. I. Coxeter, H. S. M. (Harold Scott Macdonald), 1907–2003. II. Davis, Chandler. III. Ellers, Erich W., 1928–.

QA177 .C69   2005
510—dc22   2005057055

Contents

Preface

The authors of the contributions to this volume were invited speakers at *The Coxeter Legacy—Reflections and Projections*, a conference held in Toronto in May 2004.

This collection is intended to capture the essence of the *Coxeter Legacy*—it is a mixture of survey, up-to-date information, history, storytelling, and personal memories. Each of the contributions is linked to some of Coxeter's achievements.

Donald Coxeter was able to create enthusiasm, even passion, for mathematics in people of any age, any background, any profession, any walk of life. Like Euclid, Coxeter was enchanted by Euclidean geometry—this means he was interested in the beauty, in the description, and in the exploration of the world around us, the world we live in. Although Coxeter's ideas are rooted in geometry, they often have wider implications.

In 1933, Coxeter proved that

every finite group of the form $R_i^2 = (R_i R_j)^{k_{ij}} = 1$ can be generated by reflections in the bounding primes of a spherical simplex.

The *Coxeter group* was born, its significance established, and a fruitful link between algebra and geometry discovered.

The quoted result, combined with others previously obtained by Coxeter, yields

The complete enumeration of finite groups of the form
$$R_i^2 = (R_i R_j)^{k_{ij}} = 1,$$

which is the title of a paper by Coxeter, published in 1935 in the *Journal of the London Mathematical Society*. In it, Coxeter points out that there are only very few types of irreducible Coxeter groups. These can be described conveniently by *Coxeter diagrams*, a concept he had introduced a few years earlier.

Three decades after their appearance, Coxeter groups acquired their name. The algebraic aspect of a Coxeter group W was developed by *Tits* and *Bourbaki*, always stressing the set of generating involutions, a feature that is clearly visible already in Coxeter's approach. It is advantageous to talk about a *Coxeter system* (W, S), where W is a Coxeter group and S is a set of generating involutions. Interpreting these involutions as reflections introduces the powerful tool of *visualization* that Coxeter handles masterfully.

In general, a Coxeter group may have infinitely many elements. This happens when some of the exponents k_{ij} are infinite.

Mühlherr surveys the known results on the isomorphism problem for Coxeter groups: Let W be a Coxeter group determined by a set of generators $\{R_i\}$ and a set of exponents $\{k_{ij}\}$. Let W' be defined similarly. When are these two Coxeter groups isomorphic? Mühlherr's report exhibits vibrant activity by several researchers working together on this problem.

Borovik surprises with an unusual point of view. He reveals many hidden connections, stresses the visualization, gives guidance to further research, suggests a palindromic approach to Coxeter groups, and brings to life the magical power of Coxeter's charisma. He writes, "One of the attractive features of the Coxeter Theory is that it is saturated with beautiful examples."

Ronan sheds light on the central role that Coxeter groups play in several branches of mathematics. They are connected with the root systems in groups of Lie type. Coxeter groups are used to define apartments which in turn are the essential ingredient in Tits's approach on buildings. Ronan gives a fast-moving account of the history of group theory, touching on many important developments, and brings us up to date on the research in buildings.

Kostant deals with the question of how an arbitrary (unitary) representation of $SU(2)$ decomposes under the action of a finite subgroup Γ of $SU(2)$. Coxeter elements and Coxeter diagrams are essential tools.

Kellerhals brings together a wealth of information on activities, some of which have their roots in Coxeter's work. In 1934, Coxeter classified elliptic and parabolic Coxeter groups; however, hyperbolic Coxeter groups were only partially classified. Hyperbolic Coxeter simplexes exist only in spaces of dimension 9 or less. Some bounds exist for more general situations. Some results on the density of sphere packings that are recorded here rely on a 1954 paper of Coxeter.

Regular polytopes were Coxeter's love. He had a life-time fascination with them. He shared his encyclopaedic knowledge with us in many papers, culminating in two books, *Regular Polytopes*, which went through three editions, and *Regular Complex Polytopes*. Once, after a talk at the Fields Institute in which he discussed many details of a great number of polytopes, when I admired his knowledge of them, he explained, "You see, they are all old friends." Coxeter passed this enthusiasm on to his students and admirers.

Coxeter's devotion to classical polytopes is unsurpassed. His achievements have been praised by Grünbaum, McMullen, and Ivić Weiss in the *Notices of the AMS* of November 2003. These authors also stress the role that Coxeter groups and certain of their quotient groups play as symmetry groups of polytopes.

More recently, the purely combinatorial aspect of regular polytopes has become a centre of attention. A comprehensive record on this subject has been given by *McMullen* and *Schulte* in their book *Abstract Regular Polytopes*, published in 2002.

McMullen and *Schulte* in their contribution here discuss progress on abstract regular polytopes, in particular on their faithful realizations, since the publication of their book.

Monson and *Weiss* offer a concise history of several of Coxeter's interests, objects of study, and achievements. These include regular maps, abstract polytopes, and chiral polytopes. The latter are polytopes whose automorphism groups have precisely two flag orbits with adjacent flags in distinct orbits. In their section on graphs, they tell the charming "Coxeter Graph—My Graph" story.

Wills presents an interesting account on today's knowledge of equivelar polyhedra. He formulates a number of open problems on them. An equivelar polyhedron is a polyhedral 2-manifold with planar faces embedded in Euclidean 3-space such that all faces are p-gons and all vertices are q-valent.

Khovanskii deals with the combinatorial theory of polytopes. His main result generalizes an estimate of Nikulin by giving an upper bound on the average number of l-dimensional faces on k-dimensional faces of an n-dimensional polytope simple at the edges.

Senechal tells the story of crystals that do not satisfy the crystallographic conditions; they have axes with 5-fold symmetry. She starts gently with the 5-fold symmetry of an apple core, one of Coxeter's conversation pieces. She proceeds to tilings with an axis of 5-fold symmetry, which are of necessity nonperiodic, and subsequently moves to Penrose kites and darts. Do the Coxeter groups of the noncrystallographic types H_3 and H_4 hold the key to understanding quasicrystals?

Configurations are the essence of Euclidean geometry, so it is not surprising that Coxeter embraced topics involving configurations, and his work stimulated further research.

Grünbaum gives a comprehensive account of configurations of points and lines. He conveys the excitement of the development of the concepts, of missteps, of great progress, periods of flourishing, lulls, and of the reinvigoration of the topic by a Coxeter paper. Grünbaum is largely concerned with realization and representation, i.e., loosely speaking, with the interpretation of combinatorial configurations in the real Euclidean plane.

Richter-Gebert was inspired by a proof in the book *Geometry Revisited* by Coxeter and Greitzer, where six Menelaus configurations are combined to prove the theorem of Pappus. Richter-Gebert has elevated this method to an art. Multiplying certain expressions that occur in the statements of the theorems of Ceva and Menelaus miraculously leads to a new theorem of Euclidean geometry.

Schattschneider concentrates on the interaction of Coxeter with artists and modelmakers. Preeminent is, of course, the long and fruitful friendship between Coxeter and Escher. She also mentions that Pelletier produced the beautiful 3-D shadow of the 4-D dodecahedron that adorns the ceiling above the staircase of the atrium in the Fields Institute in Toronto.

Emmer enthusiastically describes filming Escher's art with Coxeter—related to geometry, of course! Emmer includes many illuminating and revealing quotes from the correspondence on art and mathematics between Coxeter and him and also between Coxeter and Escher.

Coxeter's involvement in art and with artists earned him admiration and adoring friends in the intellectual community all over the globe. Coxeter's devotion to polytopes lives on in his students and entranced followers. Coxeter groups can arise in various subjects in applied mathematics, and they have a permanent place in some of the most demanding and fascinating branches of abstract mathematics, such as Lie algebras, algebraic groups, Chevalley groups, and Kac-Moody groups.

Coxeter's legacy is much larger than shown here. For instance, we have not mentioned his investigations into optimally dense packings of spheres. A simple search reveals that for the year 2004 there are 110 "Coxeter" entries in *Mathematical Reviews*—giving testimony to Coxeter's everlasting popularity.

Toronto, April 2005   Erich W. Ellers

Acknowledgements

We are delighted that Donald Coxeter's daughter, Mrs. Susan Thomas, kindly provided us with the photo of Donald at the beginning of this book.

We are grateful to Dr. Carl Riehm of the Fields Institute for gently guiding us through the production process and for answering all our questions pleasantly.

We wish to thank Ms Debbie Iscoe, Publication Manager at the Fields Institute, for being cheerful, understanding, and patient during the preparation of this volume.

Our thanks are also due to the Fields Institute and the Mathematics Department of the University of Toronto for their technical support.

                                                          Chandler Davis
                                                          Erich W. Ellers

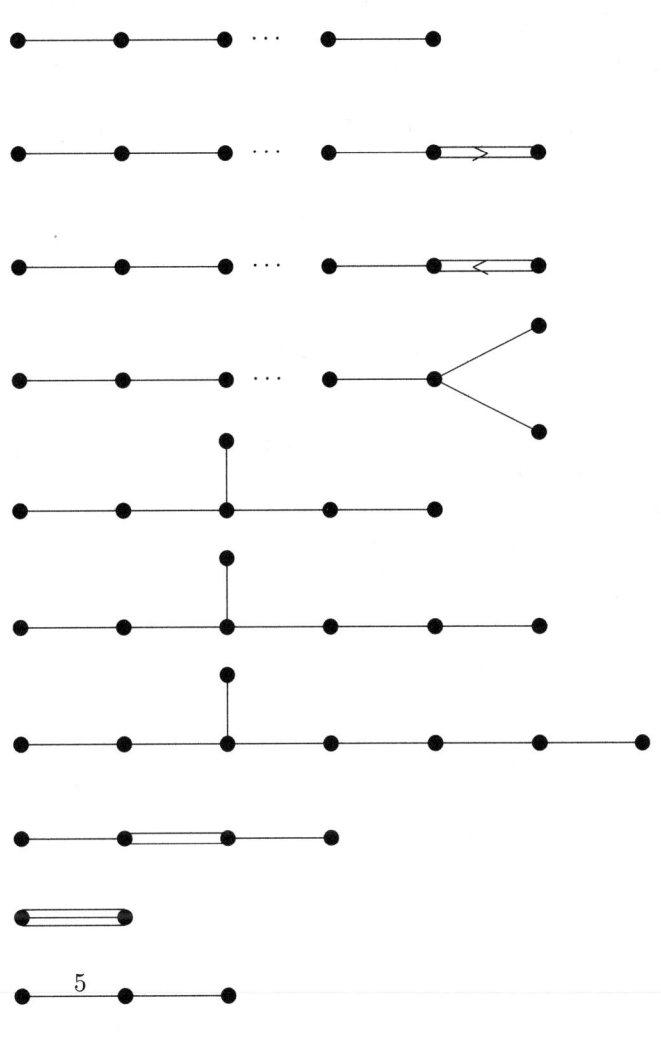

Coxeter Diagrams

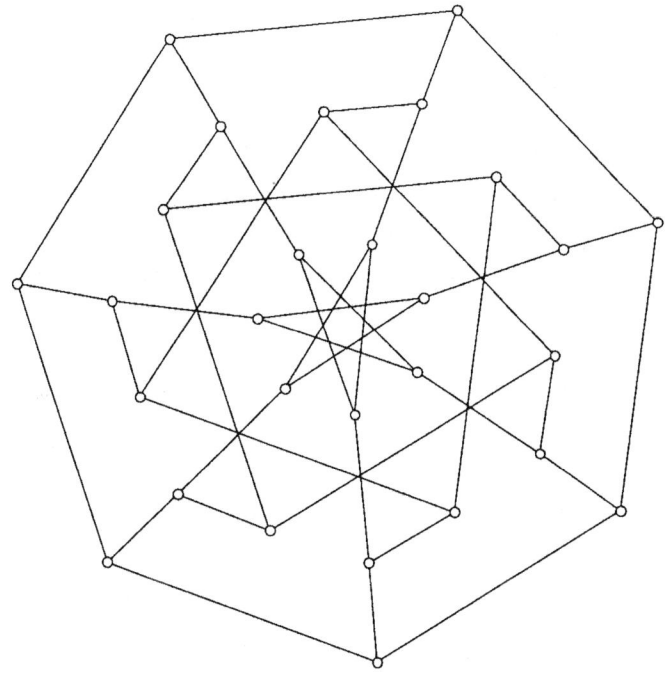

Coxeter Graph

The Isomorphism Problem for Coxeter Groups

Bernhard Mühlherr

ABSTRACT. By a recent result obtained by R. Howlett and the author considerable progress has been made towards a complete solution of the isomorphism problem for Coxeter groups. In this paper we give a survey on the isomorphism problem and explain in particular how the result mentioned above reduces it to its 'reflection preserving' version. Furthermore we desrcibe recent developments concerning the solution of the latter.

1. Introduction

Coxeter groups are important in several areas of mathematics. It is therefore a bit surprising that the isomorphism problem for those groups does not seem to have been considered before the late 1990's. The only earlier reference known to the author where this problem has been indicated is [17]. The first major contributions to it are [16] and [10]. In [16] a rigidity result is proved for a certain class of Coxeter groups. Rigidity means that the Coxeter generating sets are all conjugate. In [10] diagram twists have been introduced. Those provide non-trivial examples of non-rigid Coxeter groups. The question which Coxeter systems are rigid arises naturally. The isomorphism problem for Coxeter groups which deals with a more general question, comes to mind just as easily.

The purpose of the present paper is to give a survey of what is known at present about the isomorphism problem. The main motivation for writing this survey is provided by a recent result obtained by the author in collaboration with Bob Howlett. This result reduces the isomorphism problem to its 'reflection-preserving version'. For the solution of the latter there is a conjecture stated in [10]. Considerable progress towards a proof of this conjecture has been made in [38], and on the basis of recent work of Pierre-Emmanuel Caprace in [13] there is reasonable hope that this conjecture will be proved in the near future. Due to these facts there is now a clear picture of what the solution of the isomorphism problem should look like. In fact, at present there is a solution if one assumes that there are no irreducible spherical residues of rank 3. We state two conjectures in Section 5. The first one is known to be true for all Coxeter systems having no H_3-subsystems; the second is a refinement of Conjecture 8.1 in [10] already mentioned. Under the assumption that

---

2000 *Mathematics Subject Classification.* Primary 20F55; Secondary 51F55.

both conjectures are true, we give an algorithm for the solution of the isomorphism problem.

Two versions of the isomorphism problem. Let W be a group and let $S \subseteq W$ be a set of involutions. Then $M(S)$ denotes the square matrix $(o(ss'))_{s,s'\in S}$ where $o(w)$ denotes the order of an element $w \in W$. The matrix $M(S)$ is called the *type* of S. As the elements of S are involutions we have the following.

1. For all $s, s' \in S$ we have $o(ss') \in \mathbf{N} \cup \{\infty\}$;
2. for all $s \neq s' \in S$ we have $o(ss') = o(s's) \geq 2$;
3. for all $s \in S$ we have $o(ss) = 1$.

Hence, the matrix $M(S)$ is a symmetric square matrix with entries in the set $\mathbf{N} \cup \{\infty\}$ where all entries on the main diagonal are equal to one and all remaining entries are strictly greater than one. Such a matrix is called a *Coxeter matrix over* S.

Let (W, S) be as above. We call (W, S) a *Coxeter system* (of type $M(S)$) if $\langle S \rangle = W$ and if the relations $((ss')^{o(ss')} = 1_W)_{s,s'\in S}$ provide a presentation of W. For a given Coxeter matrix $M = (M_{ij})_{i,j\in I}$ over a set I, we define the *Coxeter group of type M* by setting $W(M) := \langle I \mid ((ij)^{m_{ij}} = 1)_{i,j\in I} \rangle$. It is a basic fact that the pair $(W(M), I)$ is a Coxeter system of type M (i.e. that $o(ij) = m_{ij}$ in $W(M)$ for all $i, j \in I$).

In this paper we will consider the isomorphism problem for finitely generated Coxeter groups. Thus, if we talk about a Coxeter system (W, S) or a Coxeter matrix M over I it is always understood that the sets S and I are finite.

Here are two versions of the isomorphism problem for Coxeter groups.

Problem 1 *Given two Coxeter matrices M and M', decide whether the groups $W(M)$ and $W(M')$ are isomorphic.*

Problem 2 *Given two Coxeter matrices M and M', find all isomorphisms from $W(M)$ onto $W(M')$.*

At first sight, Problem 1 seems to be a more natural question than Problem 2. The latter is just a more general version of the first. Roughly speaking, the solution of Problem 2 is equivalent to the solution of Problem 1 and a description of the automorphism group of $W(M)$ for any Coxeter matrix M. This is in fact the main motivation to consider Problem 2. It turns out that for certain Coxeter matrices M a good understanding of the automorphism group of the group $W(M)$ is only possible if a solution of Problem 2 is available for all Coxeter matrices M'.

Content. In Section 2 we recall some definitions, fix notation and mention some basic facts concerning Coxeter groups. In Section 3 we will consider the rigidity problem for Coxeter groups. This is an interesting special case of the isomorphism problem. In that section we will provide examples of non-rigid Coxeter systems which will play an important role later. Section 4 is devoted to explaining the results obtained in [**23**] and [**27**] and how these results reduce the isomorphism problem to its 'reflection-preserving version' which will then be treated in Section 5. In Section 6 we explain an algorithm to solve the isomorphism problem under the assumption that Conjectures 1 and 2 of Section 5 hold. Finally, in Section 7 we will make some remarks on the automorphism groups of Coxeter groups.

Remark: It was mentioned above that there is no contribution to the isomorphism problem for Coxeter groups before the late 1990's. Since then, however, there are several publications concerning this subject. For instance, Problem 1 has been solved completely in the case where M is assumed to be even (i.e. no odd entries) by P. Bahls and M. Mihalik (see [**34**] and the references given there).

In this survey paper we do not attempt to give a systematic description of all contributions to the isomorphism problem for Coxeter groups. We mention results (or consequences of them) whenever it will be convenient. However, we try to include all references on the subject in the bibliography. Thus, quite a few references will be mentioned only there.

Acknowledgement. The content of this paper is based on my talk at the conference *The Coxeter Legacy—Reflections and Projections* at Toronto in May 2004. I thank the organizers for the invitation to present this survey at this conference.

2. Preliminaries

Coxeter diagrams. With a Coxeter matrix $M = (m_{ij})_{i,j \in I}$ we associate its *diagram* $\Gamma(M)$. It is the edge-labelled graph $(I, E(M))$, where the edge-set is $E(M) := \{\{i,j\} \mid m_{ij} \geq 3\}$ and where an edge $\{i,j\} \in E(M)$ has the label m_{ij}. We do not distinguish between a Coxeter matrix and its diagram since they carry the same information. We call a Coxeter matrix *irreducible* if its associated Coxeter diagram is connected. An *irreducible component* of M is a subset J of I, which is a connected component of the diagram. A Coxeter matrix M is called *spherical* if $W(M)$ is finite. The irreducible spherical Coxeter diagrams have been classified by H.S.M. Coxeter in [**18**]; we will use the Bourbaki notation for denoting them with the exception that we denote rank 2 diagrams for the dihedral groups of order $2n$ by $I_2(n)$. Thus we have the four series $A_n, C_n = B_n, D_n$ and $I_2(n)$ and the 6 exceptional diagrams E_6, E_7, E_8, F_4, H_3, and H_4.

An isomorphism from a Coxeter diagram $M = (m_{ij})_{i,j \in I}$ onto a Coxeter diagram $M' = (m'_{ij})_{i,j \in I'}$ is a graph isomorphism which preserves the edge-labels.

Let $M = (m_{ij})_{i,j \in I}$ be a Coxeter matrix over I and let J be a subset of I. Then we put $M_J := (m_{jk})_{j,k \in J}$ and $J^\perp := \{k \in I \mid m_{kj} = 2 \text{ for all } j \in J\}$.

A Coxeter matrix M is called *right-angled* if all edge-labels of $\Gamma(M)$ are infinite; it is called *2-spherical* if there are no infinities; it is called *even* if there are no odd labels and it is called *of large type* if the diagram is a complete graph (hence if there are no 2's in M).

Coxeter systems. Let (W, S) be a Coxeter system. The set of its *reflections* is defined to be the set $S^W := \{wsw^{-1} \mid s \in S \text{ and } w \in W\}$. The *length* of $w \in W$ is the length of a shortest product of elements in S representing w; it is denoted by $l(w)$. We call (W, S) right-angled, 2-spherical, even, or of large type if this is the case for $M(S)$.

We list some facts about Coxeter systems which are important in the sequel. Facts 1 and 2 are basic and can be found in any standard reference on Coxeter groups (see [**9**] or [**29**]); Fact 3 is a non-trivial exercise in [**9**] but it follows also from the fact that the Davis-complex of a Coxeter system is CAT(0); Fact 4 is contained in [**44**]; Fact 5 can be shown by considering the geometric representation and Fact 6 is just an easy consequence of the definition of a Coxeter system.

1. If $J \subseteq S$, then $(\langle J \rangle, J)$ is a Coxeter system.
2. Let $J \subseteq S$ and $l : W \to \mathbf{N}$ be the length function of (W, S). Then the following are equivalent:

a) $(\langle J \rangle, J)$ is finite;

b) there is an element $\rho_J \in \langle J \rangle$ such that $l(\rho_J) > l(x)$ for all $x \in \langle J \rangle$ with $x \neq \rho_J$.

Moreover, if these two conditions are satisfied, then $\rho_J^2 = 1_W$.

3. If $X \leq W$ is a finite subgroup, then there exist $w \in W$ and $J \subseteq S$ such that $X^w \leq \langle J \rangle$ and such that J is a spherical subset of S (i.e. $\langle J \rangle$ finite).

4. Let $r \in W$ be an involution. Then there exist $w \in W$ and $J \subseteq S$ such that J is spherical, $w\rho_J w^{-1} = r$ and such that ρ_J is central in $\langle J \rangle$.

5. Suppose that J is a spherical subset of S such that ρ_J is central in $\langle J \rangle$. Then the normalizer of $\langle J \rangle$ in W and the centralizer of ρ_J in W coincide.

6. Let (W, S) be a Coxeter system. Then each permutation π of S which is an automorphism of $M(S)$ extends uniquely to an automorphism γ_π of W.

Let (W, S) be a Coxeter system. By Fact 6 we can identify the stabilizer of S in $\mathrm{Aut}(W)$ with the group of automorphisms of $M(S)$; this subgroup will be denoted by $\Gamma_S(W)$ and its elements are called the *graph-automorphisms* of (W, S). The group Γ_S has trivial intersection with the group $\mathrm{Inn}(W)$ of inner automorphisms. An automorphism of W will be called *inner-by-graph* if it can be written as a product of an inner automorphism and a graph-automorphism.

3. Rigidity

Let G be a group and $R \subseteq G$ a set of involutions. Recall that the Coxeter matrix $M(R)$ is called the *type* of R; the set R is called *universal* if $(\langle R \rangle, R)$ is a Coxeter system; it is called a *Coxeter generating set of G* if it is universal and $G = \langle R \rangle$.

A Coxeter matrix M is called *rigid* if for each Coxeter generating set R of $W(M)$ the Coxeter diagrams $M(R)$ and M are isomorphic. It is called *strongly rigid* if any two Coxeter generating sets of $W(M)$ are conjugate in $W(M)$.

Clearly, strong rigidity implies rigidity. If a Coxeter diagram is (strongly) rigid, then we call the corresponding Coxeter group and Coxeter system (strongly) rigid as well.

If one can show that the Coxeter diagram M of Problem 1 is rigid, then this problem is trivially solved. The answer is just that the Coxeter diagram M' has to be isomorphic to M.

Similarly, if one can show that the Coxeter diagram M is strongly rigid, then Problem 2 is solved. An isomorphism onto $W(M')$ exists if and only if M' and M are isomorphic. Moreover, the automorphism group of $W(M)$ is just the semi-direct product of the group of inner automorphisms with the group of graph-automorphisms of $W(M)$; in other words: all automorphisms of W are inner-by-graph.

There are several interesting classes of Coxeter systems which are not rigid. Before describing them we present some positive results. The first is due to D. Radcliffe [**43**].

THEOREM 3.1. *Right-angled Coxeter systems are rigid.*

Although we fixed the convention that all Coxeter systems in this paper are by definition of finite rank it is appropriate to mention that the theorem above has been generalized to right-angled Coxeter systems of arbitrary rank by A. Castella (see [15]). The next result about strong rigidity is the result of R. Charney and M. Davis already mentioned in the introduction (see [16]).

THEOREM 3.2. *Let (W, S) be a Coxeter system. If W is capable of acting effectively, properly and cocompactly on some contractible manifold, then (W, S) is strongly rigid. In particular, Coxeter groups of affine and compact hyperbolic type are strongly rigid.*

The next result is very recent. An important step towards a proof of it was already made in [28]; in the version presented here it is a consequence of the main results in [14] and [23].

THEOREM 3.3. *Suppose that (W, S) is irreducible, non-spherical, and 2-spherical, then (W, S) is strongly rigid.*

In the following we describe two ways to manipulate the generating set of a given Coxeter system in order to produce a new one whose type is possibly non-isomorphic to the type of the original one. It is conjectured (and known to be true in many special cases) that Coxeter systems are rigid up to these manipulations.

Pseudo-Transpositions. Let $k \geq 1$ be a natural number and put $n := 2(2k+1)$. We consider the dihedral group W of order $2n$ as the group of isometries preserving a regular n-gon in the euclidian plane. Let $s, t \in W$ be two reflections whose axes intersect at an angle $\frac{\pi}{n}$, and let ρ be the central symmetry. Then it is easily seen that $\{s, t\}$ and $\{s, tst, \rho\}$ are both Coxeter generating sets for W of type $I_2(n)$ and $I_2(2k + 1) \times A_1$, respectively. Thus, the dihedral group of order $2n$ is a non-rigid Coxeter group because it has two Coxeter generating sets of different types. This example is of course trivial and somewhat deceiving because one of the two Coxeter matrices is not irreducible. However, it can be used to produce more general examples by taking direct products or free products. In [27] pseudo-transpositions have been introduced in order to describe the general feature.

Let (W, S) be a Coxeter system and let $\tau \in S$. We call τ a *pseudo-transposition* if the following holds.

PT1 There is a unique $t \in S$ such that $o(\tau t) = 2(2k + 1)$ for some natural number $k \geq 1$.

PT2 For all $s \in S \setminus \{\tau, t\}$ one has $o(\tau s) \in \{2, \infty\}$ and if $o(s\tau) = 2$, then $o(st) = 2$ as well.

The following is an easy observation about pseudo-transpositions.

LEMMA 3.4. *Let (W, S) be a Coxeter system, let $\tau \in S$ be a pseudo-transposition of (W, S) and let $t \in S$ be as in the definition above. Then $S \setminus \{\tau\} \cup \{\tau t\tau, \rho_{\{\tau, t\}}\}$ is a Coxeter generating set of W.*

There is also another kind of pseudo-transpositions for Coxeter systems based on the fact that the Coxeter groups $W(C_n)$ and $W(D_n \times A_1)$ are isomorphic for odd n. They yield also non-isomorphic Coxeter generating sets in a similar way. We refer to [27] for the details.

Let (W, S) be a Coxeter system, let $\tau \in S$ be a pseudo-transposition, and let R be the 'new' Coxeter generating set as described in the lemma above. Then we call the Coxeter system (W, R) an *elementary reduction of* (W, S). A Coxeter system (W, S') will be called a *reduction of* (W, S) if it can be obtained from (W, S) by a sequence of elementary reductions. Finally, we call (W, S) *reduced*, if there are no pseudo-transpositions in (W, S). It is easy to see that each Coxeter system has a reduced reduction.

Given a Coxeter diagram M over a set I, then a Coxeter diagram M' over I' is called an *elementary reduction of* M if there is an elementary reduction of the Coxeter system $(W(M), I)$ whose type is isomorphic to M'; we call M' a *reduction of* M if M' can be obtained from M by a sequence of elementary reductions and we call M *reduced* if the system $(W(M), I)$ does not contain any pseudo-transpositions.

Clearly, any rigid Coxeter system has to be reduced in view of Lemma 3.4 above. The converse is true for even Coxeter systems; indeed, the following result is due to M. Mihalik [34] and is based on earlier work of P. Bahls [1].

THEOREM 3.5. *An even Coxeter system is rigid if and only if it does not contain any pseudo-transposition.*

Note that this result generalizes Theorem 3.1.

Twistings. In this subsection we describe twistings as they were introduced in [10] and we give some further definitions concerning them.

Let (W, S) be a Coxeter system and let $J, K \subseteq S$. We call the pair (J, K) an *S-admissible* pair if the following holds.

AD1 J is a spherical subset of S and $K \cap (J \cup J^\perp) = \emptyset$.

AD2 For all $k \in K$ and $l \in L := S \setminus (J \cup J^\perp \cup K)$ the order of kl is infinite.

An S-admissible pair (J, K) is called trivial if K or L is empty. For an S-admissible pair (J, K) we put $T_{(J,K)}(S) := J \cup J^\perp \cup K \cup \{\rho_J l \rho_J \mid l \in L\}$.

The following lemma is not too difficult to prove (see [10]).

LEMMA 3.6. *Let (W, S) be a Coxeter system and let (J, K) be an S-admissible pair. Then $T_{(J,K)}(S)$ is a Coxeter generating set of W which is contained in S^W.*

Let $(W, S), (J, K)$, and $S' := T_{(J,K)}(S)$ be as in the previous lemma. If ρ_J is central in $\langle J \rangle$, then it is easily verified that $M(S)$ is isomorphic to $M(S')$. If ρ_J is not central in $\langle J \rangle$, then $M(S)$ is not isomorphic to $M(S')$ in the generic case. The following example of such a situation was given in [37].

Example: Let (W, S) be a Coxeter system such that $S = \{s_1, s_2, s_3, s_4\}$ and such that $o(s_1 s_2) = o(s_2 s_3) = o(s_3 s_4) = 3$ and $o(s_1 s_3) = o(s_1 s_4) = o(s_2 s_4) = \infty$. We put $J := \{s_2, s_3\}$ and $K := \{s_1\}$. It follows that

$$S' := T_{(J,K)}(S) := \{s_1' := s_1, s_2' := s_2, s_3' := s_3, s_4' := s_2 s_3 s_2 s_4 s_2 s_3 s_2\},$$

that $o(s_1' s_2') = o(s_2' s_3') = o(s_2' s_4') = 3$, and that $o(s_1' s_3') = o(s_1' s_4') = o(s_3' s_4') = \infty$. Thus $M(S)$ and $M(S')$ are not isomorphic.

Let S, R be Coxeter generating sets of a group W; we call R a *twist* of S if there is an S-admissible pair (J, K) such that $R = T_{(J,K)}(S)$. It is readily verified that R is a twist of S if and only if S is a twist of R and that $S^W = R^W$ in this case. A Coxeter generating set S is called *twist-rigid* if there are no non-trivial S-admissible pairs; i.e. if there are no twists of S which are not conjugate to S in W.

Let M be a Coxeter matrix over I. A Coxeter matrix M' is called a *twist of* M if there is a twist I' of I in the Coxeter system $(W(M), I)$ such that $M(I')$ is isomorphic with M'. As before one verifies that M' is a twist of M if and only if M is a twist of M'.

We close this section with a result about strong rigidity for Coxeter groups. Obviously, if (W, S) is a strongly rigid Coxeter system, then S has to be twist-rigid.

The following theorem provides the converse under the additional assumption that all Coxeter generating sets R of W are contained in S^W. Corollary 4.2 below indicates a class of groups for which this assumption holds.

THEOREM 3.7. *Let M be a non-spherical, irreducible Coxeter diagram over I such that there is no subdiagram of type H_3. Suppose that I is a twist-rigid subset of $W(M)$ and that all Coxeter generating sets of $W(M)$ are contained in $I^{W(M)}$. Then M is strongly rigid.*

This theorem was first proved in the large-type case ($m_{ij} > 2$ for all i, j) in [**38**]; the result as it is stated above has been obtained recently by P.-E. Caprace [**13**].

4. The reduction to the restricted isomorphism problem

The restricted isomorphism problems for Coxeter groups are the following:

Problem 3: *Given a Coxeter system (W, S) and a Coxeter matrix M, decide whether there is a Coxeter generating set $R \subseteq S^W$ of W such that $M(R) = M$.*

Problem 4: *Given a Coxeter system (W, S) and a Coxeter matrix M, find all Coxeter generating sets $R \subseteq S^W$ of W with $M(R) = M$.*

In [**27**] Problems 1 and 2 of the introduction have been reduced to Problems 3 and 4, respectively. This reduction is based on the results for the *finite continuation of a reflection in a Coxeter group*, which have been obtained in [**23**]. The purpose of this section is to describe the results obtained in both references. The original motivation for the investigations in [**23**] was to find a tool to characterize reflections in abstract Coxeter groups. We first provide some examples, where an abstract Coxeter group does not determine 'its set of reflections'.

We have already seen examples, where an abstract Coxeter group has different Coxeter generating sets yielding different sets of reflections. If (W, S) is not reduced and if R is an elementary reduction of S, then $S^W \nsubseteq R^W$ and $R^W \nsubseteq S^W$. We will now obtain further examples by producing automorphisms of Coxeter groups which do not preserve reflections. There are two kinds of such automorphisms, namely s-transvections and J-local automorphisms.

s-**Transvections.** Let (W, S) be a Coxeter system and let $s \in S$. We define the odd connected component of s in the diagram $\Gamma(S)$ to be the set of all elements $t \in S$ for which there is a path from s to t such that all its edge-labels are odd. We denote the odd component of s by $\text{odd}(s)$ and we put

$$\text{eodd}(s) := \text{odd}(s) \cup \{t \in S \mid o(tt') \neq \infty \text{ for some } t' \in \text{odd}(s)\}.$$

Let J_s denote the irreducible component of $\text{eodd}(s)$ which contains s, and let K_s denote the union of all spherical irreducible components of $\text{eodd}(s)$ that do not contain s.

Let z be an element in the center of $\langle K_s \rangle$. We define the mapping $\theta_{s,z} : S \to W$ by setting $\theta_{s,z}(t) = tz$ if $t \in \text{odd}(s)$ and by setting $\theta_{s,z}(t) = t$ for the remaining $t \in S$. One readily verifies that this mapping extends to an involutory automorphism of W and that sz is not contained in S^W. Hence $\theta_{s,z}(S)$ is a Coxeter generating set of W providing a set of reflections different from S.

The involutory automorphism described above is called an *s-transvection* of the Coxeter system (W, S). In fact, the definition of an s-transvection given in [**27**] is slightly more general. This is due to particular instances which might arise when there are subsystems of type C_3. Due to these instances the formal definition of an s-transvection is somewhat involved and will be omitted here. Nevertheless, we give an example of such a C_3-transvection because—unlike for the other kinds of automorphisms—it is not an 'obvious automorphism easily seen from the diagram'.

Example Let (W, S) be a Coxeter system where $S = \{s, t, t', c\}$ such that $o(st) = o(st') = 3$, $o(ct) = o(ct') = 4$, $o(sc) = 2$ and $o(tt') = \infty$. Define $\theta : S \to W$ by setting $\theta(c) := c$, $\theta(s) := sc$, $\theta(t) := stcsts$ and $\theta(t') := st'cst's$. One verifies that θ extends uniquely to an involutory automorphism of W.

J-local automorphisms. Let (W, S) be a Coxeter system. A subset J of S is called a *graph factor* of (W, S) if J is spherical and if for all $t \in S \setminus J$ either $tj = jt$ for all $j \in J$ or $o(tj) = \infty$ for all $j \in J$.

Let J be a graph factor of (W, S) and let α be an automorphism of $\langle J \rangle$. Then it is readily verified that there is a unique automorphism of W stabilizing the subgroup $\langle J \rangle$, inducing α on it and inducing the identity on $S \setminus J$. We call such an automorphism a *J-local automorphism*.

This observation can be used to produce non-reflection preserving automorphisms. There are many examples of finite Coxeter groups, having automorphisms which are not reflection preserving. Obvious examples are the elementary abelian 2-groups. A particularly interesting example is of course the exceptional automorphism of $\mathrm{Sym}(6)$ which is the Coxeter group of type A_5.

The finite continuation of a reflection. Let (W, S) be a Coxeter system. As S is supposed to be finite and as each finite subgroup of W is conjugate to a subgroup of some spherical standard parabolic subgroup it follows that there is an upper bound for the order of any finite subgroup of W. This implies that there is for any subgroup X of W a unique maximal normal finite subgroup of X which we denote by $O_{\mathrm{fin}}(X)$.

Let $r \in W$ be an involution of W; by the result of Richardson mentioned in Section 2 (Fact 4) we know that r is conjugate to some ρ_J for some spherical subset J of S and such that ρ_J is central in $\langle J \rangle$. Now one knows that $N_W(\langle J \rangle) = C_W(\rho_J)$ (Fact 5) and hence $\langle J \rangle$ is contained in $O_{\mathrm{fin}}(C_W(\rho_J))$. These considerations show that r must be a reflection if $O_{\mathrm{fin}}(C_W(r)) = \langle r \rangle$. Hence we have found a handy criterion which ensures that a given involution of an abstract Coxeter group is a reflection for any Coxeter generating set of that group.

This idea was the starting point for the results obtained in [**23**]. It soon turned out that it is more convenient to work with the *finite continuation* $\mathrm{FC}(r)$ rather than with the group $O_{\mathrm{fin}}(C_W(r))$. This is defined to be the intersection of all maximal finite subgroups of W containing r. The main result of [**23**] is the following theorem. Its proof is based on a careful analysis of the centralizer of a reflection which had been desribed in detail in [**11**].

THEOREM 4.1. *Let (W, S) be a Coxeter system and let $s \in S$. Then $\mathrm{FC}(s)$ is known. Moreover, if $\mathrm{FC}(s) = \langle s \rangle$, then s is a reflection for each Coxeter generating set of W.*

The description of $\mathrm{FC}(s)$ may become complicated if there are subsystems of type C_3 or D_4. If this is not the case, one can describe $\mathrm{FC}(s)$ by means of the subsets J_s and K_s defined in the paragraph on s-transvections as follows.

COROLLARY 4.2. *Let (W, S) be a Coxeter system and suppose that (W, S) does not contain any subsystem of type C_3 or D_4. Let $s \in S$. If J_s is spherical, then $\mathrm{FC}(s) = \langle J_s \cup K_s \rangle$; in the remaining cases one has $\mathrm{FC}(s) = \langle \{s\} \cup K_s \rangle$. In particular, if $K_s = \emptyset$ and J_s is non-spherical, then s is a reflection for each Coxeter generating set of W.*

The reduction theorem. Let (W, S) be a Coxeter system. We call $s \in S$ FC-*centered* if $\mathrm{FC}(s) = \langle J \rangle$ for some $J \subseteq S$. A fundamental reflection might not be FC-centered if there are subsystems of type C_3 or D_4. Moreover, the group of automorphisms of W which stabilize the subset S^W is denoted by $\mathrm{Ref}_S(W)$. We are now able to state the main result of [**27**].

THEOREM 4.3. *Let (W, S) be a reduced Coxeter system. For each FC-centered $s \in S$, let T_s denote the group of all s-transvections of (W, S). For each graph factor $J \subseteq S$ let L_J denote the group of all J-local automorphisms of (W, S). Let Σ be the subgroup of $\mathrm{Aut}(W)$ which is generated by all T_s and all L_J, where s runs through the FC-centered elements of S and J runs through the set of graph factors of (W, S). Let $\tilde{\Sigma}$ be the subgroup of $\mathrm{Aut}(W)$ which stabilizes $\mathrm{FC}(s)$ for all $s \in S$. Then we have the following:*

  a) *The group $\tilde{\Sigma}$ is finite and $\Sigma \leq \tilde{\Sigma}$. In particular, Σ is a finite subgroup of $\mathrm{Aut}(W)$.*
  b) *Given a reduced Coxeter system (W', S') and an isomorphism $\alpha : W \to W'$, then there exists $\sigma \in \Sigma$ such that $\alpha(\sigma(S)) \subseteq S'^{W'}$.*
  c) *The group Σ (and hence also the group $\tilde{\Sigma}$) is a finite supplement of $\mathrm{Ref}_S(W)$ in $\mathrm{Aut}(W)$.*

Part b) of the theorem above says in particular, that if (W, S) and (W', S') are Coxeter systems that are both reduced, and if there is an isomorphism from W onto W', then there is also an isomorphism between them which maps S^W onto $S'^{W'}$. This yields the reduction of Problem 1 to Problem 3 for reduced Coxeter systems. Moreover, given any reduced Coxeter system (W, S), then its group of automorphisms can be written as $\Sigma \mathrm{Ref}_S(W)$, hence Problem 2 is reduced to Problem 4 for reduced Coxeter systems.

5. The restricted isomorphism problem

In view of the reduction result described in the previous section it suffices to solve Problems 3 and 4 in order to solve Problems 1 and 2, respectively. Thus we are led to the following question.

Question: *Let (W, S) be a Coxeter system and let $R \subseteq S^W$ be a Coxeter generating set of W. What can be said about R?*

We have to consider Coxeter generating sets whose elements are reflections in a given Coxeter system. The following is a first observation which can be shown by using the geometric representation of a Coxeter group.

LEMMA 5.1. *Let (W, S) be a Coxeter system, let $R \subseteq S^W$ be a Coxeter generating set of W and let $X \subseteq R$ be such that $\langle X \rangle$ is finite. Then there exists a subset J of S and an element $w \in W$ such that $\langle X \rangle^w = \langle J \rangle$. In particular, if $r, r' \in R$ are such that $o(rr') = n \neq \infty$, then there exist $s, s' \in S$ such that $o(ss') = n$ and such that the subgroups $\langle r, r' \rangle$ and $\langle s, s' \rangle$ are conjugate.*

Let (W, S) be a Coxeter system and let $R \subseteq S^W$ be a Coxeter generating set. We call R *sharp-angled* with respect to S if for any two reflections $r, r' \in R$ there exists $w \in W$ such that $\{r, r'\}^w \subseteq S$.

Let W be the dihedral group of order $2n$ for some natural number $n \geq 2$. We consider W as the group of automorphisms of the regular n-gon in the euclidean plane. Let $S = \{s, t\}$, where s and t are reflections whose axes intersect at an angle $\frac{\pi}{n}$. Given $r \neq r' \in S^W$, then $\{r, r'\}$ is sharp-angled with respect to S if the reflection axes of r and r' intersect at an angle $\frac{\pi}{n}$.

Angle-deformations. Let (W, S) be a Coxeter system, let $s \neq t \in S$ be such that st has finite order, let $x \in \langle s, t \rangle$ be such that $\langle s, xtx^{-1} \rangle = \langle s, t \rangle$ and put $Y := S \setminus (\{s, t\} \cup \{s, t\}^\perp)$. Let Y_s be the set of all $y \in Y$ for which there exists a sequence $y_1, \ldots, y_k = y$ in Y such that $o(sy_1), o(y_1 y_2), \ldots, o(y_{k-1} y_k)$ are finite and define Y_t similarly. We define the mapping $\delta_x : S \to W$ by setting $\delta_x(r) := r$ if $r \in S \setminus (Y_t \cup \{t\})$ and $\delta_x(r) = xrx^{-1}$ in the remaining cases. The following is easy to verify.

LEMMA 5.2. *If $Y_s \cap Y_t = \emptyset$ then δ_x extends uniquely to an automorphism of W which stabilizes the set S^W.*

If $\{s, xtx^{-1}\}$ is not sharp-angled with respect to $\{s, t\}$ and if δ_x is as above, then $\delta_x(S)$ is not sharp-angled with respect to S. We therefore call the automorphisms of the lemma above *angle-deformations*.

The following result can be obtained by using rigidity of Fuchsian Coxeter groups in a similar way as it was done in [**38**].

PROPOSITION 5.3. *Let (W, S) be a Coxeter system and suppose that there is no 3-subset J of S such that $M(J) = H_3$. Let Δ be the group generated by all angle deformations of (W, S). Given a Coxeter generating set $R \subseteq S^W$, then there exists $\delta \in \Delta$ such that $\delta(R)$ is sharp-angled with respect to S.*

In view of the previous proposition the following conjecture is known to be true for Coxeter systems having no subsystem of type H_3.

Conjecture 1: Let (W, S) be a Coxeter system and $R \subseteq S^W$ be a Coxeter generating set. Then there exists an automorphism α of W such that $\alpha(S^W) = S^W$ and such that $\alpha(R)$ is sharp-angled with respect to S.

Twist-equivalence. Let (W, S) be a Coxeter system and let $R \subseteq S^W$ be a Coxeter generating set of W. Recall that $R' \subseteq S^W$ is called a twist of R if there is an R-admissible pair (J, K) such that $R' = T_{(J,K)}(R)$. Moreover, R' is a twist of R if and only if R is a twist of R'. By taking the transitive closure we obtain an equivalence relation on the set of the Coxeter generating sets contained in S^W which is called *twist-equivalence*.

If R' is a twist of $R \subseteq S^W$, then $R' \subseteq R^W$ and R' is sharp-angled with respect to R. Hence, if R' is twist-equivalent with $R \subseteq S^W$, then $R' \subseteq R^W$ and R' is sharp-angled with respect to R. There is some evidence that the converse is also true. This is the content of the conjecture below. This conjecture is a refinement of Conjecture 8.1 in [**10**].

Conjecture 2: Let (W, S) be a Coxeter system and $R \subseteq S^W$ a Coxeter generating set of W which is sharp-angled with respect to S. Then R is twist-equivalent to S.

At present, the following two theorems are known by recent work of P.-E. Caprace. The first improves earlier results obtained in [**10**], and [**38**].

THEOREM 5.4. *Conjecture 2 holds for all Coxeter systems which do not contain an irreducible spherical subsystem of rank 3.*

THEOREM 5.5. *If (W, S) is a Coxeter system such that $M(J^\perp)$ is 2-spherical for each spherical subset J of S, Conjecture 2 holds for (W, S).*

The main tool to prove Conjecture 2 in the references above is known to the experts as 'Kac Conjugacy Theorem for root bases'. This theorem is proved in [31] for affine and compact hyperbolic groups. A proof for all Coxeter groups is given in [28].

6. The solution of Problem 1

Let M be a Coxeter diagram over a set I. Recall that M' is called a twist of M if there is a twist I' of $I \subseteq W(M)$ such that $M(I')$ is isomorphic to M'. Again, M' is a twist of M if and only if M is a twist of M' and by taking the transitive closure we obtain an equivalence relation on the set of Coxeter matrices which is called twist-equivalence as well.

The following lemma is easy to prove.

LEMMA 6.1. *Let (W, S) be a Coxeter system and let M be a Coxeter matrix. Then the following are equivalent.*

   a) *There exists a Coxeter generating set $R \subseteq S^W$ such that $M(R)$ is isomorphic to M and such that R is twist-equivalent to S.*
   b) *The matrices $M(S)$ and M are twist-equivalent.*

Using the previous lemma one obtains the following theorem, which yields the solution of Problem 3.

THEOREM 6.2. *Let (W, S) and (W', S') be Coxeter systems and suppose that Conjectures 1 and 2 hold for (W, S). Then the following are equivalent.*

   a) *$M(S)$ and $M(S')$ are twist-equivalent.*
   b) *There exists an isomorphism $\alpha : W' \to W$ such that $\alpha(S') \subseteq S^W$*

We recall that a Coxeter system (W, S) is reduced if the set S contains no pseudo-transposition, that there is a natural notion of a Coxeter system or a Coxeter matrix to be a reduction of another, and that it is always possible to produce a reduced reduction of a Coxeter system or Coxeter matrix by an easy algorithm. Now the previous theorem and Theorem 4.3 yield the following.

THEOREM 6.3. *Let M and M' be irreducible Coxeter matrices of rank at least 3 and let (W, S) be a Coxeter system of type M. If Conjectures 1 and 2 hold for (W, S), then the following are equivalent.*

   a) *The groups $W(M)$ and $W(M')$ are isomorphic.*
   b) *If M_1 is a reduced reduction of M and if M'_1 is a reduced reduction of M', then M_1 and M'_1 are twist equivalent.*

In view of Theorem 5.4 and Proposition 5.3 we have the following corollary.

COROLLARY 6.4. *Let M and M' be Coxeter matrices and suppose that M has no subdiagram of type A_3, C_3, or H_3, then the following are equivalent:*

   a) *The groups $W(M)$ and $W(M')$ are isomorphic.*
   b) *If M_1 is a reduced reduction of M and if M'_1 is a reduced reduction of M', then M_1 and M'_1 are twist equivalent.*

7. On automorphisms of Coxeter groups

The previous section shows that there is—under the hypothesis that Conjectures 1 and 2 are true—a satisfactory solution of Problem 1. Unfortunately, we cannot offer a satisfactory description of the automorphism groups of Coxeter groups under the same assumptions which would yield a solution of Problem 2 as well. In fact, the author has serious doubts whether such a handy description exists in the general case. Nevertheless there are several natural subgroups of the automorphism group of a Coxeter group which are quite well understood. In most of the 'interesting' cases, the understanding of these subgroups suffices to understand the group of automorphisms as a whole. Our discussion will be restricted to those subgroups. Before going more into the details we would like to mention that the automorphism groups of Coxeter groups had been determined in various special cases.

1. A presentation of the automorphism groups of right-angled Coxeter groups was given in [36]. This work is based on the results obtained in [45] and the latter is a far reaching generalization of the result in [30].
2. The automorphism groups of 2-spherical Coxeter groups are 'trivial' (i.e. all automorphisms are inner-by-graph) if there is no direct factor which is spherical. This result was accomplished in [14] and [23]. A 'virtual' result in this direction has been obtained already in [28] and the main tool developed there was used again in [14].
3. The automorphism groups of several classes of Coxeter groups which are 'almost spherical' have been described in [19], [20], [21] and [22]. In [22] a complete description of the automorphism groups of the irreducible spherical Coxeter groups is given.

Given an abstract Coxeter group W, then there is always a Coxeter generating set $S \subseteq W$ such that (W, S) is reduced. Thus, there is no loss of generality if we consider only reduced Coxeter systems in this section. Let (W, S) be a reduced Coxeter system. We define the following subgroups:

1. $\mathrm{Ref}_S(W) := \{\alpha \in \mathrm{Aut}(W) \mid \alpha(S^W) = S^W\}$,
2. $\mathrm{Ang}_S(W) := \{\alpha \in \mathrm{Ref}_S(W) \mid \alpha(S) \text{ sharp-angled with respect to } S\}$,
3. $\tilde{\Sigma}_S(W) := \{\alpha \in \mathrm{Aut}(W) \mid \alpha(\mathrm{FC}(s)) = \mathrm{FC}(s) \text{ for all } s \in S\}$
4. $\Gamma_S(W) := \{\alpha \in \mathrm{Aut}(W) \mid \alpha(S) = S\}$

In view of Theorem 4.3 we have $\mathrm{Aut}(W) = \tilde{\Sigma}_S(W) \mathrm{Ref}_S(W)$ and the group $\tilde{\Sigma}_S(W)$ is a finite group. Thus, there is a finite supplement of $\mathrm{Ref}_S(W)$ in $\mathrm{Aut}(W)$. There is the natural question about minimal supplements (or even complements) of $\mathrm{Ref}_S(W)$ in $\mathrm{Aut}(W)$. The example of the Coxeter group of type A_1^k shows that there are not always complements. However, a careful analysis of several special cases provides some evidence for the following conjecture.

Conjecture 3: *Let (W, S) be a reduced Coxeter system. Then there exists a subgroup $\Omega \leq \tilde{\Sigma}_S(W)$ such that $\Pi := \Omega \cap \mathrm{Ref}_S(W) \leq \Gamma_S(W)$ and such that Ω is a supplement of $\mathrm{Ref}_S(W)$ in $\mathrm{Aut}(W)$. Moreover, there is a normal 2-subgroup U of Ω and a complement L of U in Ω such that $L = L_1 \times L_2 \times \ldots L_k$ where L_i is isomorphic to $\mathrm{GL}(n_i, 2)$ for some natural number n_i for $1 \leq i \leq k$ and $\Pi \cap L_i$ is just the set of permutation matrices.*

There is a canonical candidate for the choice of the group Ω and based on this choice the validity of the conjecture is not difficult to see in several special cases. However, the arguments become somewhat involved in the general case.

Reflection-preserving automorphisms. As $\mathrm{Ref}_S(W)$ has a finite supplement, a big part of $\mathrm{Aut}(W)$ is understood if $\mathrm{Ref}_S(W)$ is understood. A first observation is that $\mathrm{Ang}_S(W)$ is a normal subgroup of finite index in $\mathrm{Ref}_S(W)$ and therefore a similar remark holds for $\mathrm{Ang}_S(W)$. We do not know whether $\mathrm{Ang}_S(W)$ always has a finite supplement in $\mathrm{Ref}_S(W)$ but we believe that there are examples where this is not the case. If there is no H_3-subdiagram, then the group $\mathrm{Ref}_S(W)$ is generated by the angle-deformations of (W, S) and $\mathrm{Ang}_S(W)$. We expect this to be true in general with a suitable definition of angle-deformations in the case where there are H_3-subdiagrams.

In the following we will consider the group $\mathrm{Ang}_S(W)$. Let

$$\mathbf{R} := \{R \subseteq S^W \mid R \text{ sharp-angled Coxeter generating set of } W \text{ with respect to } S\}$$

and call two elements $R \neq R'$ in $\mathbf{R}$ adjacent if one is a twist of the other. This yields a graph which we call $\mathbf{C}$. Conjecture 2 is equivalent to the statement that the graph $\mathbf{C}$ is connected.

We consider first the special case where $M(S)$ is even in which case Conjectures 1 and 2 are known to be true. If $M(S)$ is even, there is for each neighbor R of S in the graph $\mathbf{C}$ a canonical involution θ_R in $\mathrm{Ang}_S(W)$ which switches S and R. Setting $X := \langle \theta_R \mid R \text{ neighbor of } S \rangle$, one verifies that $\mathbf{C}$ is the Cayley graph of X with respect to this generator set and that Γ_S is a complement of X in $\mathrm{Ang}_S(W)$. It is probably possible to generalize the arguments given in [**36**] in order to give a presentation of the group $\mathrm{Ang}_S(W)$. The key ingredient of such a generalization would be the observation that the group $\mathrm{Ang}_S(W)$ is something like a 'generalized Coxeter group' as it is in the right-angled case.

Let us consider the general case under the assumption that Conjecture 2 holds. The situation becomes more complicated. The graph $\mathbf{C}$ is no longer the Cayley graph of a group but of a groupoid. We do not go into the details here. But it is worth mentioning that a similar situation occurs if one is interested in the normalizer of a parabolic subgroup in a Coxeter group. These normalizers have been described in [**8**] and [**12**] in a satisfactory way. The key observation in [**12**] is that they are finite index subgroups of a groupoid which one might call a Coxeter groupoid in view of its properties which are quite similar to those of Coxeter groups. We believe, that a presentation of $\mathrm{Ang}_S(W)$ can be given by using analogous ideas. It would be based on the observation that the graph $\mathbf{C}$ is the Cayley graph of a generalized Coxeter groupoid of which $\mathrm{Ang}_S(W)$ is a subgroup of finite index. However, a concrete description of such a presentation might become rather involved.

References

[1] Bahls, P. *Even Rigidity in Coxeter Groups*, PhD-Thesis, Vanderbilt University, (2002).

[2] Bahls, P. *A new class of rigid Coxeter groups*. Internat. J. Algebra Comput. **13** (2003), 87–94.

[3] Bahls, P. *Strongly rigid even Coxeter groups*. Preprint (2002), 29p. to appear in Topology Proc..

[4] Bahls, P. *Automorphisms of Coxeter groups*. Preprint (2003), 20p. to appear in Trans. Amer. Math. Soc..

[5] Bahls, P. *Rigidity of two-dimensional Coxeter groups*. Preprint (2003).

[6] Bahls, P. The Isomorphism Problem in Coxeter Groups, Lecture Notes in Mathematics series, Word Scientific Publishing Company, Manuscript in preparation (2004).

[7] Bahls, P. and Mihalik, M. *Reflection independence in even Coxeter groups*. Preprint (2002) to appear in Geom. Dedicata.

[8] Borcherds, R. E. *Coxeter groups, Lorentzian lattices, and K3 surfaces.* Internat. Math. Res. Notices **19** (1998), 1011–1031.

[9] Bourbaki, N. *Groupes et algèbres de Lie, Chapitres 4, 5 et 6* Hermann, Paris, 1968.

[10] Brady, N., McCammond, J., Mühlherr, B. and Neumann, W. *Rigidity of Coxeter groups and Artin groups.* Geom. Dedicata **94** (2002), 91–109.

[11] Brink, B. *On centralizers of reflections in Coxeter groups.* Bull. London Math. Soc. **28** (1996), 465–470.

[12] Brink, B. and Howlett, R. B. *Normalizers of parabolic subgroups in Coxeter groups.* Invent. Math. **136** (1999), 323–351.

[13] Caprace, P. E. PhD-thesis in preparation.

[14] Caprace, P. E. and Mühlherr, B. *Reflection rigidity of 2-spherical Coxeter groups.* Preprint (2003), 24p. to appear in *Proc. London Math. Soc.*.

[15] Castella, A. *Sur les automorphismes et la rigidite des groupes de Coxeter a angles droits.* Preprint (2004), 20p, ArXiv, math/GR0411575.

[16] Charney, R. and Davis, M. *When is a Coxeter system determined by its Coxeter group?* J. London Math. Soc. **61** (2000), 441–461.

[17] Cohen, A. M. *Coxeter groups and three related topics.* pp. 235–278 in *Generators and Relations in Groups and Geometries.* by A. Barlotti, E. W. Ellers, P. Plaumann and K. Strambach (eds.), NATO ASI Series C: Math. and Phys. Sciences – Vol. 333, Kluwer Acad. Publ., Dordrecht, 1991.

[18] Coxeter, H. S. M. *The complete enumeration of finite groups of the form $R^2 = (R_i R_j)^{k_{ij}} = 1$.* J. London Math. Soc. **10** (1935), 21–25.

[19] Franzsen, W. N. *Automorphisms of rank three Coxeter groups with infinite bonds.* J. Algebra **248** (2002), 381–396.

[20] Franzsen, W. N. *Automorphisms of Coxeter Groups,* PhD thesis, January 2001 (University of Sydney).

[21] Franzsen, W. N. and Howlett, R. B., *Automorphisms of Coxeter groups of rank three.* Proc. Amer. Math. Soc. **129** (2001), 2607–2616.

[22] Franzsen, W. N. and Howlett, R. B. *Automorphisms of nearly finite Coxeter groups.* Adv. Geom. **3** (2003), 301–338.

[23] Franzsen, W. N., Howlett R. B. and Mühlherr, B. *Reflections in abstract Coxeter groups.* Preprint (2004), 25p.

[24] Hosaka, T. *Determination up to isomorphism of right-angled Coxeter systems.* Proc. Japan Ac. **79** (2003), 33–35.

[25] Hosaka, T. *Coxeter systems with two-dimensional Davis-Vinberg complexes.* AcXiv, math.GR/0405553.

[26] Howlett, R. B. *Normalizers of parabolic subgroups of reflection groups.* J. Lond. Math. Soc.(2) **21** (1980), 62–80.

[27] Howlett, R. B. and Mühlherr, B. *Isomorphisms of Coxeter groups which do not preserve reflections.* Preprint (2004) 18p.

[28] Howlett, R. B., Rowley, P. J. and Taylor, D. E. *On outer automorphism groups of Coxeter groups.* Manuscripta Math. **93** (1997), 499–513.

[29] Humphreys, J. E. *Reflection Groups and Coxeter Groups,* Cambridge Studies in Advanced Mathematics, Vol. 29, Cambridge University Press, Cambridge, 1990.

[30] James, L. D. *Complexes and Coxeter groups—operations and outer automorphisms.* J. Algebra **113** (1988), 339–345.

[31] Kac, V.G. *Infinite-dimensional Lie Algebras,* 3rd edition, Cambridge University Press, Cambridge, 1990.

[32] Kaul, A. *Rigidity for a Class of Coxeter Groups,* PhD-Thesis, Oregon State University (2000).

[33] Kaul, A. *A class of rigid Coxeter groups.* J. London Math. Soc. **66** (2002), 592–604.

[34] Mihalik, M. *The even isomorphism theorem for Coxeter groups.* Preprint (2003), 38p, to appear in Trans. Amer. Math. Soc..

[35] Mihalik, M., Ratcliffe, J. and Tschantz, S. *On the isomorphism problem for finitely generated Coxeter groups. I, Basic matching.* Preprint (2005), 36p AcXiv, math.GR/0501075.

[36] Mühlherr, B. *Automorphisms of graph-universal Coxeter groups.* J. Algebra **200** (1998), 629–649.

[37] Mühlherr, B. *On isomorphisms between Coxeter groups.* Des. Codes Cryptogr. **21** (2000), 189.

[38] Mühlherr, B. and Weidmann, R. *Rigidity of skew-angled Coxeter groups.* Adv. Geom. **2** (2002), 391–415.

[39] Nuida, K. *On the direct indecomposability of infinite irreducible Coxeter groups and the isomorphism problem for Coxeter groups.* Preprint (2005), 30p, ArXiv, math.GR/0501276.

[40] Paris, L. *Irreducible Coxeter groups.* Preprint (2004), 13p, ArXiv, math.GR/0412214

[41] Radcliffe, D.G. *Unique Presentation of Coxeter Groups and Related Groups. PhD-Thesis, University of Wisconsin, Milwaukee,* (2001).

[42] Radcliffe, D.G. *Rigidity of right-angled Coxeter Groups.* Preprint (2002), 4p, ArXiv, math/GR9901049.

[43] Radcliffe, D.G. *Rigidity of graph products of groups.* Algebr. Geom. Topol. **3** (2003), 1079–1088 (electronic).

[44] Richardson, R.W. *Conjugacy classes of involutions in Coxeter groups.* Bull. Austral. Math. Soc. **26** (1982), 1–15.

[45] Tits, J. *Sur le groupe des automorphismes de certains groupes de Coxeter.* J. Algebra **113** (1988), 346–357.

Département de Mathématiques, U. L. B. CP 216, Bd. du Triomphe, 1050-Bruxelles, Belgium

E-mail address: bernhard.muhlherr@ulb.ac.be

Coxeter Theory: The Cognitive Aspects

Alexandre V. Borovik

ABSTRACT. Coxeter Theory is a fascinating example of a mathematical theory where the cognitive roots of mathematical thinking are directly evident. The paper contains an informal discussion, made from a mathematician's point of view, of possible cognitive aspects of the theory. We hope that it will be interesting to mathematicians and cognitive scientists alike.

Introduction

This paper is a mathematician's attempt to reflect on the explosive development of *mathematical cognition*, an emerging branch of neurophysiology which purports to locate structures and processes in the human brain responsible for mathematical thinking [14, 24]. So far the research efforts were concentrated mostly on the brain processes during quantification and counting; important as they are, these activities occupy a very low level in the hierarchy of mathematics. Not surprisingly, the remarkable achievements of cognitive scientists and neurophysiologists are mostly ignored by the mathematical community. The situation might change fairly soon, since conclusions drawn from the neurophysiological research could happen to be very attractive to policymakers in mathematical education, especially since neurophysiologists themselves do not shy away from making direct recommendations [15].

I refer the reader to the fascinating book *The Number Sense: How The Mind Creates Mathematics* by Stanislas Dehaene [24] for the first-hand account of the neurophysiological research of number sense and numerosity, admittedly some of simpler facets of mathematical thinking. In this paper, I am trying to bridge the gap between mathematics and mathematical cognition by pointing out structures and processes of mathematics which, on one hand, are sufficiently non-trivial to be interesting to a professional mathematician, while, on the other hand, are deeply integrated into some basic structures of our mind and, hopefully, lie within the reach of a cognitive scientist. I will talk mostly about visual processing (and, in particular, about our sense of symmetry), and about basic parsing procedures in language processing.

Remarkably, Coxeter Theory, that is, the theory of reflection groups and Coxeter groups as originated in seminal works of H. S. M. Coxeter [19, 20], provides a

2000 *Mathematics Subject Classification.* Primary 00A30; Secondary 00A35, 20F55.

fascinating example of a mathematical theory where we occasionally have a glimpse of the inner working of our mind. As we shall soon see, the underlying cognitive mechanisms which make Coxeter Theory so natural and intuitive are deeply rooted in both visual and verbal processing modules of our mind.

CONTENTS

Part 1. Mirrors and Reflections

1. The starting point: kaleidoscopes

One remarkable feature of Coxeter Theory is that its principal objects can be defined right on the spot in the most intuitive and elementary way. I give first an informal description:

> imagine a few (semi-transparent) mirrors in ordinary three dimensional space. Mirrors (more precisely, their images) multiply by reflecting in each other, like in a kaleidoscope or a gallery of mirrors. A *closed system of mirrors* is what we see when we look into such a kaleidoscope.

As the reader of course knows, of special interest are systems of mirrors which generate only finitely many reflected images. One of the implications of Coxeter Theory is that such finite systems of mirrors are cornerstones of modern mathematics and lie at the heart of many mathematical theories.

Of course, the theory is actually concerned with the more general case of the n-dimensional Euclidean space, with mirrors being $(n-1)$-dimensional hyperplanes rather than two dimensional planes. To that end, we give a formal definition:

a system of hyperplanes (mirrors) $\mathcal{M}$ in the Euclidean space $\mathbb{R}^n$ is called *closed* if, for any two mirrors M_1 and M_2 in $\mathcal{M}$, the mirror image of M_2 in M_1 also belongs to $\mathcal{M}$ (Figure 1).

Thus, the principal objects of Coxeter Theory can be described as *finite closed systems of mirrors*. In more general terms, the theory can be described as the geometry of multiple mirror images. This approach to Coxeter groups is well known and fully exploited, for example, in Chapter 5, §3 of Bourbaki's classical text [**11**][1], or in Vinberg's paper [**57**]. Of course, the definition of a finite closed mirror system is cryptomorphic (that is, equivalent but expressed in different language) to finite reflection groups and root systems. Its important property, however, is that it lends itself to a weakening and can be adapted to become a very efficient axiomatic of matroid theory, an important branch of combinatorics; see Section 13 for more detail.

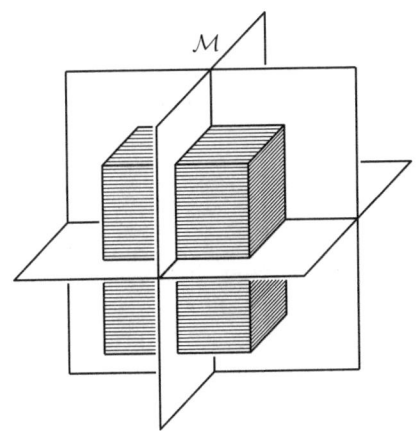

The system $\mathcal{M}$ of mirrors of symmetry of a geometric body Δ is *closed*: the reflection of a mirror in another mirror is a mirror again. Notice that if Δ is compact then all mirrors intersect at a common point.

FIGURE 1. A closed system of mirrors.

So we have two aspects, expressed in two different mathematical languages, of the same mathematical theory (I will call it *Coxeter theory*). This is not an unusual thing in mathematics. What makes the case of mirror systems / Coxeter groups interesting is that a closer look at the corresponding mathematical languages reveals their cognitive (and even neurophysiological!) aspects, much more obviously than in the rest of mathematics. In particular, as we shall soon see, the mirror system / Coxeter group alternative precisely matches the great visual / verbal divide of mathematical cognition.

---

[1]Arguably, one of the better books by Bourbaki; it even contains a drawing, which is an unexpected deviation from his usual style. See a very instructive discussion of the history of this volume by its main contributor, Pierre Cartier [**45**].

I wish to stress that, although the theory of Coxeter groups formally belongs to "higher" mathematics, the issues it raises are relevant to the teaching of mathematics at all levels, from elementary school to graduate studies. Indeed, I will be talking about such stuff as *geometric intuition*. I will also touch on the role of pictorial proofs and self-explanatory diagrams; some of them may look naïve, but, as I try to demonstrate, frequently lead deep into the heart of mathematics, see Section 6 for one of the more striking cases.

2. Image processing in humans

The mirror is, of course, one of the most powerful and evocative symbols of our culture; seeing oneself in a mirror is equated to the very self-awareness of a human being.[2] But the reason why the language of mirrors and reflections happens to be so useful in the exposition of mathematical theories lies not so much at a cultural as at a psychophysiological level.

How do people recognize mirror images? Tarr and Pinker [54] showed that recognition of planar mirror images is done by subconscious mental rotation of 180° about an appropriate axis. Remarkably, the brain computes the position of this axis!

This is how Pinker describes the effect of their simple experiment.

> So we showed ourselves [on a computer screen] the standard upright shape alternating with one of its mirror images, back and forth once a second. The perception of flipping was so obvious that we didn't bother to recruit volunteers to confirm it. When the shape alternated with its upright reflection, it seemed to pivot like a washing machine agitator. When it alternated with its upside-down reflection, it did backflips. When it alternated with its sideways reflection, it swooped back and forth around the diagonal axis, and so on. *The brain finds the axis every time.*
> [**42**, pp. 282–283]

Interestingly, the brain is doing exactly the same with misorientated three-dimensional shapes, *provided they have the same orientation* and can be identified by a rotation [52]. The interested reader may wish to take any computer graphics package which allows animation and see it for himself.[3] It is really difficult to avoid the conclusion that the classical Euler's Theorem:

> If an orientation-preserving isometry of the affine Euclidean space $\mathbb{AR}^3$ has a fixed point then it is a rotation around an axis.

is hardwired into our brains.

The illusion of rotation disappears when the brain faces the problem of identification of three-dimensional mirror images of *opposite orientation*; of course, they can still be identified by an appropriate rotation, but, this time, in four-dimensional space. The environment which directed the evolution of our brain never provided our ancestors with four-dimensional experiences.

---

[2]Gregory [**30**] is a comprehensive survey of the cultural and psychological significance of mirrors.

[3]To reproduce Tarr's experiments, I was using PAINTSHOP PRO, with 3-dimensional images produced by XARA, both software packages picked from the cover disk of a computer magazine.

Human vision is a solution of an ill-posed inverse problem of recovering information about three-dimensional objects from two-dimensional projections on the retinas of the eyes. Pinker stresses that this problem is solvable only because of the many assumptions about the nature of the objects and the world in general built into the human brain or acquired from previous experiences.[4]

The algorithm of identification of three-dimensional shapes is only one of many modules in the immensely complex system of visual processing in humans. It is likely that various modules are implemented as particular patterns of connections between neurons. It is natural to assume that different modules developed at different stages of evolution of humans' ancestors [53]. The older ones were likely to be more primitive and, probably, involved relatively simple wiring diagrams. But since they had adaptive value, they were inherited and acted as constraints in the evolution of later additions to the system, new modules which happened to process the outputs of, and interact with, the pre-existent modules. At every stage, evolution led to the development of an algorithm for solving a very special and narrow problem.

The "flipping" algorithm for the recognition of mirror images of a flat object and the closely related (and possibly identical) "rotation" algorithm for the identification of misoriented three-dimensional objects provide rare cases when we can glimpse the inner workings of our mind. Observe, however, that the algorithms are solutions of relatively simple mathematical problems with a very rigid underlying mathematical structure, namely, the group of rotations of the three-dimensional Euclidean space. There is no analogue of Euler's Theorem for four-dimensional space!

The reader has possibly noticed that I prefer to use the term "algorithm" rather than "circuit", emphasizing the strong possibility that the same algorithms can be implemented by different circuit arrangements if some of the arrangements became impossible as the result, say, of trauma, especially during the early stages of a child's development.[5]

Studies of compensatory developments are abundant in the literature. When I was looking for some recent studies, my colleague David Broomhead directed me to paper [29], a case study of a young woman who cannot make eye movements since birth but had surprisingly normal visual perception. It is surprising because the so-called saccadic movements of eyes are crucial for tracing the contours of objects. The woman compensates for the lack of eye movement by quick movements of her head which follow the usual patterns of saccadic movements. I quote the paper: "Her case suggests that saccadic movements, of the head or the eye, form the *optimal sampling method* for the brain". The italics are mine, since I find the choice of words very attractive: some aspect of the inner working of the brain is described as a mathematical procedure, which raises some really interesting metamathematical

---

[4]Jody Azzouni [4, p. 125] made a subtle comment on pictorial proofs: they work only because we impose many assumptions on diagrams admissible as part of such proofs. As he put it,

> We can conveniently stipulate the properties of *circles* and take them as mechanically recognizable because there are no *ellipses* (for example) in the system. Introduce (arbitrary) *ellipses* and it becomes impossible to tell whether what we have drawn in front of us is a *circle* or an *ellipse*.

It is likely that his remark would not surprise cognitive psychologists; they believe that this is what our brain is doing anyway.

[5]Compare Vandervert [56].

questions. I plan to write more on that elsewhere; meanwhile, I refer the interested reader to papers on mathematical models of eye movement [1, 12].

3. A small triumph of visualisation: Coxeter's proof of Euler's Theorem

If you need convincing that visualization might work in learning, teaching and doing mathematics, there is no better example than the proof of Euler's Theorem as it is given by Coxeter [21, p. 36]; I quote it *verbatim*. Remember that Coxeter's book was first published in 1948, hence was written for readers who were likely to have taken a standard course of Euclidean geometry and therefore had reasonably well-trained geometric imagination.

> In three dimensions, a congruent transformation that leaves a point **O** invariant is the product of at most three reflections: one to bring together the two x-axes, another for the y-axes, and a third (if necessary) for the z-axes.
>
> Since the product of three reflections is opposite, a direct transformation with an invariant point **O** can only be the product of reflections in *two* planes through **O**, i.e., a rotation.

I add just a few comments to facilitate the translation into modern mathematical language: a *congruent transformation* is an isometry; a *direct transformation* preserves the orientation, while an *opposite* one changes it. Coxeter refers to the fact that the product of two mirror reflections is a rotation about the line of intersection of mirrors. It is something that everyone has seen in a tri-fold dressing table mirror; the easiest way to prove the fact is to notice that the product of two reflections leaves invariant every point on the line of intersection of mirrors.[6]

We humans are blessed with a remarkable piece of mathematical software for image processing hardwired into our brains. Coxeter made the full use of it, and expected the reader to use it, in his lightning proof of Euler's Theorem. The perverse state of modern mathematics teaching is that "geometric intuition", the skill of solving geometric problems by looking at (simplified) two- and three-dimensional models is mostly expelled from classroom practice.

4. Mathematics: interiorization and reproduction

David Mumford [39, p. 199] paraphrased Davis and Hersh [22], to say that mathematics is

> *the study of mental objects with reproducible properties.*[7]

---

[6]We accept that the reader has every right to insist that the best way to prove Euler's Theorem is by reduction to algebra and eigenvalues of a three dimensional orthogonal matrix. But is that simpler than Coxeter's proof?

[7]Mumford continues:

I love this definition because it doesn't try to limit mathematics to what has been called mathematics in the past but really attempts to say why certain communications are classified as math, others as science, others as art, others as gossip. Thus reproducible properties of the physical world are science whereas reproducible mental objects are math. Art lives on the mental plane (the real painting is not the set of dry pigments on the canvas nor is a symphony the sequence of sound waves that convey it to our ear) but, as the post-modernists insist, is reinterpreted in new contexts by each appreciator. As for gossip, which includes the vast majority of our thoughts, its essence is its relation to a unique local part of time and space.

Thus learning mathematics has at least two intertwined aspects:

- Interiorization of other people's mental objects.
- The development of reproduction techniques for your own mental objects.

There is a natural hierarchy of reproduction methods. A partial list includes: proof; algorithm; symbolic and graphic expression. I wish to clarify that reproduction is more than communication: you have to be able to reproduce your own mental work for yourself.

Interiorization is less frequently discussed; for our purposes, we mention only that it includes visualization of abstract concepts; transformation of formal conventions into psychologically acceptable "rules of the game"; development of subconscious "parsing rules" for the processing of strings of symbols (most importantly, for reading mathematical formulae). At a more mundane level, you cannot learn an advanced technique of symbolic manipulation without first polishing your skills in more routine computations to the level of almost automatic perfection. Interiorization is more than understanding; to handle mathematical objects, one has to imprint at least some of their functions at the subconscious level of one's mind. My use of the term "interiorization" is slightly different from the understanding of this word, say, by Weller et al. [**58**]. I put emphasis on subconscious, neurophysiological components of the process.

Some mathematical activities are of synthetic nature and can be used as means of both interiorization and reproduction. A really remarkable one is the generation and discussion of examples. Really useful examples can be loosely divided into two groups: "typical", generic examples of the theory, or, on the contrary, "simplest possible", almost degenerate examples, which emphasize the limitations and the logical structure of the theory. Of course, one of the attractive features of the Coxeter Theory is that it is saturated with beautiful examples of both kinds; I discuss some "exceptional" cases in Section 6.

Proof, being the highest level of reproduction activity, has an important interiorization aspect: as Yuri Manin stresses, a proof becomes such only after it is *accepted* [**37**, pp. 53–54]. Manin describes the act of acceptance as a social act; however, the importance of its personal, psychological component could hardly be overestimated.

Visualization is one of the most powerful interiorization techniques. It anchors mathematical concepts and ideas into one of the most powerful parts of our brain, the visual processing module. Returning to Coxeter Theory, I want to point out that finite reflection groups allow an approach to their study based on a systematic reduction of complex geometric configurations to much simpler two- and three-dimensional special cases. Mathematically it is expressed by Coxeter's theorem:

a finite reflection group is a Coxeter group,

that is, all relations between elements are consequences of relations between *pairs* of generating reflections. But a pair of mirrors in the n-dimensional Euclidean space is no more sophisticated a configuration than a pair of lines on the plane, and all the properties of the former can be deduced from that of the latter. *This provides a metamathematical explanation why visualization is so effective in the theory of finite reflection groups.*

5. How to draw an icosahedron on a blackboard

My understanding of visualization as an interiorization technique leads me to believe that drawing pictures is an important way of facilitating mathematical work. This means that pictures have to be treated as mathematical objects, and, consequently, be *reproducible.*

I have to emphasize the difference between *drawings* or *sketches* which are supposed to be reproduced by the reader or student, and more technically sophisticated illustrative material (I will call it *illustrations*), especially computer-generated images designed for visualization of complex mathematical objects. It would be foolish to impose the restrictions on the technical perfection of illustrations. [8] I believe that *drawings* should be intentionally made very simple, almost primitive. Mathematical pictures represent *mental* objects, not the real world! In words of one of the leading geometers of our time, William Thurston, people

> do not have a very good built-in facility for *inverse vision*, that is, turning an internal spatial understanding back into a two-dimensional image. Consequently, mathematicians usually have fewer and poorer figures in their paper and books than in their heads. [**55**, p. 164]

Mathematical pictures should not instil an inferiority complex in the reader who has not attempted to draw anything since his or her halcyon days at the elementary school; they should act as an invitation to the reader to express his or her own mental images.

Figure 2 illustrates possibly the most effective way of drawing an icosahedron, so simple that it is accessible to the student with very modest drawing skills.[9] First we mark symmetrically positioned segments in an alternating fashion on the faces of the cube (left), and then connect the endpoints (right). The drawing actually provides a proof of the existence of the icosahedron: varying the lengths of segments on the left cube, it is easy to see from the continuity principles, that, at certain length of the segments, all edges of the inscribed polytope on the right become equal. Moreover, this construction helps to prove that the group of symmetries of the resulting icosahedron is as big as it should be.

Figure 2 works as a proof because it is produced by "inverse vision". To draw it, you have to run, in your head, the procedure for construction of the icosahedron. And, of course, the continuity principles used are self-evident – they are part of the same mechanisms of perception of motion which glue, in our mind, cinema's 24 frames a second in a continuous motion.

I hope that you agree with me that Figure 2 deserves to be treated as a mathematical statement.

Of course, construction of the icosahedron is the same as construction of the finite reflection group H_3; it can be done by means of linear algebra – which leads to rather nasty calculations, or by means of representation theory – which requires some knowledge of representation theory. It also can be done by quaternions – which is nice and beautiful, but requires the knowledge of quaternions. The graphical

---

[8]However, one should be aware of the danger of excessive details; as William Thurston stresses, "words, logic and detailed pictures rattling around can inhibit intuition and associations." [**55**, p. 165]

[9]This construction of the icosahedron is adapted from the method of H. M. Taylor [**34**, pp. 491–492].

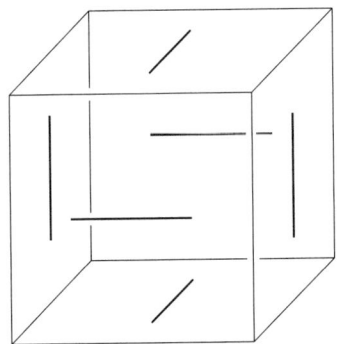 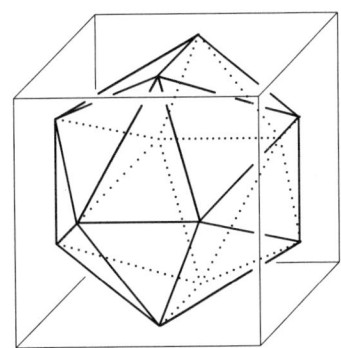

FIGURE 2. A self-evident construction of an icosahedron.

construction is the simplest; using computer jargon, it is WYSIWYG ("What You See Is What You Get") mode of doing mathematics, which deserved to be used at every opportunity.

6. Self-explanatory diagrams

Self-explanatory diagrams are virtually expunged from modern mathematics. I believe they might be useful, maybe not as formal tools for use in proofs, etc., but as means of metamathematical discussion of the structure and interrelations of mathematical theories.

Figure 3 (taken from [9]) is one example: the isomorphism of the root systems D_3 (shown on the left inscribed into the unit cube $[-1,1]^3$) and A_3 is not immediately obvious, but the corresponding mirror systems coincide most obviously. The mirror system D_3 (the system of mirrors of symmetry of the cube) is shown in the middle by tracing the intersections of mirrors with the surface of the cube, and, on the right, by intersections with the surface of the tetrahedron inscribed in the cube. Comparing the last two pictures we see that the mirror system of type D_3 is isomorphic to the mirror system of the regular tetrahedron, that is, to the system of type A_3.

As we shall soon see, this isomorphism has far-reaching implications.

Indeed, at the level of complex Lie groups the isomorphism $D_3 \simeq A_3$ becomes the rather mysterious isomorphism between the 6-dimensional orthogonal group $SO_6(\mathbb{C})$ and $\frac{1}{2}SL_4(\mathbb{C})$, the factor group of the 4-dimensional special linear group $SL_4(\mathbb{C})$ by the group of scalar matrices with diagonal entries ± 1 (or, if you prefer to work with spinor groups, between $D_3(\mathbb{C}) = \mathrm{Spin}_6(\mathbb{C})$ and $A_3(\mathbb{C}) = SL_4(\mathbb{C})$).

This is not yet the end of the story. The compact form of $SL_4(\mathbb{C})$ is SU_4, hence the embedding

$$SU_4 \hookrightarrow \mathrm{Spin}_6(\mathbb{C})$$

features prominently in the representation theory of SU_4, and hence in the SU_4-symmetry formalism of theoretical physics.

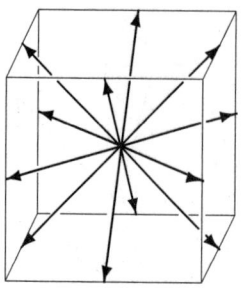 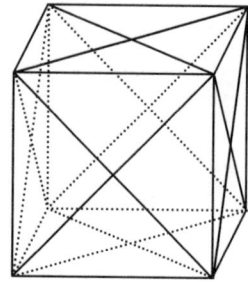 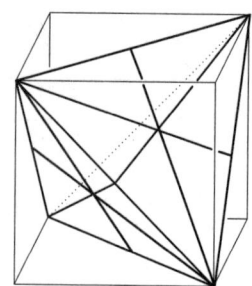

FIGURE 3. The mirror system of type D_3 is the same as the mirror system of type A_3.

But the underlying reason for the isomorphisms is still ridiculously elementary: the tetrahedron can be inscribed into the cube.

Because of their truly fundamental role in mathematics, even simplest diagrams concerning finite reflection groups (or finite mirror systems, or root systems—the languages are equivalent) have interpretations of cosmological proportions. Figure 4 is even more instructive. It is a classical case of the *simplest possible example* as discussed in Section 4. For example, it is the simplest rank 2 root system, or the simplest root system with a non-trivial graph automorphism; the latter, as we shall see in a minute, has really significant implications.

Figure 4 also demonstrates that the root system $D_2 = \{ \pm\epsilon_1 \pm\epsilon_2 \}$ is isomorphic to $A_1 \oplus A_1$. At the level of Lie groups, this isomorphism plays an important role in the description of the structure of 4-dimensional space-time of special relativity, namely, it yields the structure of the Minkowski group, that is, the group of isometries of the 4-dimensional space-time of special relativity theory with the metric given by quadratic form

$$x^2 + y^2 + z^2 - t^2.$$

Indeed, the isomorphism of root systems $D_2 \simeq A_1 \oplus A_1$ leads to the isomorphisms

$$\mathrm{Spin}_4(\mathbb{C}) \simeq \mathrm{SL}_2(\mathbb{C}) \times \mathrm{SL}_2(\mathbb{C})$$

and

$$\mathrm{SO}_4(\mathbb{C}) \simeq \mathrm{SL}_2(\mathbb{C}) \otimes \mathrm{SL}_2(\mathbb{C})$$

(the tensor product of two copies of $\mathrm{SL}_2(\mathbb{C})$, each acting on its canonical 2-dimensional space $\mathbb{C}^2$). The Minkowski group is a real form of $\mathrm{SO}_4(\mathbb{C})$). Hence it is the group of fixed points of some involutory automorphism τ of $\mathrm{SO}_4(\mathbb{C})$). What is this automorphism τ? Let us look again at the quadratic form

$$x^2 + y^2 + z^2 - t^2;$$

it ought to be a real form of the complex quadratic form

$$z_1^2 + z_2^2 + z_3^2 + z_4^2,$$

but lost the symmetric pattern of coefficients. One can see that this means that τ swaps the two copies of $\mathrm{SL}_2(\mathbb{C})$ in $\mathrm{SL}_2(\mathbb{C}) \otimes \mathrm{SL}_2(\mathbb{C})$ and therefore has to be

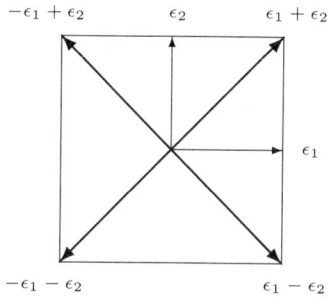

$-\epsilon_1 + \epsilon_2 \qquad \epsilon_2 \qquad \epsilon_1 + \epsilon_2$

ϵ_1

$-\epsilon_1 - \epsilon_2 \qquad\qquad \epsilon_1 - \epsilon_2$

FIGURE 4. The isomorphism of root systems $D_2 = \{\pm\epsilon_1 \pm \epsilon_2\}$ and $A_1 \oplus A_1$.

the symmetry between the two diagonals of the square in Figure 4. Being an involution, τ fixes pointwise the "diagonal" subgroup in $\mathrm{SL}_2(\mathbb{C}) \otimes \mathrm{SL}_2(\mathbb{C})$ isomorphic to $\mathrm{PSL}_2(\mathbb{C})$. (It is $\mathrm{PSL}_2(\mathbb{C})$ rather than $\mathrm{SL}_2(\mathbb{C})$ because its center $\langle -\mathrm{Id} \otimes -\mathrm{Id} \rangle$ is killed in the tensor product.) Hence the Minkowski group is isomorphic to $\mathrm{PSL}_2(\mathbb{C})$.

Part 2. Words and Brackets

7. Parsing

So far I have emphasized the role of visualization in mathematics, and its power of persuasion. Here I will try to relate the visual and symbolic aspects of mathematics, and touch on limitations of visualization.

Indeed, visualization works perfectly well in the naive geometric theory of finite reflection groups, while you do not venture into the more general and stunningly beautiful theory of (infinite) Coxeter groups. Being basic and truly fundamental mathematical objects, Coxeter groups also provide an example of a theory where links of mathematical teaching/learning to cognitive psychology lie exposed. Besides the power of geometric interpretation and visualization, the theory of Coxeter groups very much relies on manipulation of words in canonical generators (chains of consecutive reflections, in the case of reflection groups) and provides one of the best examples of the effectiveness of the language metaphor in mathematics.[10]

Of course, one can be tempted to try to link the psychology of symbolic manipulation in mathematics with the conjecture, vigorously promoted by Steven Pinker [43], that humans have an innate facility for parsing human language. Basically, parsing is the introduction of structure into strings of symbols (phonemes, letters, etc.) We are parsing everything we read or hear; here is an example from Pinker's book [43, pp. 203–205]:

Remarkable is the rapidity of the motion of the wing of the hummingbird.

---

[10]For a further development of the language metaphor, see, for example, the book [26] which discusses applications of the theory of formal languages to group theory.

To make the sense of the phrase, we have to mentally bracket the linked words, making something like

[Remarkable is [the rapidity of [the motion of [the wing of [the hummingbird]]]]].

A sentence might have a different bracket pattern, just compare

[Remarkable is [the rapidity of [the motion]]]

and

[[The rapidity that [the motion] has] is remarkable].

Some patterns are harder to deal with than others:

[[The rapidity that [the motion that [the wing] has] has] is remarkable],

while some come close to incomprehensible, even if the sentence conveys the same message:

[[The rapidity that [the motion that [the wing that [the hummingbird] has] has] has] is remarkable].

Different human languages have different grammars, resulting in different parsing patterns. The grammar is not innate; Pinker emphasizes that innate is the human capacity to generate parsing rules. Generation of parsing patterns is a part of language learning (and infants are extremely efficient in it). Also, it is a part of interiorization of mental objects of mathematics, especially when they are represented by strings of symbols.

Cognitive scientists are very much attracted to case studies of "idiots savants", autistic persons with disproportionate, in comparison with their low general IQ, ability to handle arithmetic or calendrical calculations. As Snyder and Mitchell formulated it [46],

> ... savant skills for integer arithmetic ... arise from an ability to access some mental process which is common to us all, but which is not readily accessible to normal individuals.

What are these "hidden" processes? In one of the extreme cases (mentioned by Butterworth [15]), a severely autistic young man was unable to understand speech but could handle factors and primes in numbers. This suggests that certain mathematical actions are related not so much to language itself, but to the parsing facility, one of the components of the language system; an autistic person might have difficulty in handling language for reasons unrelated to his parsing ability, for example for his incapacity to recognize the source of speech communication as another person.[11] But, in order to achieve such feats as "doubling 8 388 628 up to 24 times to obtain 140 737 488 355 328 in several seconds" [46, p. 589], an autistic person still has to be able to input into his brain the numbers given, inevitably, as strings of phonemes or digits.

I dare to suggest that the parsing mechanisms of human brain is the key to the understanding of low-level arithmetic and formula processing.

Moving several levels up the hierarchy of mathematical processes, we have a fascinating idea in the theory of automatic theorem proving: *rippling*, a formalization of a common way of mathematical reasoning where "formulae are manipulated in a

---

[11]See Baron-Cohen et al. [5] about the failure, in autistic children, of the ability to conceive of mental states of others.

way that increases their similarities by incrementally decreasing their differences"
[**13**, p. 13]. This is facilitated by differentiating the formula in parts which have
to be preserved and parts which have to be changed. Again, we see that in order
to understand how humans use rippling in mathematical thinking (and whether
they actually use it), we have to understand how our brain parses mathematical
formulae.

8. Number sense and grammar

I turn to another remarkable story from cognitive psychology, which links mechanisms of language processing to mastering arithmetic.

When infants learn to speak (in English) and count, there is a distinctive period,
of 5-6 months, in their development, when they know words *one, two, three, four*,
but can correctly apply only the numeral "one", when talking about a single object;
they apply words "two, three, four", apparently at random, to any collection of more
than one object. Susan Carey [**16**] calls the children at this stage *one-knowers*. The
most natural explanation is that they react to the formal grammatical structures
of the adults' speech: *one doll*, but *two dolls, three dolls*. At the next stage of
development, they suddenly start using the numerals *two, three, four, five* correctly.
Chinese and Japanese children become one-knowers a few months later – because
the grammar of their languages has no specific markers for singular or plural in
nouns, verbs, and adjectives.

When the native language is Russian, the "one-knower" stage is replaced by
"two-three-four knower" stage, when children can differentiate between three categories of quantities: single object sets, the sets of two, three or four objects (without
further differentiation between, say, two or three objects), and sets with five or more
objects. This is happening because morphological differentiation of plural forms in
the Russian language goes further than in English.

Well, when I heard about special plural forms of two, three or four nouns
in a lecture by Susan Carey at the *Mathematical Knowledge 2004* conference in
Cambridge, I was mildly amused because it made no sense to me, a native Russian
speaker. Still, I started to write on note paper:

one doll	одна кукл**А**
two doll**S**	две кукл**Ы**
three doll**S**	три кукл**Ы**
four doll**S**	четыре кукл**Ы**
five doll**S**	пять кук**ОЛ**
⋮	⋮
ten doll**S**	десять кук**ОЛ**

I was startled: yes, Susan Carey was right! I was using, all my life, the morphological rules for forming plurals without ever paying any attention to them,
subconsciously. But, apparently, an infant's brain is tuned exactly at picking the
rules: it is easier for the child to associate the number of objects with the morphological marker in the noun signifying the object than with the word *one* or *two*.[12]
In learning numbers, the grammar precedes words!

---

[12]See a detailed discussion of plurality marking in Sarnecka et al. [**44**].

9. Palindromes and mirrors

To demonstrate the role of parsing and other word processing mechanisms in doing mathematics, let us briefly describe Coxeter groups in terms of words, intentionally using as low level "non-mathematical" terminology as possible.

We work with an alphabet $\mathbb{A}$ consisting of finitely many letters, which we denote $a, b, \ldots$, etc. A *word* is any finite sequence of letters, possibly empty (we denote the empty word ϵ). Notice that we have infinitely many words. To impose an algebraic structure onto the amorphous mass of words, we proclaim that some of them are equivalent (or synonymous) to other words; we shall denote the equivalence of words V and W by writing $V \equiv W$. We demand that concatenation of words preserves equivalence: if $U \equiv V$ then $UW \equiv VW$ and $WU \equiv WV$: if *mail* is the same as *post* then *mailroom* is the same as *postroom*. We denote the language defined by the equivalence relation $\equiv$ by $\mathcal{L}_{\equiv}$.

So far all that was just the proverbial "general nonsense". It is remarkable how little we have to add in order to create an extremely rigid, crystal-like structure of a Coxeter group. To that end, we say that a word is *reduced* if it is not equivalent to any shorter word. Now we introduce just two axioms which define *Coxeter languages*:

Deletion Property: If a word $W = a_1 \cdots a_k$ is *not* reduced, then W is equivalent to a word

$$a_1 \cdots a_{i-1} a_{i+1} \cdots a_{j-1} a_{j+1} \cdots a_k$$

obtained from $W = a_1 \cdots a_k$ by deleting some two letters a_i and a_j.

(Of course, it may happen that the new word is still not reduced, in which case the process continues in the same fashion, two letters at a time.)

Reflexivity: The words like aa obtained by doubling a letter are not reduced (hence equivalent to the empty word, by the Deletion Property); *aardvark* is not a reduced word.

Actually, a Coxeter language is exactly a Coxeter group, but I intentionally ignore this crucial fact and formulate everything in terms of words and languages.

I can now give a (rather straightforward and simple) reformulation of a classical theorem of 20th century algebra, due to Coxeter and Tits. My formulation is a bit of a caricature and invented specifically for the purposes of the present paper.

To emphasize the language aspects, I make a *palindrome*, that is, a word that reads the same backwards as forwards, the central object of the theory. When talking about Coxeter languages, I will make an extra technical assumption that *palindromes are reduced and non-empty*. Now the Coxeter-Tits Theorem becomes a theorem about representations of palindromes by mirrors.

The Palindrome Representation Theorem. Assume that a Coxeter language $\mathcal{L}_{\equiv}$ contains, up to equivalence, only finitely many palindromes.[13] Then:

- There exists a finite closed system $\mathcal{M}$ of mirrors in a finite-dimensional Euclidean space $\mathbb{R}^n$ such that the mirrors in $\mathcal{M}$ are in one-to-one correspondence with the classes of equivalence of palindromes.

---

[13]Without this finiteness assumption, the Palindrome Representation Theorem is still true if we accept mirrors in non-Euclidean spaces.

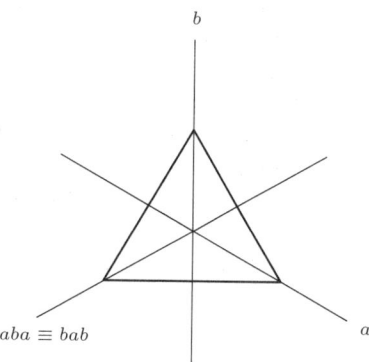

FIGURE 5. The Palindrome Representation Theorem: The three mirrors of symmetry of the equilateral triangle correspond to the palindromes a, b and aba. Together with the equivalences $aa \equiv bb \equiv \epsilon$ (the empty word), the equivalence $aba \equiv bab$ warrants that the corresponding Coxeter language does not contain any other palindromes.

- Moreover, if M_1 and M_2 are mirrors and P_1, P_2 their palindromes, then the palindrome associated with the reflected image of the mirror M_1 in the mirror M_2 is equivalent to $P_2 P_1 P_2$.
- Finally, every closed finite system of mirrors in the Euclidean space $\mathbb{R}^n$ can be obtained in that way from the system of palindromes in an appropriate Coxeter language.

The interested reader may find all the necessary ingredients of a proof of this result in Chapters 5 and 7 of [10]. It involves, at some point, the following equivalence [9, Exercise 11.8]:

$$a_1 \cdots a_l \equiv a_l^{a_{l-1} \cdots a_1} \cdot a_{l-1}^{a_{l-2} \cdots a_1} \cdots a_2^{a_1} \cdot a_1,$$

where the group conjugation $b^{a_k \cdots a_1}$ can be viewed simply as an abbreviation for the symmetric (or palindromical) expression

$$a_1 \cdots a_k \cdot b \cdot a_k \cdots a_1;$$

The identity expresses an arbitrary word as the concatenation of palindromical words; its proof consists of rearrangement of brackets and cancellation of doubled letters $a_i a_i$ whenever they appear. Proofs like that is one of the many reasons why, in order to master the theory of Coxeter groups expressed in a "linguistic" manner, the novice reader has to develop the ability to manipulate the imaginary mental brackets with a rapidity comparable only with the remarkable rapidity of the motion of the wing of the hummingbird.

My "palindrome" formulation of the Coxeter-Tits Theorem is one of many manifestations of *cryptomorphism*, the remarkable capacity of mathematical concepts and facts for translation from one mathematical language to another. I emphasize that, in this paper, I adopted a "local", "microscopic" viewpoint. Although the "palindrome theory" is of little "global" value for mathematics in general, it is sufficiently amusing and demonstrates some interesting "local" features of mathematics.

I stress again that I invented the palindrome formulation of the Representation Theorem specifically for the present paper. When afterwards I made a standard search on Google and MathSciNet, I was pleased to discover that my formulation appeared to be new.

I was pleasantly surprised to find more than a hundred papers on palindromes produced by computer scientists. The set of all palindromic words in a given alphabet is one of the simplest examples of a language which can be generated only by a device with some kind of memory, say, with a stack or push-down storage which works on the principle "last come – first go", like bullets in a handgun clip. It makes palindromes a very attractive test problem in the study of the complexity of word processing, for example, for comparing the two concepts of complexity: space-complexity, measured by the amount of memory required, and time-complexity. The difference between the two complexities is deeply philosophical: we can re-use space, but, unfortunately, cannot re-use time. I was particularly fascinated to learn that palindromes are recognizable by Turing machines working within sublogarithmic space constraints [51]. Hence, in this particular problem it is possible to overwrite and re-use the memory.

Maybe, it is exactly the necessity to engage – and re-use – the memory that turns palindromes into such popular and addictive brainteasers.

10. Parsing, continued: do brackets matter?

The balance of interiorization/reproduction is crucial for any serious discussion of what is actually happening in teaching and learning mathematics, and it is very worrying that this cognitive core is so frequently missing from the professional discourse on mathematical education. This is especially true for the discussion of merits of computer-assisted learning of mathematics, where the use of technology changed the cognitive content of standard elementary routines which for centuries served as building blocks for learning of mathematics.

And here is a small case study, concerned mostly with the basic parenthesizing of the "wing of the hummingbird" sort, like in Section 7. For some years I taught courses in mathematical logic based on two well-known software packages: SYMLOG [41] and TARSKI'S WORLD [6] (reviews: [8, 35]). SYMLOG used a DOS command line interface which was extremely poor even by the standards of its time, while TARSKI'S WORLD very successfully exploited the graphic user interface of Apple and Windows for the visualization of one of the key concepts of logic, a model for a set of formulae, see [7] for the discussion of the underlying philosophy. Also, TARSKI'S WORLD made a very clever use of games for explaining another key concept, the validity of a formula in an interpretation (although the range of interpretations was limited [35]). However, when it came to a written test, students taught with SYMLOG made virtually no errors in composition of logical formulae, while those taught with TARSKI'S WORLD very obviously struggled with this basic task. The reason was easy to find: SYMLOG's very unforgiving interface required retyping the whole formula if its syntax had not been recognized, while TARSKI'S WORLD's

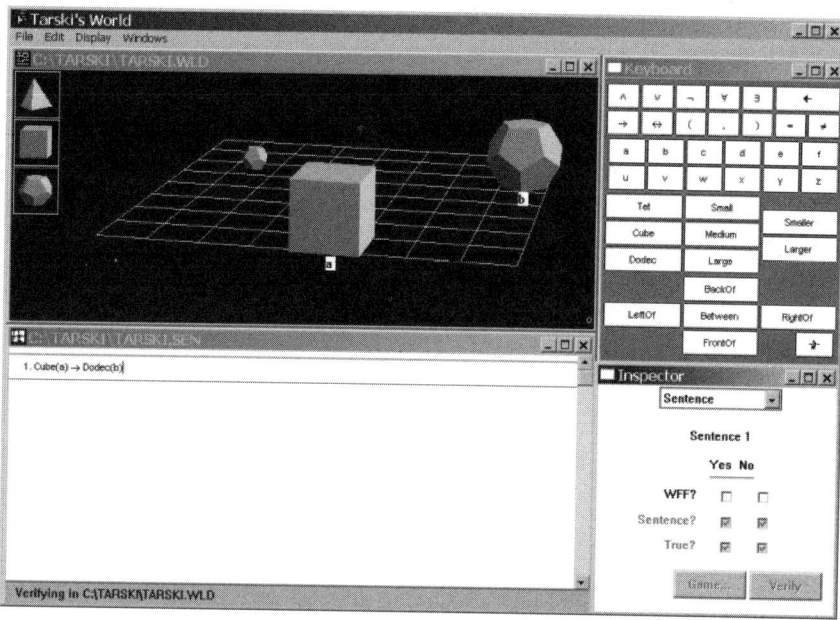

FIGURE 6. A screen shot of TARSKI'S WORLD.

user-friendly formula editor automatically inserted matching brackets. Although TARSKI'S WORLD's students had no difficulty with rather tricky logic problems when they used a computer, their inability to handle formulae without a computer was alarming. Indeed, in mathematics, the ability to reproduce your mental work has to be media-independent. Relieving the students of a repetitive and seemingly mindless task led them to lose a chance to develop an essential skill.

It is appropriate to mention that, besides visualization, there is another mode of interiorization, namely *verbalization*. Indeed, we much better understand those things which we can describe in words. In naive terms, typing a command is like saying a sentence, while clicking a mouse is equivalent to pointing a finger in conversation. The reader would probably agree that, when teaching mathematics, we have to make our students speak. The tasks of opening and closing matching pairs of brackets, however dull and mundane they are, activate the deeply rooted neural mechanisms for generation of parsing rules, and are crucial for the interiorization of symbolic mathematical techniques.

11. The mathematics of bracketing and Catalan numbers

> We have not begun to understand the relationship
> between combinatorics and conceptual mathematics.
>
> Jean Dieudonné [25]

The parsing examples we have considered so far were of a special kind, *binary parenthesizing*; I do not want to venture into anything more sophisticated because

even placing parentheses in an expression made by repeated use of a binary operation, like

$$a + b + c + d$$

is already an immensely rich mathematical procedure. In various disguises, it appears throughout the entire mathematics. There is no better example than Richard
Stanley's famous collection of 66 problems on Catalan numbers [**48**, Exercise 6.19,
pp. 219–229][14]. I quote a couple of examples.

The number of various ways to parenthesize the sum of $n + 1$ numbers,

$$a_1 + a_2 + \cdots + a_n + a_{n+1}$$

is called the *n-th Catalan number* and is denoted C_n; it can be shown that

$$C_n = \frac{1}{n+1} \cdot \binom{2n}{n}.$$

For example, when $n = 3$, we have 5 ways to place the brackets in $a + b + c + d$:

$$a + (b + (c + d)),\ a + ((b + c) + d),\ (a + (b + c)) + d,\ (a + b) + (c + d),\ ((a + b) + c) + d$$

(following the usual convention, I skip the outmost pair of brackets).

Remarkably, when you count ways to triangulate a convex $(n+2)$-gon by $n - 1$
diagonals which touch each other only at their endpoints, you come to the same
result:

The mysterious coincidence is resolved as soon as we treat drawing diagonals
as taking the sums of $n + 1$ vectors

$$\vec{a} + \vec{b} + \vec{c} + \vec{d}$$

going along the $n + 1$ sides of the $(n + 2)$-gon , with the last side (the base of the
polygon) representing the sum:

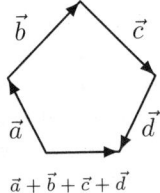

$$\vec{a} + \vec{b} + \vec{c} + \vec{d}$$

Now the one-to-one correspondence between parenthesizing the vector sum and
drawing the diagonals becomes self-evident:

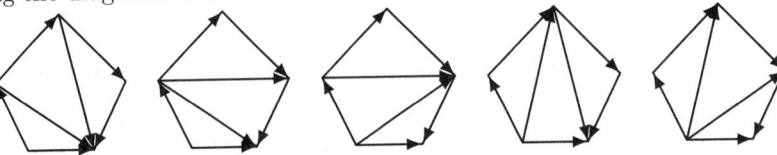

$\vec{a} + (\vec{b} + (\vec{c} + \vec{d}))$ $\vec{a} + ((\vec{b} + \vec{c}) + \vec{d})$ $(\vec{a} + (\vec{b} + \vec{c})) + \vec{d}$ $(\vec{a} + \vec{b}) + (\vec{c} + \vec{d})$ $((\vec{a} + \vec{b}) + \vec{c}) + \vec{d}$

---

[14]Solutions can be found on Internet [**49**].

And here is another class of combinatorial objects which are also counted by Catalan numbers. Take a graph paper with a square grid, and assume that the unit (smallest) squares have length 1. A *Dyck path* is a path in the grid with steps $(1, 1)$ and $(1, -1)$. I claim that the number of Dyck paths from $(0, 0)$ to $(2n, 0)$ which never fall below the coordinate x-axis $y = 0$ is, again, the Catalan number C_n. I give here the list of such paths for $n = 3$, arranged in a natural one-to-one correspondence with the patterns of parentheses in $a + b + c + d$:

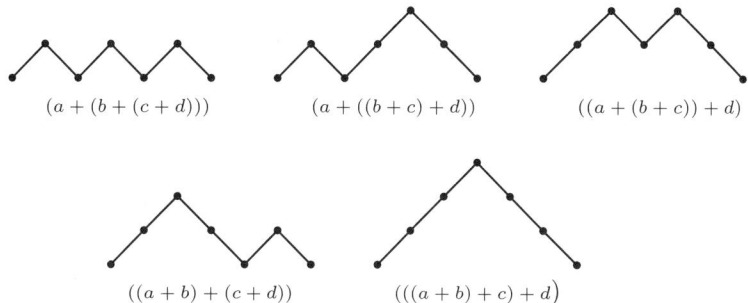

$$(a + (b + (c + d)))$$ $$(a + ((b + c) + d))$$ $$((a + (b + c)) + d)$$

$$((a + b) + (c + d))$$ $$(((a + b) + c) + d)$$

Is the rule self-evident to you? Notice that I added, for your convenience, the exterior all-embracing pairs of parenthesis, they are usually omitted in algebraic expressions.

One more example is concerned with n nonintersecting chords joining $2n$ points on the circle (or on an oval):

Again, there are

$$C_n = \frac{1}{n + 1} \cdot \binom{2n}{n}$$

different ways to draw the chords. Can you find a one-to-one correspondence between the 5 chord diagrams and the 5 ways to parenthesize the sum $a + b + c + d$?[15]

Richard Stanley [48, pp. 219–229] has a list of 66 similar problems, each, of course, having Catalan numbers as the answer! He makes a wry comment that, ideally, the best way to solve the problems is to construct directly the one-to-one correspondences between the 66 sets involved, $66 \cdot 65 = 4290$ bijections in all!

This is still not the end of the story: the striking influence of a seemingly mundane structure, grammatically correct parenthesizing, can be traced all the way to the most sophisticated and advanced areas of modern mathematics research. A brief glance at Stasheff's associahedra (Figure 7) suggests that they live in the immediate vicinity of the Coxeter Theory. Actually, generalized associahedra can be defined for any finite Coxeter group (Stasheff's associahedra [50] being related, of course, to the symmetric group Sym_n viewed as the Coxeter group of type A_{n-1}); for some recent results see, for example, [27].

---

[15]I can give you a hint: the chords link opening parentheses "(" with the plus symbol "+" last applied in the sum enclosed by the matching closing parenthesis.

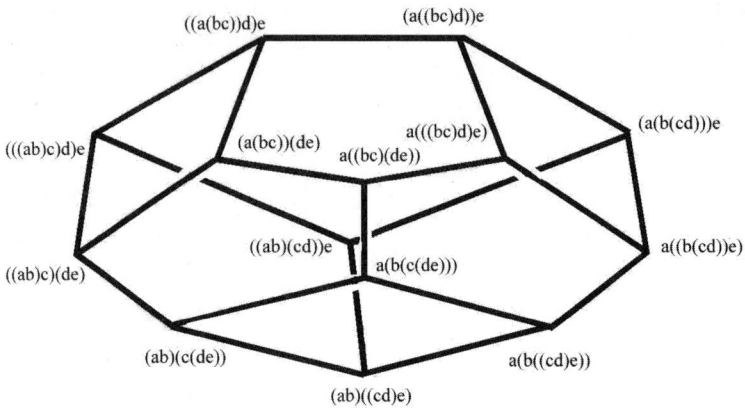

FIGURE 7. Stasheff's associahedron: binary parenthesizings of n symbols can be arranged as vertices of a convex $(n-2)$-gon, with two vertices connected by an edge if the corresponding parenthesizings differ by change in position of exactly one pair of brackets.

12. The mystery of Hipparchus

It appears that the importance of parsing was appreciated by mathematicians and philosophers from truly ancient times. The following fragment from Plutarch, a famous Greek biographer and philosopher of 2nd century A.D., remained a mystery for centuries:

> Chrysippus says that the number of compound propositions that can be made from only ten simple propositions exceeds a million. (Hipparchus, to be sure, refuted this by showing that on the affirmative side there are 103,049 compound statements, and on the negative side 310,952.)

Here Plutarch refers to two prominent thinkers of Classical Greece: philosopher Chrysippus (c. 280 B.C.–207 B.C.) and astronomer Hipparchus (c. 190 B.C.–after 127 B.C.). Only in 1994 did David Hough notice that 103,049 is the number of arbitrary (non-binary) parenthesizings of 10 symbols, that is, the number of all possible expressions like

$$(xxxx)((xx)(xx)xx).$$

This suggests that, for Chrysippus and Plutarch, "compound" propositions were built from "simple" propositions simply by bracketing.

The mathematics and history of Hipparchus' number is discussed in detail in a paper by Richard Stanley [47]. The number of parenthesizings of n symbols is known as the *Schröder number* $s(n)$; the first 11 values of Schröder numbers are

$$1, 1, 3, 11, 45, 197, 903, 4279, 20793, 103049, 518859.$$

In 1998, Laurent Habsieger, Maxim Kazarian and Sergei Lando [**31**] suggested a very plausible explanation of the second Hipparchus number, of compound statements on "negative side". They observe that

$$\frac{s(10) + s(11)}{2} = 310,954$$

and, assuming a slight arithmetic or copying error in Plutarch's text, suggest to interpret the compound statements on the "negative side" as parenthesizings of expressions

$$\text{NOT } x_1 x_2 \cdots x_{10}$$

under the following convention: the negation NOT is applied to all the simple propositions included in the first brackets that include NOT. That means that parenthesizings

$$[\text{NOT } [P_1] \cdots [P_k]]$$

and

$$[\text{NOT } [[P_1] \cdots [P_k]]]$$

give the same result, and most of the negative compound propositions can be obtained in two different ways. The only case which is obtained in an unique way is when one only takes the negation of x_1. Therefore twice the number of negative compound propositions equals the total number of parenthesizings on a string of 11 elements

$$\text{NOT } x_1 x_2 \cdots x_{10}$$

plus the total number of parenthesizings on a string of 10 elements

$$(\text{NOT } x_1) x_2 \cdots x_{10}.$$

This, indeed, provides the value $(s(10) + s(11))/2 = 310,954$.

Nowadays, the thinkers of Classic Antiquity do not enjoy the same authority and revered status as they had up to the 18th century. Armed with the machinery of enumerative combinatorics, we may look condescendingly at the fantastic technical achievement of Hipparchus (he had at his disposal just basic arithmetic and only rudimentary algebraic notation). But I find it highly significant that ancient Greek philosophers, as soon as they started to think about the logical structure of human thought, identified the problem of parsing and attempted to treat it mathematically.

13. Combinatorics as non-parametric mathematics

> The belief that all simple (having no continuous moduli) objects in the nature are controlled by the Coxeter groups is a kind of religion.
>
> V. Arnold [**3**]

I mentioned in Section 1 that the "kaleidoscope" definition of mirror systems can be weakened and made into an axiomatics of matroid theory. For that, we should not demand that our mirror system is closed and use instead an appropriate convexity property, making the whole theory even more geometrical in its spirit. This approach to matroid theory and its immediate and obvious generalization is developed in the book [**10**]. It is truly remarkable how naturally Coxeter Theory arises from the simplest, genuinely basic structures of combinatorics.

Many eloquent speeches were made, and many beautiful books written in explanation and praise of the incomprehensible unity of mathematics. In most cases, the unity was described as a cross-disciplinary interaction, with the same ideas being fruitful in seemingly different mathematical disciplines, and the technique of one discipline being applied to another. The *vertical* unity of mathematics, with many simple ideas and tricks working both at the most elementary and at rather sophisticated levels, is not so frequently discussed – although it appears to be highly relevant to the very essence of mathematical practice. I discuss matroids mostly because I wish to emphasize the "vertical" integrity of mathematics, link the "local" and the "global" viewpoints.

Combinatorics studies structures on a finite set; many of the most interesting of these arise from elimination of continuous parameters in problems from other mathematical disciplines. Joseph Kung [**36**] characterized the corresponding areas of combinatorics as *non-parametric mathematics*.

For example, graphs appear in real life optimization problems as, say, sets of cities (vertices of the graph) connected by roads (edges of the graph) of certain length. Combinatorics looks at the structure left after we ignore the lengths of the roads (which are continuous parameters in the original problem), as well as all topographical considerations, etc.. The combinatorial structure of the graph determines many important features of the original parametric problem. If we work on an optimal delivery problem it does matter, for example, whether our graph is connected or disconnected.

Matroid [**40, 59**] is a combinatorial concept which arises from the elimination of continuous parameters from one of the most fundamental notions of mathematics: that of linear dependence of vectors.

Indeed, let E be a finite set of vectors in a vector space $\mathbb{R}^n$. Vectors $\alpha_1, \ldots, \alpha_k$ are linearly dependent if there exist real numbers $c_1, \ldots, c_k$, not all of them zero, such that $c_1 \alpha_1 + \cdots + c_k \alpha_k = 0$. In this context, the coefficients $c_1, \ldots, c_k$ are continuous parameters; what properties of the set E remain after we decide never to mention them? The solution was suggested by Hassler Whitney [**60**] in 1936. He noticed that the set of linearly independent subsets of E has some very distinctive properties. In particular, if $\mathcal{B}$ is the set of *maximal* linearly independent subsets of E, then, by a well known result from linear algebra, it satisfies the following *Exchange Property*:

> *For all $A, B \in \mathcal{B}$ and $a \in A \smallsetminus B$ there exists $b \in B \smallsetminus A$, such that $A \smallsetminus \{a\} \cup \{b\}$ lies in $\mathcal{B}$.*

Whitney introduced the term *matroid* for a finite structure consisting of a set E with a distinguished collection $\mathcal{B}$ of subsets satisfying the Exchange Property. The origin of the word 'matroid' is in 'matrix': this is what is left of a matrix if we are interested only in the pattern of linear dependences of its column vectors. Since the Gaussian elimination procedure is about linear dependence, matroids naturally describe its combinatorial "skeleton".

Matroids arise in many areas of mathematics, including combinatorics itself. For example, when we take the set E of edges of a connected graph together with the collection $\mathcal{B}$ of its maximal trees, they happen to form a matroid. Moreover, the validity of the Exchange Property is almost self-evident and can be established by a simple combinatorial argument. However, there are deeper reasons why a matroid arises: it can be shown that the edges of a graph can be represented by vectors in such a way that linearly dependent sets of edges are exactly those containing closed cycles. The cohomological nature of the last observation should be apparent to everyone familiar with algebraic topology. This should not be surprising, since cohomology (with coefficients in $\mathbb{Z}$) is itself a classical example of elimination of continuous parameters and therefore belongs to non-parametric mathematics.

The work of three generations of mathematicians confirmed that matroids, indeed, capture the essence of linear dependence. Since linear dependence is a ubiquitous and really basic concept of mathematics, it is not surprising that the concept of matroid has proven to be one of the most pervasive and versatile in modern combinatorics.

It is the idea strongly promoted by Israel Gelfand that even such a simple object as a finite set should be endowed with some extra structure, and that the most fundamental structure on a finite set – even in the absence of any other structures – is provided by its symmetric group acting on it. The symmetric group already lurks between the lines of the Exchange Property in the form of transpositions (a, b) responsible for the exchange of elements.

The symmetric group Sym_n is the simplest example of a finite Coxeter group (or, equivalently, a finite reflection group). It can be interpreted geometrically as the group of symmetries of the regular $(n-1)$-dimensional simplex in $\mathbb{R}^n$ with the vertices

$$(1, 0, \ldots, 0), (0, 1, 0, \ldots, 0), \ldots, (0, \ldots, 0, 1)$$

resulting from permutation of vertices. In terminology of Coxeter groups and root systems, Sym_n has notation A_{n-1}.

We can replace the symmetric group with the reflection group C_n – the group of symmetries of another Platonic solid in $\mathbb{R}^n$, the n-cube $[-1, 1]^n$. Then we get a very natural generalisation of matroids, called *symplectic matroids*. We usually refer to matroids (in Whitney's classical sense) as *ordinary matroids*, to distinguish them from the more general symplectic matroids and from even more general Coxeter matroids.

Symplectic matroids are related to the geometry of vector spaces endowed with bilinear forms, although in a more intricate way than ordinary matroids to ordinary vector spaces: they come from isotropic subspaces of the space. The most interesting class (which we call Lagrangian) comes from Lagrangian subspaces, that is, isotropic subspaces of maximal possible dimension. Furthermore, Sym_n is naturally embedded in the group of symmetries of the n-cube, because we can make Sym_n permute the coordinate axes without changing their orientation; this action obviously preserves the n-cube $[-1, 1]^n$. Thus ordinary matroids can be also understood as symplectic (and, moreover, Lagrangian) matroids, the latter becoming the most natural generalisations of the former.

A general definition of Coxeter matroid can be best done in terms of collections of mirrors; unlike the case of finite reflection groups, we do not require from matroids that their systems of mirrors are *closed*, but still want the system of mirrors to have some nice geometric properties. Here is the definition:

> Let Δ be a convex polytope. For every edge of Δ, take the hyperplane that cuts the edge in its midpoint and is perpendicular to the edge and imagine this hyperplane being a semitransparent mirror (see Figure 8); denote by $\mathcal{M}$ the resulting collection of mirrors. Now mirrors from $\mathcal{M}$ multiply by reflecting in other

mirrors, as in a kaleidoscope. We end up with the *closed* system of mirrors $\mathcal{M}^*$; if it contains only finitely many mirrors, we call Δ a *Coxeter matroid polytope*; this concept is cryptomorhically equivalent to that of a Coxeter matroid.

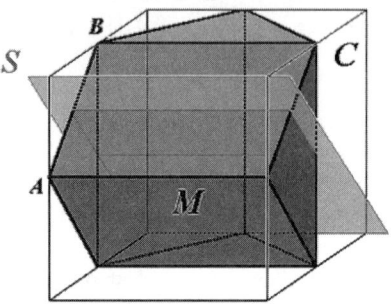

FIGURE 8. A Coxeter matroid polytope. After similarly drawing mirrors of symmetries for every edge, and reflecting mirrors in mirrors, we end up with the full system of mirrors of symmetries of the cube. Hence this matroid polytope corresponds to a symplectic matroid, that is, Coxeter matroid for the group C_3. (Drawing by Maria Borovik.)

Essentially, Coxeter matroids are n-dimensional kaleidoscopes which generate only finitely many mirror images.

However, this is only one of many possible definitions of Coxeter matroids. Being really basic mathematical structures, Coxeter matroids (and particularly the ordinary matroids, related to the symmetric group Sym_n viewed as the reflection group A_{n-1}) are notorious for their ability to cryptomorph and reappear in dozens of disguises which, at the first glance, appear to have nothing in common. This universality and ubiquity comes from the really elementary nature of Coxeter matroids: they are elementary particles, if you wish, of mathematical theories stripped of continuous parameters.

So far the story was strictly "local". It is interesting to follow Arnold [2] and retell it from the global point of view. I quote Arnold:

> Linear algebra is essentially the theory of the special roots system A_n. The basic facts of linear algebra (like the eigenvalues and the Jordan block theory) can be reformulated in terms of the roots, making the statements meaningful for other root systems. These new systems miraculously occur to be correct (while suitably modified). The theories of the root systems B_k, C_k, D_k (corresponding to the Euclidean and symplectic spaces geometry) are from this point of view *rather the sisters than the daughters* of the usual vector space geometry.

Arnold describes several ways of informal generalisation of huge tracts of mathematics, of which *symplectisation*, move from the root systems of type A_n to C_n, is a prominent case. He stresses that symplectisation is acting

> not on such small things, as points, functions, varieties, categories or functors, but on the whole of mathematics.

Also, according to Arnold, symplectisation (as well as complexification, quaternisation, etc.) is not confined to the finite-dimensional domain, and becomes even more interesting when applied to infinite-dimensional objects, see [2] for many colourful details.

However, it is Coxeter matroids for reflection groups other than A_n, especially symplectic matroids (for C_n), which provide the combinatorial underpinning for some of the "sister theories" of classical mathematics.

Acknowledgements

I thank my children Sergey and Maria, who corrected my English (further errors introduced by me are not their responsibility). I am grateful to my wife Anna, co-author and the harshest critic of our textbook on finite reflection groups [9]; neither [9], nor this paper would ever have appeared without her.

Many ideas in the present paper can be traced to the book *Coxeter Matroids* [10], and I am grateful to my co-authors Israel Gelfand and Neil White.

David Corfield, Chandler Davis, Erich Ellers, and Ray Hill carefully read and corrected the whole or parts of the paper.

My thanks are due to David Corfield for lively discussion of the philosophy of mathematics, to Satyan Devadoss for comments on the role of illustrations in mathematical texts, and to Paul Ernest, Reuben Hersh and Ann Kostant for encouraging comments.

An invitation to the conference *The Coxeter Legacy: Reflections and Projections* at the University of Toronto had a considerable influence on my work on this paper, and I am most grateful to its organizers.

While working on the paper, I used, on several occasions, facilities of Mathematisches Forschungsinstitut Oberwolfach and The Fields Institute for Research in Mathematical Sciences.

Parts of the paper were written during my visit to University Paris VI in January 2004 on invitation from Michel LasVergnas, and I use this opportunity to tell Janette and Michel LasVergnas how enchanted I was by their hospitality.

During our conversation in Paris, late Paul Moszkowski put forcefully the case for the development of the theory of Coxeter groups without reference to geometry and pointed me to his remarkable paper [38].

References

[1] O. Akman, D. S. Broomhead and R. A. Clement, Mathematical models of eye movements, Mathematics Today, 39 no. 2 (2003) 54–59.

[2] V. I. Arnold, Polymathematics: is mathematics a single science or a set of arts? In **Mathematics: Frontiers and Perspectives** (V. I. Arnold et al., eds.). Amer. Math. Soc., 2000, pp. 403–416.

[3] V. Arnold, First steps of local contact algebra, Samizdat.

[4] J. Azzouni, Proof and ontology in Euclidean mathematics, in **New Trends in the History and Philosophy of Mathematics** (T. H. Kjeldsen, S. A. Pedersen and L. M. Sonne-Hanse, eds.). University Press of Southern Denmark, 2004, pp. 117–133.

[5] S. Baron-Cohen, A. M. Leslie and U. Frith, Does the autistic child have a "theory of mind"? Cognition, 21 (1985) 37-46.

[6] J. Barwise and J. Etchemendy, **The Language of First-Order Logic: Including the IBM-compatible Windows Version of Tarski's World 4.0**. Stanford, 1992.

[7] J. Barwise and J. Etchemendy, Computers, visualization, and the nature of reasoning, in **The Digital Phoenix: How Computers are Changing Philosophy** (T. W. Bynum and J. H. Moor, eds.). Blackwell, 1998, pp. 93–116.

[8] G. Boolos, Review of Jon Barwise and John Etchemendy, Turing's World and Tarski's World, J. Symbolic Logic 55 (1990) 370–371.

[9] A. V. Borovik and A. Borovik, **Mirrors and Reflections**, Birkhäuser, Boston, 2004.

[10] A. V. Borovik, I. M. Gelfand and N. White, **Coxeter Matroids**, Birkhäuser, Boston, 2003.

[11] N. Bourbaki, **Groupes et Algebras de Lie, Chap. 4, 5, et 6**. Hermann, Paris, 1968.

[12] D. S. Broomhead, R.A. Clement, M.R. Muldoon, J.P. Whittle, C. Scallan and R.V. Abadi, Modelling of congenital nystagmus waveforms produced by saccadic system abnormalities, Biological Cybernetics, 82 no. 5 (May 2000) 391–399.

[13] A. Bundy, D. Basin, D. Hutter, A. Ireland, **Rippling: Meta-Level Guidance for Mathematical Reasoning**, in preparation.

[14] B. Butterworth, **The Mathematical Brain**. Papermac, London, 1999.

[15] B. Butterworth, Mathematics and the Brain: Openning address to the Mathematics Association, Reading: April 3rd 2003, www.mathematicalbrain.com/pdf/.

[16] S. Carey, Bootstrapping and the origin of concepts, Dedalus, Winter 2004 59–68.

[17] D. Corfield, **Towards a Philosophy of Real Mathematics**. Cambridge University Press, 2003 (ISBN 0521817226).

[18] H. S. M. Coxeter, **Introduction to Geometry**, Joh Wiley & Sons, New York, London, 1961 and later eds.

[19] H. S. M. Coxeter, Finite groups generated by reflections and their subgroups generated by reflections, Proc. Camb. Phil. Soc. 30 (1934) 466–482.

[20] H. S. M. Coxeter, The complete enumeration of finite groups of the form $R_i^2 = (R_i R_j)^{k_{ij}} = 1$, J. London Math. Soc. 10 (1935) 21–25.

[21] H. S. M. Coxeter, **Regular Polytopes**. Methuen and Co., London, 1948.

[22] P. Davis and R. Hersh, **The Mathematical Experience**. Birkhäuser, Boston, 1980.

[23] S. Dehaene, E. Spelke, P. Pinet, R. Stanescu and S. Tsivkin, Sources of mathematical thinking: behavioral and brain–imaging evidence, Science 284 (7 May 1999) 970–974.

[24] S. Dehaene, **The Number Sense**. Penguin Books, 2001.

[25] J. Dieudonné, **A Panorama of Pure Mathematics: As seen by N. Bourbaki**. Academic Press, New York, 1982.

[26] D. B. A. Epstein, with J. W. Cannon, D. F. Holt, S. V. F. Levy, M. S. Paterson and W. P. Thurston, **Word Processing in Groups**. Jones and Bartlett, Boston-London, 1992.

[27] S. Fomin and A. Zelevinsky, Y-systems and generalized associahedra, Adv. Math. 158 (2003) 977–1088.

[28] H. Freudenthal, **Mathematics as an Educational Task**. Reidel, Dordrecht, 1973.

[29] I. J. Gilchrist, V. Brown and J. Findlay, Saccades without eye movements, Nature 390 no. 3 (1997) 130–131.

[30] R. Gregory, **Mirrors in Mind**. W. H. Freeman, 1997, 302 pp.

[31] L. Habsieger, M. Kazarian and S. Lando, On the second number of Plutarch, Amer. Math. Monthly 105 (1998), 446.

[32] J. Hadamard, **The Psychology of Invention in the Mathematical Field**. Dover, New York, 1945.

[33] G. H. Hardy, **A Mathematician's Apology** (with a foreword by C. P. Snow). Cambridge University Press, 1967.

[34] T. L. Heath, **The Thirteen Books of Euclid's Elements, with Introduction and Commentary**. Cambridge: at the University Press, 1908.

[35] W. Hodges, Review of J. Barwise and J. Etchemendy, Tarski's World and Turing's World, Computerised Logic Teaching Bulletin 2 (1) (1989) 36–50.

[36] J. P. S. Kung, Combinatorics and nonparametric mathematics, Annals of Combinatorics 1 (1997) 105–106.

[37] Yu. Manin, **Provable and Unprovable**. Sovetskoe Radio, Moscow, 1979 (in Russian).

[38] P. Moszkowski, Longueur des involutions et classification des groupes de Coxeter finis, Séminaire Lotharingien de Combinatoire, B33j (1994).

[39] D. Mumford, The dawning of the age of stochasticity, in **Mathematics: Frontiers and Perspectives** (V. I. Arnold et al., eds.). Amer. Math. Soc., 2000, pp. 197–218.

[40] J. G. Oxley, **Matroid Theory**, Oxford University Press, Oxford, 1992.

[41] F. D. Porturaro and R. E. Tully, **Logic with Symlog: Learning Symbolic Logic by Computer**. Prentice Hall, 1994.

[42] S. Pinker, **The Language Instinct**. Penguin Books, 1995.

[43] S. Pinker, **How the Mind Works**. Penguin Books, 1999.

[44] B. W. Sarnecka, V. G. Kamenskaya, T. Ogura, Y. Yamana and J. B.Yudovina, Language as Lens: Plurality Marking and Numeral Learning in English, Japanese, and Russian, Proceedings of the Boston University Conference on Language Development, pp. 494–505, 2004.

[45] M. Senechal, The continuing silence of Bourbaki—An interview with Pierre Cartier, June 18, 1997. The Mathematical Intelligencer 1 (1998) 22–28.

[46] A. W. Snyder and D. J. Mitchell, Is integer arithmetic fundamental to mental processing?: The mind's secret arithmetic, Proc. R. Soc. Lond. B 266 (1999) 587–592.

[47] R. P. Stanley, Hipparchus, Plutarch, Schröder and Hough, Amer. Math. Monthly 104 (1997), 344–350.

[48] R. P. Stanley, **Enumerative Combinatorics**. Vol. 2, Cambridge University Press, 1999.

[49] R. P. Stanley, Solutions to Exercises on Catalan and Related Numbers, http://www-math.mit.edu/ rstan/ec/catsol.pdf.

[50] J. D. Stasheff, Homotopy associativity of H-spaces, Trans. Amer. Math. Soc. 108 (1963), 275–292.

[51] A. Szepietowski, **Turing Machines with Sublogarithmic Space**. Lecture Notes in Computer Science, 843. Springer-Verlag, Berlin, 1994.

[52] M. J. Tarr, Rotating objects to recognise them: A case study on the role of viewpoint dependency in the recognition of three dimensional shapes, Psychonomic Bulletin and Review 2 (1995) 55–82.

[53] M. J. Tarr and M. J. Black, A computational and evolutionary perspective on the role of representation in vision, Computer Vision, Graphics, and Image Processing: Image Understanding 60 (1994) 65–73.

[54] M. J. Tarr and S. Pinker, Mental rotation and orientation-dependence in shape recognition, Cognitive Psychology 21 (1989) 233–282.

[55] W. P. Thurston, On proof and progress in mathematics, Bulletin AMS 30 no. 2 (1994) 161–177.

[56] L. R. Vandervert, From idiots savants to Albert Einstein: A brain-algorithmic explanation of savant and everyday performance, New Ideas in Psychology 14 no. 1 (March 1996) 81–92.

[57] E. B. Vinberg, Калейдоскопы и группы отражений, Математическое Просвещение, 3 no. 7 (2003) 45–63.

[58] K. Weller, A. Brown, E. Dubinsky, M. McDonald and C. Stenger, Intimations of infinity, Notices AMS 51 no. 7 (2004) 741–750.

[59] D. J. A. Welsh, **Matroid Theory**. Academic Press, London a.o., 1976.

[60] H. Whitney, On the abstract properties of linear dependence, Amer. J. Math. 57 (1935) 509–533.

SCHOOL OF MATHEMATICS, PO BOX 88, THE UNIVERSITY OF MANCHESTER, SACKVILLE STREET, MANCHESTER M60 1QD UK

E-mail address: borovik@manchester.ac.uk

From Galois and Lie to Tits Buildings

Mark Ronan

Introduction

This paper is an essay on how the development of group theory led to the discovery of various families of simple groups, and how these in turn led to the theory of buildings. In outline the story is this. Galois first used the term "group" in the technical sense, and found the first simple groups. Jordan, in his famous Treatise published in 1870, promoted Galois' work and put the theory of groups on a firm foundation. At this time groups were treated as groups of permutations, but other aspects of group theory were soon on the way. Lie visited Paris in 1870 as a graduate student, and went on to create the theory of continuous transformation groups. Killing came to such groups independently, and in 1888 found the classification of the simple Lie groups, using semisimple complex Lie algebras (families A through G). Cartan refined this classification in 1894, correcting some errors in the proofs, and it is now known as the Killing-Cartan classification.

The classical families (A through D) soon led to groups over fields other than the real or complex numbers, and a comprehensive study was published by Dickson in 1901. Later he dealt with E_6 and G_2, but progress on the others did not occur until after the Second World War. Tits was working on the problem, as was Chevalley who was a more established mathematician at that time. Chevalley succeeded in 1955, and his paper was soon followed by variations due to Steinberg, Tits, Suzuki, and Ree (see Section 3). During this time Tits was gradually developing the theory of buildings, and his book [**T8**] in 1974 produced a fully fledged theory that has since found enormous uses. In Section 4 some of Tits' early work on buildings is mentioned, and in Section 5 the contents of his book on buildings of spherical type [**T8**] are discussed. Finally in Section 6 a later approach to buildings, also due to Tits, is mentioned, and we return at the end to the construction of the exceptional groups of Lie type using Building Theory.

1. Group theory in nineteenth century France

The study of groups started with groups of permutations, and the first advances were made by Lagrange (1736–1813). In a paper in 1770 entitled, "Reflections on the algebraic resolution of equations" [**L**], Lagrange treated the roots to an algebraic equation as abstract, formal objects that could be permuted among themselves.

2000 *Mathematics Subject Classification.* Primary 01A60; Secondary 51B25, 51E24.

The coefficients of an equation are symmetric functions of its roots, and Lagrange hoped to use permutations to explain why polynomials of low degree were solvable by radicals, whereas polynomials of higher degree were not. His idea was new, and although he was not able to push it to a conclusion, it inspired others, particularly Ruffini and Abel.

For equations of degree less than 5, formulas had already been found. A formula for solving those of degree 2 was first discovered by the Babylonians in about 1800 BC. It then took more than 3000 years before equations of degrees 3 and 4 were solved (in the early sixteenth century in Italy [**Cd**]), but there the development stopped. No-one could find a formula for solving equations of degree 5, or any higher degree, and in 1799 Gauss wrote:

> Since the works of many geometers left very little hope of ever arriving at the resolution of the general equation algebraically, it appears increasingly likely that this resolution is impossible and contradictory.

That same year, Paolo Ruffini (1765–1822), a professor of clinical medicine and applied mathematics at Modena, published a "proof" of the fact that there could be no formula for solving equations of degree 5, nor any higher degree. This was wonderful work, but his proof was long—two volumes covering 516 pages—and very hard to follow. Some people distrusted his methods and, although no-one was able to refute them, the work was never fully accepted. Ruffini became dismayed at the lack of appreciation, and in 1810 he submitted a new paper on this subject to the French Academy of Sciences. The referees failed to respond in a timely manner, so Ruffini withdrew the paper, and the Secretary responded politely to him:

> Your referees needed very considerable work to give their approval, or to refute your proof. You know how precious time is to realise how reluctant most geometers are to occupy themselves for a long time with the works of each other, and ... they would have to be moved by quite a powerful motive to enter the lists against a geometer so learned and so skillful.

Ruffini was, of course, cracking a really major problem, and was certainly on the right lines, though it now appears that his work contained an important gap. The matter was not settled until a quarter century later in 1824, when Abel (1802–1829) produced a convincing proof, and it was published in Crelle's Journal in 1826 [**A**].

The question of *which* equations were solvable by radicals, and which were not, was still open. Abel was closing in on a method for dealing with this problem when tuberculosis carried him off in 1829 at the age of 26, but the stage was now set for Évariste Galois (1811–1832) who died even younger. He was shot in a duel at the age of 20, and died the next day. The night before the duel he wrote a long letter explaining his ideas. The story is well known and I shall not dwell on it.

For any given polynomial with rational coefficients, Galois considered those permutations of its roots that preserved algebraic relations between them. The group of all such permutations—the *Galois group* of the polynomial—was what he studied, and he was the first to use the term group in its modern technical sense. When the Galois group of a polynomial has a composition series consisting of prime cyclic groups (in other words when it is a *solvable* group) the equation can be solved by radicals, otherwise not.

According to the fundamental theorem of algebra, first proved by Gauss in 1815, every equation has solutions in the complex numbers. According to Ruffini and Abel there are equations of degree 5 whose solutions cannot be expressed in terms of radicals, so not all finite groups are solvable. This implies that the prime cyclic groups were not the only simple groups. Galois knew this very well, and wrote:

> The smallest number of permutations an indecomposable [i.e., simple] group can have, when this number is not prime, is $5 \cdot 4 \cdot 3$.

This simple group of permutations having size $5 \cdot 4 \cdot 3$ (i.e., 60) is the alternating group Alt_5. For many equations of degree 5 or more Alt_5 appears as a subgroup of the Galois group, and hence there is no general formula expressing the solutions in terms of radicals. Galois realised that the alternating groups Alt_n, for $n \geq 5$, formed a family of simple groups, and he also found another family: the special linear groups of degree 2—the groups $\mathrm{PSL}_2(p)$ where p is a prime greater than 3.

Although the problem of finding all finite simple groups would remain unresolved for over a hundred years, the idea of permutation groups had caught on, and a sustained and systematic treatment was needed. The person who rose to the occasion was Camille Jordan (1838–1922). He was an engineer by profession, like his father, though he followed the career of a mathematician, and lectured at the École Polytechnique and the Collège de France. His courses were widely attended, and his mathematical work covered a broad spectrum. He wrote the definitive text on analysis, and in 1870 he published his *Traité des Substitutions* [**J**] (Treatise on Permutations) which became the standard reference for group theory for the next thirty years.

In this Treatise, Jordan gave a definitive account of Galois' work, and showed that every finite group has a composition series. His later theorem with Hölder showed that any two composition series give the same collection of simple groups (with the same multiplicity). Jordan's work finally gave group theory a foundation on which future generations of mathematicians could build. His exposition attracted a wide audience and his fame spread well beyond France. Foreign students attended his lectures and two among them, Felix Klein (1849–1925) from Germany, and Sophus Lie (1842–1899) from Norway later produced ideas leading group theory in new directions.

2. Lie groups

A driving force behind Lie's early work was to do for differential equations what Galois had done for algebraic equations. The situation is of course quite different: an algebraic equation has a finite number of solutions, but a differential equation has a continuum of solutions, each depending on the choice of initial conditions. Like Galois, Lie wanted to consider all the solutions together, and this led him to examine groups of continuous transformations.

Lie was born in 1842 in Norway (thirty years later than Galois) and was the sixth and last child of a Lutheran pastor. After attending university in his home town of Oslo (called Christiania at that time), he took up private teaching. In 1868 when he was 25 years old he got hold of some papers on geometry by mathematicians in France and Germany. This changed his life, and the next year he went to Berlin. There he met another young mathematician, Felix Klein. Lie and Klein formed a very productive relationship because Klein loved hearing of new things to fit into his big picture of mathematics, while Lie loved to pursue his own idiosyncratic ideas. The result was that Lie explained his ideas to Klein; Klein reacted to them; Lie responded to Klein's reactions, and a very useful discussion developed.

In early 1870 the two of them went to Paris, where they met, among others, Jordan whose magnificent Treatise on permutations (mentioned in the previous section) was just rolling off the presses. The visit was very stimulating, but ended suddenly in mid-July when the Franco-Prussian war broke out. Lie and Klein left Paris. Klein returned immediately to Germany, and Lie left for Norway, but decided on a very roundabout route by first hiking to Italy. He loved hiking and had a clever technique for keeping his clothes dry in the rain—he removed them, put them in his backpack and carried on. Unfortunately, Lie was arrested, and then imprisoned on suspicion of being a German spy (the Franco-Prussian war had now started). His mathematical papers were taken to be coded messages, "lines" and "spheres" being interpreted as "infantry" and "artillery", and it took the intervention of a young French mathematician, Gaston Darboux to get him released.

Lie finally got back to Norway in December, completed an excellent doctoral thesis, and applied for a position that had recently opened up in Sweden. This prompted the Norwegian National Assembly to establish a new chair for him, and he accepted. It was a resounding vote of confidence in his abilities, and Lie quickly made a big name for himself. He had a very geometric turn of mind, and those trained in the traditional methods of differential equations found it difficult to follow him. Initial incomprehension like this occurred with Galois' work, but unlike Galois, Lie lived to promote his methods and encourage students. He was a cheerful optimistic man, although he was struck by a sudden depression later in life and entered a mental hospital. We shall return to that later, but let us first see how things developed.

*   *   *

After Lie assumed his new chair at the university, he continued his research programme with vigour. He called his new groups "finite, continuous groups", the word finite referring to finite dimensionality. We now call them Lie groups. One should realize that while Lie and Klein were using geometry, they did not take it as their starting point. As Klein wrote in a letter to Max Noether in 1870, "... we do not think of the geometrical configuration as given, and ask about the transformations; rather we consider the system of transformations as given, and ask about the geometrical configurations ...". This was analogous to Galois' attitude; he used groups of permutations, but that was not his starting point. He started with equations. And so did Lie, differential equations in his case, and this led him to groups of continuous transformations.

In the meantime a German high school teacher, Wilhelm Killing, who was five years younger, was pursuing similar ideas on groups of transformations. He wrote a long essay in 1884, sent it to Felix Klein (who now occupied the chair of geometry in Leipzig), and received an immediate response informing him of Lie's work. Killing then wrote to Lie in Norway, requesting copies of his papers but got no response. He had no access to the papers he needed, which were published in Norwegian journals. Killing wrote again to Klein, who wrote to Lie, and the next year papers were sent, but only on loan, and when Killing asked to keep them longer he got no response. Being a very sincere man, he felt he had to return them before fully digesting the results.

As for Lie, he felt a bit cut off in Norway, and in September 1884, Klein dispatched a young German mathematician from Leipzig to help him out. The young German was named Friedrich Engel, and he returned home next summer with a huge manuscript. The following year, 1886, Klein moved from Leipzig to a chair in Göttingen (where he built up a magnificent school of mathematics), and Lie was persuaded to leave his beloved Norway and take Klein's old chair at Leipzig. It seemed an excellent move for Lie who could now work with Engel on the massive book he was writing. Students could easily come from other parts of Germany to work with Lie, and since he was well known in Paris, and was very much at home with their mathematical attitudes, they sent some of their best students to Leipzig.

Meanwhile, Killing got into communication with Engel who, as soon as he read the long 1884 essay, wrote back to assure Killing that he had indeed independently "discovered transformation groups in Lie's sense". Killing responded two days later to urge Lie and Engel to get on and publish their results, "[This] will, I hope, induce you and Lie to publish more rapidly. Naturally I do not want to enter into competition over this theory, but ... I have been led to results that at least until now have not been published."

Killing soon went to visit Lie in Leipzig, but it seems the two of them did not hit it off. Lie's reaction is recorded in a letter to Klein, "Killing was just here. He has some really nice ideas. In many other respects, however, he does not make a solid impression." Their relationship did not improve, and animosity developed later as Lie's mental health broke down, but let us first see where Killing came from and what he achieved.

He began university studies in 1865 in north-western Germany at a university (in Münster) that had no mathematicians; the subject was taught by observational astronomers. His fellow students were also a disappointment, and Killing recalled that they, "showed almost no interest whatsoever in science itself; they wished (with very few exceptions) to study only what was needed for the examinations." After four semesters he abandoned Münster and went to Berlin which was the centre of mathematics in Germany at the time. Killing appears to have been a good-hearted man because he interrupted his studies one year to teach, sometimes as much as thirty-six hours a week, at a school in a small town where his father was mayor. The school had been threatened with closure and Killing taught all subjects. He then resumed his studies in Berlin, got his doctorate and went back into teaching. In 1880 he became professor at an academic training establishment for Catholic clergymen.

From this unpromising position, Killing wrote his 1884 essay, and then started pursuing the idea of classifying the simple Lie groups over the complex numbers. He was aware of the linear and orthogonal groups (types A, B, and D) and wondered whether there were any more. He was unaware of the symplectic groups (type C) at this stage, and although Engel suggested to him that there might be a third type, he appears to have forgotten Engel's comments and failed to find type C until later. In 1887, Killing got the concept of a semi-simple Lie algebra, and on the 23rd May that year he discovered the G_2 Lie algebra. By the following year he had found essentially all simple Lie algebras, and in rapid succession he wrote a series of three papers on this classification, and sent them to Klein for publication in Mathematische Annalen. The third paper was sent in October 1888 and published the next year.

Killing had discovered the families A through G of simple Lie groups. He had worked very swiftly and knew only too well that some of the theoretical analysis was inadequate. He wrote to Klein that, "If I said I was satisfied with the stated results I would be lying. I have attempted on many occasions to find an error in the proof, but so far without success. ... I thought it best to publish the results with the proof as quickly as possible, since only then can a serious thorough examination take place; and it is of the greatest importance to me that complete certainty about these questions be achieved as quickly as possible."

The importance of Killing's work was immediately clear to Lie, who wrote that it "contains results of the greatest importance, if only everything is correct ...." Later he wrote to Klein that, "Killing has done beautiful research. If, as I believe, the results are correct, he has performed an outstanding service. Generally speaking, now the theory of transformation groups ... will reign over vast areas of mathematics."

It is really unfortunate that Killing, working as he did at a school for the training of future clerics, had no research students. He had no junior mathematicians to clean up after him, and tie up some of the loose ends, and he regretted it, "If only I had students in mathematics, I would allow many excursions into structure. The groups of ranks 4 to 8 provide easy material for seminar works."

Lie on the other hand had plenty of students in his new position at Leipzig, but he found the teaching to be a very different matter from his easy experiences in Norway. As he explained to a Norwegian friend, "While in Norway I hardly spent five minutes a day on preparing the lectures; in Germany I have to spend an average of three hours. The language is always a problem and above all, the competition implies that I have to give eight to ten lectures a week." The pressure took its toll, and in 1889, the year Killing published the third of his classification papers, Lie had a nervous breakdown and was admitted to a mental hospital. He never fully recovered.

*   *   *

In the meantime, Killing's discomfort with his proofs was well founded. The results were correct, but an error in the first paper contaminated the other two papers, and a new approach was needed. Killing left the matter to others and returned to his main love in mathematics, which was the foundations of geometry. That was what had driven him to study groups of continuous transformations in the first place, and he was now offered the chair of mathematics in Münster (the university where he had first been a student).

He was finally free to pursue his original research programme, but it is an extraordinary thing in life that creative people often do their greatest work when circumstances are at their most difficult, and this seems to be the case with Killing. In his new comfortable position at Münster, he published a book on the foundations of geometry, but as Engel wrote, "Killing's latest opus on the foundations of geometry contains real nonsense." Nevertheless, Engel strongly supported Killing for a book prize because he knew how important this work was for its author, and how it had led him to his great work on the classification of Lie groups.

*   *   *

As Killing faded out of the picture, two problems remained. One was to fix up the theoretical analysis that his results were based on, the other was to construct all the groups in the families Killing classified, to verify that they all exist.

It remained to Élie Cartan (1869–1951), a young graduate student in Paris to put these things right. Cartan was the son of a miner, plucked from obscurity by an inspector of schools. This gentleman encouraged one of the teachers to give the boy special coaching and he won a full scholarship to a fine boarding school. From there he went from strength to strength and in 1888 entered the École Normale (the same place Galois had been a student, and now the premier establishment in France for mathematics).

Lie's theory was now all the rage in Paris, and as Émile Picard wrote to Lie, "You have created a theory of major importance that will be counted among the most remarkable mathematical works of the second half of this century." By 1893 Picard could write again that, "Paris is becoming a centre for groups; it is all fermenting in young minds, and one will have an excellent wine after the liquors have settled a bit."

In 1892 Cartan returned to Paris from a year's military service, and took rooms with another student who had just returned from studying in Leipzig. This student told Cartan about Lie's groups and Killing's classification. Cartan became fascinated and decided to devote his doctoral thesis to this topic. In 1893 Lie himself visited Paris, having partly recovered and left the mental hospital. Cartan was thrilled to meet the great man. He completed his doctoral thesis the next summer, filling the gaps in Killing's work, and confirming the table of simple Lie groups (or rather, Lie algebras) that Killing had found.

Cartan was just the man for the job. He had a great talent for abstract structural reasoning and this helped him clarify and develop Killing's ideas. Some of the technical details were renamed, and new details were added. For example the "Killing form" originated with Cartan, whereas "Cartan subalgebras" and "Cartan integers" originated with Killing (see [H, p. 207]). The result is now known as the Killing–Cartan classification.

In 1894, the year Cartan completed the classification, the Norwegian National Assembly established a chair for Lie. This was very welcome because he wanted to return to his homeland. But his wife and daughters had friends in Leipzig and were reluctant to leave, so Lie stayed on, until by 1898 he was suffering from cerebral anaemia, and it was clearly time to go. The family returned to Norway. Lie delivered some lectures in the autumn of that year but soon had to give up, and was reduced to holding seminars in his home. They too soon had to stop, and Lie died in February 1899.

3. Finite groups of Lie type

Finite analogues of Lie groups, obtained by replacing the real or complex numbers by finite fields, came to fruition in the work of Leonard Eugene Dickson (1874–1954). In 1901 he published his first book [D], which was on the classical groups (families A, B, C, and D) and contained a clear exposition of finite fields. Dickson's method was to use the fact that the classical groups act as linear groups preserving quadratic forms or sesquilinear forms. For a vector space over a finite field these forms can be constructed and classified, and this yields all the classical groups.

Dickson was the very first student to do a Ph.D. at the University of Chicago. He was born in the American mid-west (in Iowa) in 1874, but his family moved to Texas and he graduated from the University of Texas with a masters degree in 1894. At the University of Chicago he obtained a Ph.D. in 1896, and immediately went to Europe where he spent time in Paris and Leipzig. In Paris he met the young Élie Cartan, who had completed his doctorate two years earlier, and in Leipzig he met Lie and Engel. After a year away he returned to America, first to California for a year, then to Texas, and finally to the University of Chicago where he remained for the rest of his career, from 1900 to 1939. In that period he published eighteen books, hundreds of research articles, and supervised no fewer than fifty-five Ph.D. students.

In Dickson's time finite classical groups were not entirely new. Jordan, in his Treatise of 1870, had already dealt with the A-family (the groups PSL_n), and Galois himself had constructed the A_1 family (the groups PSL_2), as I mentioned earlier.

Dickson's methods for the families B, C, and D, which involved quadratic and sesquilinear forms, did not extend to the non-classical Lie groups (families E, F, and G), and although Dickson later dealt with E_6 and G_2, it was only after the Second World War that finite groups of all exceptional types were constructed. In the early 1950s two alternative ways forward were being pioneered: one geometric, the other algebraic. The geometric method was due to Jacques Tits in Belgium, who had obtained his doctorate in 1949. The algebraic method was undertaken by Claude Chevalley in France, an established mathematician, twenty-one years older than Tits and a founding member of the Bourbaki school. Chevalley was the first to succeed, and in 1955 he produced his famous paper "On certain simple groups" [Ch1]. Chevalley showed that for each simple Lie algebra over the complex numbers, one could define a group over an arbitrary field. These groups became known as *Chevalley groups*.

In the finite case, Chevalley obtained, for each finite field, analogues of the complex semisimple Lie groups in families A through G. Some of the finite classical groups were still missing however. For example the finite unitary groups were not Chevalley groups in the strict sense because the unitary groups $\mathrm{SU}_n(\mathbf{C})$ are not complex Lie groups, but are real forms of $\mathrm{SL}_n(\mathbf{C})$. Their definition uses a field automorphism of $\mathbf{C}$, namely complex conjugation, and in order to obtain unitary groups over a finite field one takes a quadratic extension of the field and uses a similar conjugation. Steinberg [St] used this idea, along with diagram automorphisms of order 2, and produced the finite unitary groups, a second class of orthogonal groups (acting on a $2n$-dimensional vector space having a form of Witt index $n-1$), and a "twisted" version of E_6. They are sometimes denoted ^2A_n (unitary), ^2D_n (orthogonal), and $^2E_6$. Tits also dealt with $^2E_6$, and did a similar thing for groups of type D_4 using a cubic field extension and an order 3 symmetry of the Dynkin diagram [T4], producing twisted groups $^3D_4$.

At about this time, Michio Suzuki discovered a new family of finite simple groups $Sz(2^n)$, where n is an odd number greater than 1. He found these in the process of determining all simple groups in which the involution centralizers are 2-groups [Su1, Su2]. Rimhak Ree then observed that Suzuki's family could be obtained as subgroups of the orthogonal groups $B_2(2^n)$. It had already been observed that three subfamilies of Chevalley groups, namely $B_2(2^n)$, $F_4(2^n)$, and $G_2(3^n)$ for n odd, possess an extra outer automorphism, owing to a degeneracy in one of the root group commutator relations, and Ree [**Re1, Re2**] used this, along with a variation on Steinberg's method, to produce three families of groups, $^2B_2(2^n)$, $^2F_4(2^n)$, and $^2G_2(3^n)$, for n odd. The first of these was the family of Suzuki groups; the others were new. These variations of Chevalley groups produced by Steinberg, Tits, Suzuki, and Ree were originally called *twisted Chevalley groups*, though they and the ordinary Chevalley groups are now usually subsumed under the moniker of *groups of Lie type*. For a good article on these groups, see the article by Curtis [**Cur**] or the later book by Carter [**Car**].

Here now are the diagrams referred to above.

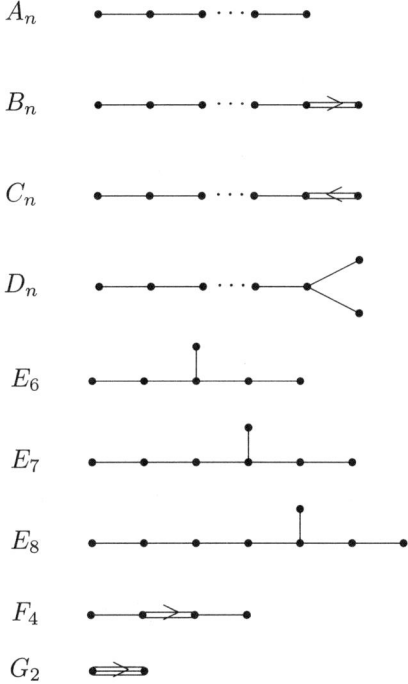

Each "twisted Chevalley group" involves a diagram automorphism, which in the case of the Suzuki and Ree groups ignores the arrow on the B_2, F_4, and G_2 diagrams. I shall come back to twisted Chevalley groups later in discussing the classification of buildings, but it is now time to turn to the work of Tits, and see how Building Theory developed.

4. Early work on buildings

The original motivation for Tits' work was to understand the geometry of the exceptional Lie groups (types E, F, and G) over the complex numbers in a way that would lead to analogues over any field. As Tits himself wrote in a paper [**T11**] published in 1980:

> Originally, the theory of buildings was aimed primarily at understanding the exceptional Lie groups from a geometrical point of view. The starting point was the observation that it is possible to associate with each complex analytic semisimple group a certain well-defined geometry, in such a way that the basic properties of

the geometries thus obtained and their mutual relationships can
be easily read from the Dynkin diagrams of the corresponding
groups.

As an example, take the A_n case, which as I mentioned above was dealt with by
Jordan and Dickson. The simple group G is $\mathrm{PSL}_{n+1}(\mathbf{C})$ and its natural geometry
is the projective space of dimension n over $\mathbf{C}$. The diagram for A_n has n nodes,
and in terms of the Lie algebra and the root system, from which the diagram is
obtained, each node corresponds to a unique conjugacy class of maximal (parabolic)
subgroups of G. Reading from left to right in the diagram, groups belonging to
the first node are stabilizers of points in the projective space, those belonging to
the second node are stabilizers of lines, and those belonging to the kth node are
stabilizers of $(k-1)$-dimensional subspaces. To put this another way, let V be an
$(n+1)$-dimensional vector space over $\mathbf{C}$; then the k-dimensional subspaces of V
belong to the kth node of the diagram.

Now construct a simplicial complex as follows. The vertices are the proper
non-trivial subspaces of V, and the simplexes are defined as those sets of vertices
forming a nested sequence of subspaces. This simplicial complex has dimension $n-1$
because a maximal nested sequence of subspaces has n members. The maximal-
dimensional simplexes are called chambers, and the whole complex is a building of
type A_n. This construction works over an arbitrary field, and the problem is to do
the same thing for all types of semisimple Lie groups. This was the problem Tits
set himself.

In general, for an arbitrary complex semisimple Lie group, the conjugacy classes
of maximal parabolic subgroups are in bijective correspondence with the nodes of
the diagram. This fact, proved by two Russian mathematicians, Morozov [**Mo**]
and Karpelevic [**Ka**] many years earlier, yields a geometry for each type of complex
semisimple Lie group, analogous to the projective space for groups of type A_n. But
the problem is how to obtain such geometries over arbitrary fields.

Tits had the idea of building up from lower rank geometries to higher rank
ones, and in explaining his earlier ideas later [**T11**, Sect.1], he makes the following
observation, which I have slightly reworded:

(*) Let M be a Dynkin diagram, and Γ the geometry for the corresponding
semisimple Lie group. Let x be a vertex of Γ, and $\Gamma(x)$ the residual geometry
comprising all simplexes of Γ distinct from but incident with x. If x belongs to the
node $\nu(x)$ of the diagram M, then $\Gamma(x)$ belongs to the diagram obtained from M
by deleting $\nu(x)$ and all strokes containing it. [Note that the objects in $\Gamma(x)$ are
in bijective correspondence with those in the link of x in Γ, which is a simplicial
complex.]

Tits goes on to make a second observation, namely that one can build up from
rank 2 geometries. Since type G_2 can only occur as a direct factor in higher-rank
groups, this leaves only three rank 2 diagrams:

$$A_1 \times A_1 \qquad \bullet \quad \bullet$$

$$A_2 \qquad \bullet\!\!-\!\!\!-\!\!\bullet$$

$$B_2 \qquad \Longrightarrow$$

Each of these can be realised as a bipartite graph. The first is a complete bipartite graph whose set of vertices is the direct sum of two projective lines. The second is the projective plane: the vertices are the points and the lines of the plane, and two vertices are joined when the point is on the line. The third is a quadric in 4-dimensional projective space, the vertices being the points and lines of this quadric. In [T2] and [T1] Tits gave examples showing how to obtain information on higher-rank geometries from these three rank 2 residual geometries, called rank 2 *residues*, and looking back at this later [T11] he writes:

> Since the three geometries of rank 2 described above have obvious analogues over an arbitrary field k, it was natural, taking those as building blocks, to try to associate with every Dynkin diagram a "geometry over k", using (*) as an axiom. This programme was started in [T2] with the additional aim of defining, in that geometrical way, analogues over arbitrary fields of the complex semisimple Lie groups. Shortly afterwards, Chevalley's fundamental paper [Ch1] appeared. As a result, the geometric approach lost much of its value as a way of defining new groups, but a technical basis was provided for a more efficient construction of the geometries in question [T3].

At the same time, the theory of algebraic semisimple groups over an arbitrary field was being developed ([Ch2], [T5], [BT1]). This led to groups of rational points that would provide what I referred to earlier as "twisted Chevalley groups" and their geometries (see e.g. [T6] and [T7]). I shall leave the story here and refer the reader to Tits [T11] for more details of this earlier approach. Let us now move on to the theory of buildings as it appeared in a well-developed form in [T8].

5. Tits' 1974 book on buildings

In his book "Buildings of Spherical Type and Finite BN-Pairs", Tits approaches buildings from the point of view of apartments (see below), and to define an apartment he uses Coxeter groups. A *Coxeter group* is a group W defined by generators s_i, $i \in I$, of order 2, and relations $(s_i s_j)^{m_{ij}} = 1$, where m_{ij} is an integer≥ 2, or ∞, for each pair of generators s_i and s_j. The possible use of ∞ merely means that in this case $s_i s_j$ generates an infinite cyclic group. The matrix $M = (m_{ij})$, where we take $m_{ii} = 1$, Tits calls a *Coxeter matrix*.

An apartment of type M he defines as a "flag complex", which means that every vertex has a *type i* in I, no two vertices of the same type lie in a common simplex, and any set of mutually adjacent vertices form a simplex. There are n different types of vertices, where n is the cardinality of I, which is assumed finite, and the type of a simplex is the set of its types of vertices. Every simplex is a face of a simplex of type I, and the simplexes of type I have maximum dimension $n - 1$ and are called chambers. The simplexes of codimension 1 (i.e. dimension $n - 2$) are called *panels*.

Tits' book [T8] deals mainly with buildings of spherical type, which means that the apartments are triangulations of a sphere (all apartments in the same building are isomorphic). A building is *spherical* precisely when the Coxeter group is finite, in which case it can be obtained as a group W generated by n reflections, $s_1, \ldots, s_n$ in $\mathbb{R}^n$. The fundamental domains for W in $\mathbb{R}^n$, when intersected with a sphere centered at the origin, give a triangulation of that sphere that is called its *Coxeter complex*.

5.1. Diagrams. Each Coxeter group belongs to a *Coxeter diagram*, obtained from the Coxeter matrix M above. There is one node for each $i \in I$, and node i is joined to node j by an edge labelled with m_{ij}, unless $m_{ij} = 2$, in which case there is no edge. As a matter of convention, when $m_{ij} = 4$ the edge and its marking are often replaced by a double edge. The connected components of a Coxeter diagram give Coxeter subgroups whose direct product is the Coxeter group of the whole diagram. There is therefore no loss in considering only connected diagrams.

The connected diagrams for finite Coxeter groups of rank 3 are almost the same as the Dynkin diagrams shown above. However the arrows are deleted (so a B_n Coxeter diagram is the same as one of type C_n), and there are two Coxeter diagrams that are not Dynkin diagrams, namely H_3 and H_4.

$$H_3 \quad \bullet \!\!\!-\!\!\!\overset{5}{-}\!\!\!\bullet$$

$$H_4 \quad \bullet \!\!\!-\!\!\!-\!\!\!\bullet \overset{5}{-}\!\!\!\bullet$$

5.2. A definition of buildings. In the definition of buildings that he gives in [**T8**], Tits uses the term complex to refer to a flag complex in the sense given above; it is called *thick* if every panel is a face of at least three chambers, and *thin* if every panel is a face of exactly two chambers. Coxeter complexes are thin. Tits defines a building as a complex Δ along with a set of subcomplexes, called apartments. Here is the definition given in [**T8**, p.38], very slightly reworded:

(B1) Δ is thick;

(B2) The apartments are thin chamber complexes;

(B3) Any two simplexes of Δ belong to an apartment;

(B4) If two apartments Σ and Σ' have in common two simplexes A and A', then there exists an isomorphism of Σ onto Σ' which leaves invariant A, A', and all their faces.

5.3. Automorphisms of spherical buildings. All apartments in a given building are isomorphic, and the *diagram* for the building is the Coxeter diagram for an apartment in the building. The major work in [**T8**] is to classify spherical buildings that are irreducible (meaning the diagram is connected), and of rank≥ 3. This is achieved by first showing the existence of a large automorphism group. In order to state the main theorems on automorphisms we need some notation. Let Δ and Δ' denote two spherical buildings that are irreducible and of rank≥ 3. For a simplex A of Δ or Δ', we let $E_i(A)$ denote the set of all chambers having a face of codimension i with A.

THEOREM 5.1. [**T8**, 4.1.1] *Let Σ be an apartment of Δ and C a chamber of Σ. Then any isomorphism from Δ to Δ' is entirely determined by its restriction to $E_1(C) \approx \Sigma$.*

THEOREM 5.2. [**T8**, 4.1.2] *Let C and C' be chambers of Δ and Δ', respectively. Then every adjacency preserving bijection from $E_2(C)$ to $E_2(C')$ extends to an isomorphism from Δ onto Δ'.*

THEOREM 5.3. [**T8**, 4.16] *Let Σ and Σ' be apartments in Δ and Δ', respectively, containing chambers C and C', respectively. Let α be an automorphism from Σ to*

Σ', and γ an adjacency-preserving mapping from $E_2(C)$ to $E_2(C')$ which coincides with α on $\Sigma \cap E_2(C)$. Then α and γ are the restrictions to Σ and $E_2(C)$ of an isomorphism from Δ onto Δ'.

Theorem 5.2 shows that the local structure determines the global structure, and it is essential for the classification. Theorem 5.3 can be used to show that the A_2 residues (residual geometries associated to A_2 subdiagrams) are Moufang planes. This implies that in the finite case they are coordinatized by a finite field; in general they must be coordinatized by a field (possibly non-commutative) or a Cayley division algebra. When the diagram contains an A_3 subdiagram (which is the case for all diagrams of rank ≥ 3 that are not of type $B_3 = C_3$ or F_4) then a classical theorem on projective geometry implies that the A_2 residues are coordinatized by a field (possibly non-commutative)—see [**T8**, p.96].

5.4. The classification of spherical buildings.
Tits' classification of thick spherical buildings that are irreducible and of rank ≥ 3 can be delineated into three cases.

(i) *Diagrams with only single bonds.* In this case the diagram is one of: A_n, D_n, E_6, E_7, or E_8. In the A_n case one is dealing with projective space and there is one building of type A_n for each field (possibly non-commutative). In the other cases there is a D_4 subdiagram, and this implies that the field must be commutative (see [**T8**, p.99]). This settles the single bond case, except for verifying the existence of a building of each of these types for every commutative field. For types A_n and D_n this is a simple question of linear algebra, but for types E_6, E_7, and E_8, Tits appeals to the existence of algebraic groups of these types (see [**T8**, p.100]), one for each commutative field. At the end of this paper I shall show how we later found a way of constructing the buildings without appealing to the existence of the groups (the groups then appear as a by-product).

(ii) *Diagrams with a double bond and an A_3 subdiagram.* In this case the diagram is $B_n = C_n$, with $n \geq 4$, and the A_{n-1} subdiagram is coordinatized by a field (possibly non-commutative). This is a tough case to classify. Tits treats all buildings of type $B_n = C_n$ as "polar spaces" (see [**T8**, 7.1]). When the rank is at least 4 (which is the case when there is an A_3 subdiagram) the polar space is "embeddable" (see [**T8**, 8.5.2]), and Tits classifies them using quadratic, pseudo-quadratic, and sesquilinear forms of Witt index n on a vector space that may be infinite-dimensional (the pseudo-quadratic forms are needed for non-commutative fields of characteristic 2)—see [**T8**, Ch.8].

(iii) *Diagrams of type $B_3 = C_3$ and F_4.* In these cases the A_2 residues can be Moufang planes that are not Desarguesian, meaning that they are coordinatized by a Cayley division algebra. Tits proves that the C_3 case is determined by the Cayley division algebra along with a quadratic norm ([**T8**, 9.1]). In the F_4 case there are five sub-cases, four of which arise from algebraic groups (of types F_4, E_6, E_7, and E_8)—see [**T8**, p.202].

(iv) *Diagrams of types H_3 and H_4.* There are no thick buildings of these types. This is proved by Tits [**T9**] using the concept of a "Moufang" building (see [**T8**, p.274]). The idea is that the group of automorphisms that is yielded by application of Theorem 5.3 contains "root groups" and by using these root groups, Tits is able to eliminate thick buildings of types H_3 and H_4.

5.5. Groups of Lie type. Groups of Lie type were mentioned at the end of Section 3. They are obtained from algebraic groups G over an algebraically closed field $\bar{k}$ by Galois descent to a subfield k. The descent process may or may not change the diagram, in the sense that the group $G(k)$ over the field k is naturally associated with a building, and the diagram for this building may or may not be the same as the diagram for the original group G. For example if $\bar{k}$ is the complex numbers C and k is the real numbers $\mathbb{R}$, then the group $SL_6(\mathbf{C})$ has three real forms, namely $SL_6(\mathbb{R})$, $SU_6(\mathbf{C})$, and $SL_3(\mathbf{H})$, where $\mathbf{H}$ denotes the quaternions. The building for the first of these groups has the same diagram as $SL_n(\mathbf{C})$, namely A_5, for the second group the building is vacuous (the group is the compact form of $SL_6(\mathbf{C})$ and is anisotropic), and for the third group the building is the flag complex of a quaternionic plane and has type A_2. The twisted Chevalley groups, mentioned in Section 3, have diagrams that are different from the algebraic groups that they are derived from. For example $^2E_6$ belongs to the F_4 diagram.

The building can be obtained from the group by using (B, N)-pairs, where B and N are subgroups (B stabilizes a chamber, and N an apartment)—see [**T8**, p.39].

5.6. Rank 2 spherical buildings. Rank 2 spherical buildings, also known as *generalized m-gons*, have as their Coxeter group the dihedral group of order $2m$. They cannot be classified like spherical buildings of rank ≥ 3 because the automorphism group can be far more limited—Theorem 5.3 fails—but assuming thickness, and finiteness, m must be 2, 3, 4, 6, or 8. This is the famous Feit-Higman theorem.

THEOREM 5.4. [**FH**] *For a finite thick generalized m-gon, m must be 2, 3, 4, 6, or 8.*

If finiteness is abandoned then there are wild examples for any m (see, e.g., [**R1**]). However, an extra condition can be imposed that then yields a similar result to the Feit-Higman theorem. The term Moufang was used earlier in this section in connection with the non-existence of thick H_3 and H_4 buildings (the term refers to the existence of certain root group automorphisms). When a generalized m-gon is Moufang (which precludes the $m = 2$ case which is reducible), then again one finds that m must be 3, 4, 6, or 8. This was proved independently by Tits and Weiss.

THEOREM 5.5. [**T10**], [**W**] *For a Moufang generalized m-gon, m must be 3, 4, 6, or 8.*

More recently a complete classification of Moufang m-gons has been achieved—see [**TW**].

6. Beyond spherical buildings

The first step beyond spherical buildings is the theory of affine buildings, where the apartments are tilings of Euclidean space. I shall not dwell on these, but merely refer the reader to my book [**R2**] and the references given there. Affine buildings were first constructed by Iwahori and Matsumoto [**IM**], and then a complete theory was worked out by Bruhat and Tits [**BT2**]. A classification of the irreducible ones of rank 4 was later given by Tits [**T13**].

Affine buildings of rank 3 can be wild (see [**R1**]), but there are some intriguing ones arising as universal covers of certain sporadic group geometries. The original discovery of one of these inspired Tits to give a new approach to buildings. He called it a "local approach" in the title of his paper [**T12**], and it is reminiscent of his original ideas, described earlier in Section 4. Tits' local approach uses the notion of a chamber as the basic ingredient from which buildings are formed. As he writes in [**T12**]:

> So far, buildings have always been described as incidence geometries or simplicial complexes. The results of the present paper find their simplest expression in a somewhat more abstract framework, that of *chamber systems*.

A spherical building can be treated as a chamber system by taking two chambers to be adjacent if they have a face of codimension 1 in common. This implies that the chambers differ in a single vertex; if that vertex has type i then the chambers are said to be i-adjacent. For a discussion of buildings as chamber systems, see [**T12**], [**T13**] or [**R2**]. A particularly good way of defining buildings is to treat them as chamber systems and use a W-valued distance between chambers, where W is the Coxeter group for the building concerned—see [**T15**]. Almost all the recent work on buildings uses this approach.

This use of chambers as the basic ingredient is particularly important when subgroups of W corresponding to proper subdiagrams are infinite. The idea is to assign faces to a chamber only for spherical residues (those for which the subdiagram has a finite Coxeter group). The idea of creating "Coxeter complexes" in this way was pioneered by Moussong [**Mou**] and Davis [**Da**]. Here is an example to give the flavour. Take a rank 5 Coxeter group with the following generators s_1, s_2, s_3, s_4, s_0, and relations $(s_i s_{i+1})^2 = 1$, where $i \in \mathbf{Z}/(5)$. The only proper spherical residues are those of types $\{i\}$, and $\{i, i+1\}$; a chamber can be realised as a right-angled pentagon, the residues of type $\{i\}$ being its edges, and those of type $\{i, j\}$ being its vertices. A suitable Coxeter complex can then be realised as the hyperbolic plane tiled by right-angled pentagons. Moufang buildings of these types, and others, have recently been constructed by Rémy and the author [**RR**] and used to construct lattices acting on hyperbolic buildings.

Another extremely useful reason for treating buildings as chamber systems having a W-valued distance function between chambers is the theory of twin buildings. A twin building is a pair of buildings, both of the same type, having a W-valued "codistance" between chambers of one and chambers of the other. They were introduced with the original purpose of dealing with Kac-Moody groups, as constructed in [**T14**]. I refer the reader to the original paper by Tits [**T15**] and a careful discussion of the local properties in [**R3**]. Twin buildings have properties analogous to spherical buildings; there is a Local-to-Global theorem [**MR**], similar to Theorem 5.2 above, and this leads to a classification—see [**Mu**] for recent work.

I want to conclude this paper by describing another use of the local approach to buildings. As I mentioned earlier, in Section 5, Tits' classification of spherical buildings in [**T8**] employed the theory of algebraic groups in order to verify the existence of buildings of type F_4, E_6, E_7, and E_8. This was unsatisfactory because one of Tits' original motivations in developing his geometric ideas was to create analogues of the exceptional Lie groups over arbitrary fields. If the buildings could be constructed independently of the groups, then Theorem 5.3 could be used to create the groups. A way of constructing these buildings as chamber systems was developed by Tits and the author [**RT**], and I quote from the introduction to that paper.

> ... it is perhaps worth remarking that one of the initial moti-vations for the theory of buildings, at a time when Chevalley's fundamental "Tôhoku paper" had not yet appeared, was the search for a geometric way of obtaining analogues of the excep-tional Lie groups. The present paper is, therefore, particularly satisfying (to one of us at least!) in that it does give a short, in fact almost trivial, proof of the existence of a building of type, say E_8 over an arbitrary field ...

The construction worked by considerably refining an earlier construction that the author gave in [**R1**]. The idea in [**RT**] was to construct a building by starting at one chamber and working outwards. A problem arose when there were rank 3 residues of irreducible spherical type (i.e. of type A_3 or B_3), and the refinement that was needed to take care of such residues we called a *blueprint*; it showed how to complete the rank 2 residues so that one created A_3 and B_3 buildings. A blueprint that gave rise to a building we called *realisable*, and the building was said to conform to it. We proved the following theorem.

THEOREM 6.1. [**RT**] *A blueprint is realisable if its restriction to each rank 3 subdiagram is realisable. In this case there is a unique building which conforms to it.*

It is a simple matter, using this theorem, to construct buildings of type E_8 by knowing how to patch together two A_2 blueprints to obtain the blueprint for an A_3 building. The details are given in [**RT**].

References

[A]   N. H. Abel, Crelle's Journal 1826.

[BT1]   F. Bruhat and J. Tits, *Groupes réductifs*, Publ. Math. I.H.E.S. **27** (1955) 55–151.

[BT2]   F. Bruhat and J. Tits, *Groupes réductifs sur un corps local, I. Données radicielles valuées*, Publ. Math. I.H.E.S. **41** (1972) 5–252.

[Cd]   G. Cardano, *Ars Magna*: see, The Great Art, or the Rules of Algebra: tr. R.MClennon, in D.E.Smith (ed.) *A Source Book in Mathematics*, Dover pp. 203–206.

[Car]   R. W. Carter, *Simple groups of Lie Type*, Wiley, London and New York 1972.

[Ch1]   C. Chevalley, *Sur certains groupes simples*, Tôhoku Math. J. (2), **7** (1955) 14–66.

[Ch2]   C. Chevalley, *Classification des groupes de Lie algébriques I, II*, Séminaire E.N.S., 1956-58, mimeographed.

[Cur]   C. W. Curtis, *Chevalley groups and related topics*, in: Finite Simple Groups (eds. Powell and Higman), 135–189, Academic Press, London 1971.

[Da]   M. Davis, *Buildings are CAT(0)*, in: Geometry and Cohomology in Group Theory (eds. P. Kropholler et al.), LMS Lecture Notes Series **252** (1997) 108–123.

[D]   L. E. Dickson, *Linear groups with an exposition of the Galois field theory*, Teubner, Leipzig 1901. Reprinted, Dover, New York 1958.

[FH]   W. Feit and G. Higman, *The nonexistence of certain generalized polygons*, J. Algebra **1** (1964) 114–131.

[H]   T. Hawkins, *Emergence of the Theory of Lie Groups*, Springer 2000.

[IM]   N. Iwahori and H. Matsumoto, *On some Bruhat decomposition and the structure of the Hecke ring of a p-adic Chevalley group*, Publ. I.H.E.S. *25* (1965) 5–48.

[J]   C. Jordan, *Traité des Substitutions et des Équations Algébriques*, Paris 1870.

[Ka]   F. Karpelevic, *O nepoluprostych maksimalnyh podalgebrah poluprostych algr Li*, Dokl. Akad. Nauk SSSR **76** (1951) 775–778.

[L]   J-L. Lagrange, *Réflexion sur la Résolution Algébrique des Équations*, Mem. Acad. Roy. Soc. 1770/71; in *Oeuvres* (eds. J.A. Serret and G. Darboux), Paris 1867-1892.

[Mo]   V. Morozov, *O nepoluprostych podgruppah prostych grupp*, Thesis, Kazan 1943.

[Mou]   G. Moussong, *Hyperbolic Coxeter Groups*, Ph.D. Thesis, Ohio State Univ. 1988.

[Mu]   B. Mühlherr, *Twin Buildings*, in: Tits Buildings and the Model Theory of Groups (ed. K. Tent), London Math. Soc. Lecture Notes **291** (2002) 103–117.

[MR]   Mühlherr, B. and Ronan, M., *Local to Global Structure in Twin Buildings*, Inventiones math. **122** (1995) 71–81.

[Re1]   R. Ree, *A family of simple groups asociated with the simple Lie algebra F_4*, Amer. J. Math. **83** (1961) 410–420.

[Re2]   R. Ree, *A family of simple groups asociated with the simple Lie algebra G_2*, Amer. J. Math. **83** (1961) 432–462.

[RR]   B. Rémy and M. Ronan, *Topological groups of Kac-Moody type, right-angled twinnings and their lattices*, preprint 2004.

[R1]   M. Ronan, *A Construction of Buildings with no Rank 3 Residues of Spherical Type*, in Lecture Notes in Math. **1181** (Buildings and the Geometry of Diagrams, Como 1984) Springer-Verlag, 1986 , 242–248.

[R2]   M. Ronan, *Lectures on Buildings*, Academic Press 1989.

[R3]   M. Ronan, *Local Isometries of Twin Buildings*, Math. Zeitschrift**234** (2000) 435–455.

[RT]   M. Ronan and J. Tits, *Building Buildings*, Math. Annalen **278** (1987) 291–306.

[St]   R. Steinberg, *Variations on a theme of Chevalley*, Pacific J. Math. **9** (1959) 875–891.

[Su1]   M. Suzuki, *Finite groups with nilpotent centralizers*, Trans. Amer. Math. Soc. **99** (1961) 425–470.

[Su2]   M. Suzuki, *On a class of doubly transitive groups: I, II*, Ann. of Math. **75** (1962) 105–145; **79** (1964) 514–589.

[T1]   J. Tits, *Les groupes de Lie exceptionels et leur interprétation géométrique*, Bull. Soc. Math. Belg. **8** (1956) 48–81.

[T2]   J. Tits, *Sur la géométie des R-espaces*, J. Math. Pure et Appl. **36** (1957) 17–38.

[T3]   J. Tits, *Sur les analogues algébriques des groupes semi-simples complexes*, Colloque d'Algèbre Supérieure du C.B.R.M., Bruxelles 1956, (1957) 261–289.

[T4]   J. Tits, *Sur la trialité at certains groupes qui s'en déduisent*, Publ. Math. I.H.E.S. **2** (1959) 14–60.

[T5]   J. Tits, *Sur la classification des groupes algébriques semi-simples*, C.R. Acad. Sci. Paris **249** (1959) 1438–1440.

[T6]   J. Tits, *Groupes algébriques semi-simples et géométries associées*, Proc. Coll. Algebraical and Topological Foundations of Geometry, Utrecht 1959, Pergamon Press, London (1962) 175–192.

[T7]   J. Tits, *Classification of algebraic semi-simple groups*, in: Proc. Symp. Pure Math. vol.9 (Algebraic Groups and Discontinuous Subgroups, Boulder 1965), 33–62, Amer. Math. Soc. 1966.

[T8]   J. Tits, *Buildings of Spherical Type and Finite BN-Pairs*, Lecture Notes in Mathematics **386**, Springer Verlag 1974.

[T9]   J. Tits, *Endliche Spiegelungsgruppen, die als Weylgruppen auftreten*, Inventiones math. **45** (1977) 283–295.

[T10]   J. Tits, *Non-existence de certains polygones généralisés*, I, II, Inventiones math. **36** (1976) 275–284; **51** (1979) 267–269.

[T11]   J. Tits, *Buildings and Buekenhout Geometries*, in: Finite Simple Groups II (ed. M.J. Collins), 309–320, Academic Press 1980.

[T12]   J. Tits, *A local approach to buildings*, in: Davis, C. et al. (eds.), The Geometric Vein—The Coxeter Festschrift. Springer Verlag, Berlin 1981, 519–547.

[T13]   J. Tits, *Immeubles de type affine*, in: Buildings and the Geometry of Diagrams (Como 1984), Lecture Notes in Mathematics **1181** 159–190, Springer Verlag 1986.

[T14]   J. Tits, *Uniqueness and Presentation of Kac-Moody Groups over Fields*, J. Algebra **105** (1987) 542–573.

[T15]   J. Tits, *Twin Buildings and Groups of Kac-Moody Type*, in: Liebeck and Saxl (eds.), Groups, Combinatorics and Geometry (Durham 1990), London Math. Soc. Lecture Note Ser. **165** 249–286, Cambridge Univ. Press 1992.

[TW]   J. Tits and R. Weiss, *Moufang Polygons*, Springer-Verlag 2002.

[W]   R. Weiss, *The Nonexistence of certain Moufang Polygons*, Inventiones math. **51** (1979) 261–266.

MATHEMATICS DEPARTMENT, UNIVERSITY OF ILLINOIS AT CHICAGO, CHICAGO, IL 60607 USA
E-mail address: ronan@uic.edu

The Coxeter Element and the Branching Law for the Finite Subgroups of $SU(2)$

Bertram Kostant

0. Introduction

0.1. Let Γ be a finite subgroup of $SU(2)$. The question we will deal with in this paper is how an arbitrary (unitary) irreducible representation of $SU(2)$ decomposes under the action of Γ. The theory of McKay assigns to Γ a complex simple Lie algebra $\mathfrak{g}$ of type $A - D - E$. The assignment is such that if $\widetilde{\Gamma}$ is the unitary dual of Γ, we may parametrize $\widetilde{\Gamma}$ by the nodes (or vertices) of the extended Coxeter-Dynkin diagram of $\mathfrak{g}$.

Let $\ell = rank\,\mathfrak{g}$ and let $I = \{1, \ldots, \ell\}$. Let $I_{ext} = I \cup \{0\}$. The set of nodes may be identified with a set of simple roots of the affine Kac-Moody Lie algebra associated with $\mathfrak{g}$ and are indexed by I_{ext}. We can then write $\Gamma = \{\gamma_i\}$, $i \in I_{ext}$. Let $\Pi = \{\alpha_i\}$, $i \in I$, be the set of simple roots of $\mathfrak{g}$ itself. One has that γ_0 is the trivial 1-dimensional representation of Γ and, for $i \in I$,

$$dim\,\gamma_i = d_i \tag{0.1}$$

where

$$\psi = \sum_{i \in I} d_i\,\alpha_i \tag{0.2}$$

is the highest root. For proofs and details about the McKay correspondence see e.g. [**G-S,V**], [**M**], and [**St**].

0.2. The unitary dual of $SU(2)$ is indexed by the set $\mathbb{Z}_+$ of nonnegative integers and will be written as $\{\pi_n\}$, $n \in \mathbb{Z}_+$ where

$$dim\,\pi_n = n + 1. \tag{0.3}$$

Our problem is the determination of $m_{n,i}$ where $n \in \mathbb{Z}_+$, $i \in I_{ext}$ and

$$m_{n,i} = \text{multiplicity of } \gamma_i \text{ in } \pi_n | \Gamma.$$

2000 *Mathematics Subject Classification.* Primary 22E40, 22F55, 20F29.

Research supported in part by NSF grant DMS-0209473 and in part by the KG&G Foundation.

It is resolved with the determination of the formal power series

$$m(t)_i = \sum_{n=0}^{\infty} m_{n,i}\, t^n. \tag{0.4}$$

To do this one readily notes that it suffices to consider only the case where $\Gamma = F^*$ and F^* is the pullback to $SU(2)$ of a finite subgroup F of $SO(3)$. This eliminates only the case where Γ is a cyclic group of odd order and $\mathfrak{g}$ is of type A_ℓ where ℓ is even. For the remaining cases the Coxeter number h of $\mathfrak{g}$ is even and we will put

$$g = h/2. \tag{0.5}$$

Also for the remaining cases there is a special index $i_* \in I$. If $\mathfrak{g}$ is of type D or E then α_{i_*} is the branch point of the Coxeter-Dynkin diagram of $\mathfrak{g}$. If $\mathfrak{g}$ is of type A_ℓ then α_{i_*} is the midpoint of the diagram (recalling that ℓ is odd).

If $i = 0$ the determination of $m(t)_0$ is classical and is known from the theory of Kleinian singularities. In fact there exist positive integers $a < b$ such that

$$m(t)_0 = \frac{1 + t^h}{(1 - t^a)(1 - t^b)}. \tag{0.6}$$

The numbers a and b in Lie theoretic terms are given in

Theorem 0.1. *One has $a = 2d_{i_*}$ and b is given by the condition that*

$$\begin{aligned} a\,b &= 2\,|F^*| \\ &= 4\,|F|. \end{aligned}$$

It remains to determine $m(t)_i$ for $i \in I$.

Proposition 0.2. *If $i \in I$ there exists a polynomial $z(t)_i$ of degree less than h such that*

$$m(t)_i = \frac{z(t)_i}{(1 - t^a)(1 - t^b)}. \tag{0.7}$$

The problem is now to determine the polynomial $z(t)_i$. This problem was solved in [K] using the orbits of a Coxeter element σ on a set of roots Δ for $\mathfrak{g}$. In the present paper we will put the main result of [K] in a simplified form. See Theorem 1.13 in the present paper. Also the present paper makes explicit some results that are only implicit in [K]. For example introducing $\widetilde{\Pi}$ (see (1.10)) and making the assertions in Remark 1.10 and Theorems 1.8, 1.9, 1.11 and 1.12.

The set Π generates a system, Δ_+, of positive roots. The highest root $\psi \in \Delta$ defines a certain subset $\Phi \subset \Delta_+$ of cardinality $2h-3$. Because of its connection with a Heisenberg subalgebra of $\mathfrak{g}$ this subset is referred to as the Heisenberg subsystem of Δ_+. The new formulation explicitly shows how the polynomials $z(t)_i$ arise from the intersection

$$\text{(orbits of the Coxeter element } \sigma) \cap \text{(the Heisenberg subsystem } \Phi). \tag{0.8}$$

The polynomials $z(t)_i$ are listed in [K]. The special case where $\mathfrak{g}$ is of type E_8 is also given in the present paper (see Example 1.7). Unrelated to the Coxeter element the polynomials $z(t)_i$ are also determined in Springer [Sp]. They also appear in another context in Lusztig [**L1**] and [**L2**]. Recently, in a beautiful result, Rossmann [**R**], relates the character of γ_i to the polynomial $z(t)_i$.

1. The main result - Theorem 1.13

1.1. Proofs of the main results stated here are in [**K**].
Let F be a finite subgroup of $SO(3)$ and let

$$F^* \subset SU(2) \tag{1.1}$$

be the pullback of the double covering

$$SU(2) \to SO(3).$$

The unitary dual $\widehat{SU(2)}$ of $SU(2)$ is represented by the set $\{\pi_n\}$, $n \in \mathbb{Z}_+$, where if $S(\mathbb{C}^2)$ is the symmetric algebra, then

$$\pi_n : SU(2) \to S^n(\mathbb{C}^2)$$

is the $(n+1)$-dimensional representation defined by the natural action of $SU(2)$ on $\mathbb{C}^2$. We are ultimately interested in determining how the restriction $\pi_n|F^*$ decomposes into irreducible representations of the finite subgroup F^*, for any n, and relating this determination to the structure of the simple Lie algebra corresponding to F^* by the McKay correspondence. We now recall this correspondence.

Let $\mathfrak{g}$ be a complex simple Lie algebra and let $\mathfrak{h}$ be a Cartan subalgebra of $\mathfrak{g}$. Let $\ell = rank\,\mathfrak{g}$, and if $\mathfrak{h}'$ is the dual space to $\mathfrak{h}$, let $\Delta \subset \mathfrak{h}'$ be the set of roots for the pair $(\mathfrak{g}, \mathfrak{h})$. Let W, operating in $\mathfrak{h}'$, be the Weyl group. Let Π be the set of simple positive roots with respect to a choice, Δ_+, of positive roots. If $I = \{1, \ldots, \ell\}$ we will write $\Pi = \{\alpha_i\}$, $i \in I$. We may regard Π as the set of vertices (or nodes) of the Coxeter-Dynkin diagram associated with $\mathfrak{g}$. The extended Coxeter-Dynkin diagram has an additional node α_0.

The McKay correspondence assigns to F^* a complex simple Lie algebra $\mathfrak{g} = \mu(F^*)$ of type $A - D - E$. The assignment has a number of properties: (1), the unitary dual $\widehat{F^*}$ may be parametrized by indices of the nodes of the extended Coxeter-Dynkin diagram of $\mathfrak{g}$. In particular $card\,\widehat{F^*} = \ell + 1$ and we can write $\widehat{F^*} = \{\gamma_i\}$, $i \in I_{ext} = I \cup \{0\}$. Next (2), γ_0 is the trivial 1-dimensional representation, and if $i \in I$, then

$$dim\,\gamma_i = d_i$$

where

$$\psi = \sum_{i=1}^{\ell} d_i \alpha_i$$

is the highest root in Δ. In addition (3), if γ is the two-dimensional representation defined by (1.1) and A is the $(\ell+1) \times (\ell+1)$ matrix defined so that

$$\gamma_i \otimes \gamma = \sum_{j=0}^{\ell} A_{ij} \gamma_j, \tag{1.2}$$

then C is the Cartan matrix of the extended Coxeter-Dynkin diagram of $\mathfrak{g}$ where

$$C_{ij} = 2\delta_{ij} - A_{ij}.$$

1.2. Returning to our main problem, for $i \in I_{ext}$ and $n \in \mathbb{Z}_+$, let

$$m_{n,i} = \text{multiplicity of } \gamma_i \text{ in } \pi_n | F^*$$

and introduce the generating formal power series

$$m(t)_i = \sum_{n=0}^{\infty} m_{n,i} \, t^n.$$

If $i = 0$, the determination of $m(t)_i$ is classical and is known from the theory of Kleinian singularities. That is, in this case

$$m_{n,0} = dim \left(S^n(\mathbb{C}) \right)^{F^*}.$$

In fact, let h be the Coxeter number of $\mathfrak{g}$ so that

$$\ell(h+1) = dim \, \mathfrak{g}.$$

Then there exist positive integers $a < b$ such that

$$m(t)_0 = \frac{1 + t^h}{(1 - t^a)(1 - t^b)}. \tag{1.3}$$

To define the numbers a and b in Lie theoretic terms one notes that $\mu(F^*)$ is of type D, E, or A_ℓ where ℓ is odd. In any of these cases there is a special index $i_* \in I$. If $\mu(F^*)$ is of type D or E, then α_{i_*} is the branch point of the Coxeter-Dynkin diagram of $\mathfrak{g}$. If $\mu(F^*)$ is of type A_ℓ, then α_{i_*} is the midpoint of the diagram (recall that ℓ is odd in this case).

Theorem 1.1. *One has $a = 2d_{i_*}$ and b is given by the condition that*

$$\begin{aligned} a\,b &= 2\,|F^*| \\ &= 4\,|F|. \end{aligned} \tag{1.4}$$

See Lemma 5.14 in [**K**]. The cases under consideration are characterized by the condition that h is even. We put $g = h/2$. The parity of g will play a later role.

Remark 1.2. One proves (see Lemma 5.7 in [**K**]) that b may also be given by

$$b = h + 2 - a \tag{1.5}$$

so that b, as well as a, is even.

The following table lists the various cases under consideration. In the table, Δ_n is the dihedral group of order $2n$.

F	$\mathfrak{g}$	a	b	h	g
$\mathbb{Z}_n$	A_{2n-1}	2	$2n$	$2n$	n
Δ_n	D_{n+2}	4	$2n$	$2n+2$	$n+1$
Alt_4	E_6	6	8	12	6
Sym_4	E_7	8	12	18	9
Alt_5	E_8	12	20	30	15

Proposition 1.3. *There exists a unique partition*

$$\Pi = \Pi_1 \cup \Pi_2 \tag{1.6}$$

such that if $k = 1, 2$ and $\alpha_i, \alpha_j \in \Pi_k$ where $i \neq j$, then α_i is orthogonal to α_j. Furthermore all the roots in Π_2 are orthogonal to the highest root ψ, or equivalently the root α_0 is orthogonal to all the roots in Π_2.

One has the disjoint union $I = I_1 \cup I_2$ where, if $k \in \{1, 2\}$, $\Pi_k = \{\alpha_i \mid i \in I_k\}$.

Remark 1.4. It is immediate from (1.2) that if $A_{ij} \neq 0$ and $i \in I_k$, then j is in the complement of I_k in I. Then γ_i descends to a representation of F (i.e., $\gamma_i(-1) = 1$) if and only if $k = 2$. In particular

$$m_{n,i} = 0 \text{ if } n \text{ and } k \text{ have opposite parities where } \alpha_i \in \Pi_k. \tag{1.7}$$

If $i \in I$ let $s_i \in W$ be the reflection defined by α_i so that s_i commutes with s_j if $i, j \in I_k$, $k \in \{1, 2\}$. Put $\tau_k = \prod_{i \in I_k} s_i$. Then

$$\begin{aligned} \tau_1^2 &= \tau_2^2 \\ &= identity. \end{aligned}$$

One defines a Coxeter element $\sigma \in W$ by putting

$$\sigma = \tau_2 \tau_1. \tag{1.8}$$

Remark 1.5. Every element in W is contained in a dihedral subgroup of W. Since, as one knows, the centralizer of a Coxeter element is the cyclic group (necessarily of order h) generated by the Coxeter element, a dihedral group containing the Coxeter element is unique. It is clear that τ_1 and τ_2 are in the dihedral group containing σ and, in fact, are in the complementary coset of the cyclic group generated by σ.

As an extension of (1.3) one knows (see (5.7.2) in [**K**]) that for any $i \in I$ there exists a polynomial $z(t)_i$ of degree less than h such that

$$m(t)_i = \frac{z(t)_i}{(1 - t^a)(1 - t^b)} \tag{1.9}$$

so that $m(t)_i$ is known as soon as one knows the polynomial $z(t)_i$.

Remark 1.6. Note that by (1.6) and evenness of a and b (Remark 1.2) one must have that the only powers of t which have a nonzero coefficient, are odd if $i \in I_1$, and even if $i \in I_2$.

Example 1.7. Consider the case where F is the icosahedral group so that $\mu(F^*) = E_8$. In the listing of $z(t)_i$ below we will replace the arbitrary index i by the more informative $\{d_i\}$. Since there exist in certain cases two distinct $i, j \in I$ such that $dim\, \gamma_i = dim\, \gamma_j$, we will write $\{d_j\}$ for j when the "distance" of α_j to α_0 is greater than the "distance" of α_i to α_0. Note that $d_{i_*} = 6$.

$$
\begin{aligned}
z(t)_{\{2\}} &= t + t^{11} + t^{19} + t^{29} \\
z(t)_{\{3\}} &= t^2 + t^{10} + t^{12} + t^{18} + t^{20} + t^{28} \\
z(t)_{\{4\}} &= t^3 + t^9 + t^{11} + t^{13} + t^{17} + t^{19} + t^{21} + t^{27} \\
z(t)_{\{5\}} &= t^4 + t^8 + t^{10} + t^{12} + t^{14} + t^{16} + t^{18} + t^{20} + t^{22} + t^{26} \\
z(t)_{\{6\}} &= t^5 + t^7 + t^9 + t^{11} + t^{13} + 2\,t^{15} + t^{17} + t^{19} + t^{21} + t^{23} + t^{25} \\
z(t)_{\underline{\{4\}}} &= t^6 + t^8 + t^{12} + t^{14} + t^{16} + t^{18} + t^{22} + t^{24} \\
z(t)_{\underline{\{2\}}} &= t^7 + t^{13} + t^{17} + t^{23} \\
z(t)_{\underline{\{3\}}} &= t^6 + t^{10} + t^{14} + t^{16} + t^{20} + t^{24}
\end{aligned}
$$

We now modify Π by defining

$$
\widetilde{\Pi} = \{\beta_i \mid i \in I\} \tag{1.10}
$$

where $\beta_i = \alpha_i$ if $i \in I_1$ and $\beta_i = -\alpha_i$ if $i \in I_2$. Let $Z \subset W$ be the cyclic group generated by the Coxeter element σ. Recall $(h+1)\,\ell$ so that

$$
\operatorname{card} \Delta = h\,\ell. \tag{1.11}
$$

We have shown that σ has ℓ orbits in Δ, each with h-elements, and that each orbit contains a unique element of $\widetilde{\Pi}$. That is, one has

Theorem 1.8. *For any $i \in I$ the σ-orbit $Z \cdot \beta_i$ has h elements and one has the disjoint union*

$$
\Delta = \sqcup_{i=1}^{\ell} Z \cdot \beta_i. \tag{1.12}
$$

This result is readily proved using (6.9.2) in [**K**].

For any $i \in I$ let $(Z \cdot \beta_i)_+ = \Delta_+ \cap Z \cdot \beta_i$. One has (see (0.5))

$$
\Delta_+ = g\,\ell. \tag{1.13}
$$

Theorem 1.9. *For any $i \in I$ one has $\operatorname{card}(Z \cdot \beta_i)_+ = g$ and the disjoint union*

$$
\Delta_+ = \sqcup_{i \in I} (Z \cdot \beta_i)_+. \tag{1.14}
$$

From (5.6.2) in [**K**] we deduce that (see (0.5))

$$
\alpha_{i_*} \in \Pi_2 \text{ if } g \text{ is even and } \alpha_{i_*} \in \Pi_1 \text{ if } g \text{ is odd.} \tag{1.15}
$$

Let κ be the long element of the Weyl group. One has (see Lemma 4.9 in [**K**]) the following result of Steinberg:

$$
\sigma^g = \kappa \tag{1.16}
$$

so that $\kappa \in Z$.

Remark 1.10. Recall that ψ is the highest root. It is a consequence of (5.6.2) in [**K**] that one has ψ and β_{i_*} are in the same σ- orbit. In fact, if g is odd, then

$$
\begin{aligned}
\sigma^{\frac{g-1}{2}}(\psi) &= \beta_{i_*} \\
&= \alpha_{i_*}
\end{aligned} \tag{1.17}
$$

and if g is even, then

$$\sigma^{\frac{g}{2}}(\psi) = \beta_{i_*}$$
$$= -\alpha_{i_*}. \tag{1.18}$$

One easily has that σ^g commutes with τ_1 and τ_2 so that, for $k \in \{1, 2\}$,

$$\sigma^g(\Pi_k) = -(\Pi_k). \tag{1.19}$$

Furthermore, since $\kappa(\psi) = -\psi$ one has that

$$\sigma^g(\alpha_{i_*}) = -\alpha_{i_*} \tag{1.20}$$

so that in any case

$$\psi \text{ and } \alpha_{i_*} \text{ lie in the same } \sigma\text{-orbit.} \tag{1.21}$$

1.3. We come now to the main result—the determination of $z(t)_i$ in terms of the orbit structure of σ on Δ. For any $\varphi \in \Delta_+$ let $i_\varphi \in I$ be defined so that (by Theorem 1.9)

$$\varphi \in (Z \cdot \beta_{i_\varphi})_+. \tag{1.22}$$

But then there exists $k_\varphi \in \{1, 2\}$ such that

$$i_\varphi \in I_{k_\varphi}. \tag{1.23}$$

The following result follows from (6.9.2) in [**K**].

Theorem 1.11. *Let $\varphi \in \Delta_+$. Then there exists a unique positive integer $n(\varphi)$ where $1 \le n(\varphi) \le h$ with the same parity as k_φ such that if $k_\varphi = 1$ then*

$$\sigma^{\frac{n(\varphi)-1}{2}}(\varphi) = \beta_{i_\varphi}. \tag{1.24}$$

If $k_\varphi = 2$ then

$$\sigma^{\frac{n(\varphi)}{2}}(\varphi) = \beta_{i_\varphi}. \tag{1.25}$$

One also has (see Remark 6.10 in [**K**])

Theorem 1.12. *For any $i \in I_1$ the map*

$$(Z \cdot \beta_i)_+ \to \{0, 1, \ldots, g - 1\}, \qquad \varphi \mapsto \frac{n(\varphi) - 1}{2} \tag{1.26}$$

is a bijection and for any $i \in I_2$ the map

$$(Z \cdot \beta_i)_+ \to \{1, \ldots, g\}, \qquad \varphi \mapsto \frac{n(\varphi)}{2} \tag{1.27}$$

is a bijection.

Let (φ, φ') be the restriction to Δ of the W-invariant bilinear form on $\mathfrak{h}'$ induced by the Killing form on $\mathfrak{g}$. Let $\Phi = \{\varphi \in \Delta \mid (\psi, \varphi) > 0\}$. One easily has that $\Phi \subset \Delta_+$. Obviously $\psi \in \Phi$. One has

$$\operatorname{card} \Phi = 2h - 3. \tag{1.28}$$

Because of its connection with a Heisenberg subalgebra of $\mathfrak{g}$ we refer to Φ as the Heisenberg subsystem of Δ_+. For $i \in I$ let $\Phi^i = \Phi \cap (Z \cdot \beta_i)_+$. Our main result is

Theorem 1.13. *Let* $i \in I - \{i_*\}$. *Then*

$$z(t)_i = \sum_{\varphi \in \Phi^i} t^{n(\varphi)}. \tag{1.29}$$

Furthermore

$$card\, \Phi^i = 2d_i. \tag{1.30}$$

In addition all the coefficients of $z(t)_i$ *are either* 1 *or* 0 *so that*

$$z(1)_i = 2\, d_i. \tag{1.31}$$

For $i = i_*$ *one has*

$$z(t)_{i_*} = 2\, t^g + \sum_{\varphi \in \Phi^{i*},\, \varphi \neq \psi} t^{n(\varphi)}. \tag{1.32}$$

In addition the coefficient of t^g *is 2 and all the other coefficients of* $z(t)_{i_*}$ *are either* 0 *or* 1. *One also has*

$$\begin{aligned} z(1)_{i_*} &= 2\, d_{i_*} \\ &= a. \end{aligned} \tag{1.33}$$

Finally

$$z(t)_{i_*} = t^{g-a+2} + t^{g-a+4} + \cdots + t^{g-2} + 2\, t^g + t^{g+2} + \cdots + t^{g+a-4} + t^{g+a-2}. \tag{1.34}$$

Theorem 1.13 combines Theorem 6.6 and Lemma 6.14 in [**K**]. We note also that the expression (1.32) for $z(t)_{i_*}$ in Theorem 1.13 follows from the proof of Theorem 6.6 in [**K**] (see especially (5.8.1) in [**K**]).

References

[G-S,V]   Gonzalez-Sprinberg, G.; Verdier, J.-L. *Construction géométrique de la correspondance de McKay.* Ann. Sci. Ecole Norm. Sup. **16**, no.3 (1983), 410–449.

[K]   Kostant, B. *The McKay correspondence, the Coxeter element and representation theory.* In *Élie Cartan et les mathématiques d'aujourd'hui* (Lyon 1984). Astérisque, hors série, (1985), 209–255.

[L1]   Lusztig, G. *Some examples of square integrable representations of semisimple p-adic groups.* Trans. AMS **277** (1983), 153–215.

[L2]   Lusztig, G. *Subregular nilpotent elements and bases in K-theory.* Canad. J. Math. **51** (6) (1999), 1194–1225.

[M]   McKay, J. *Graphs, singularities and finite groups.* Proc. Symp. Pure Math. **37** (1980), 183–186.

[R]   Rossmann, W. *McKay's correspondence and characters of finite subgroups of SU(2). Noncommutative Harmonic Analysis, in honor of Jacques Carmona,* Prog. in Math. **220** (2004), Birkhäuser, 441–458.

[Sp]   Springer, T. *Poincaré series of binary polyhedral groups and McKay's correspondence.* Math. Ann. **278** (1985), 587–598.

[St]   Steinberg, R. *Finite subgroups of* SU_2, *Dynkin diagrams and affine Coxeter elememts.* Pac. J. Math. **118** (1985), 587–598, Preprint 1982.

DEPARTMENT OF MATHEMATICS, MASSACHUSETTS INSTITUTE OF TECHNOLOGY, CAMBRIDGE, MA 02139 USA

E-mail address: kostant@math.mit.edu

Hyperbolic Coxeter Groups and Space Forms

Ruth Kellerhals

ABSTRACT. We give a short survey of hyperbolic Coxeter groups and their quotient spaces serving as important prototypes for hyperbolic space forms. We present some volume extremality and structural results for hyperbolic space forms.

Acknowledgement. We would like to thank the organisers for the invitation to speak at the conference *The Coxeter Legacy: Reflections and Projections* and to report on some aspects of the fundamental work of H.S.M. Coxeter.

1. Hyperbolic space forms

1.1. Standard hyperbolic space. Let H^n denote the standard hyperbolic n-space, that is, the (up to isometry) unique simply connected Riemannian manifold of constant sectional curvature -1. Its isometry group $Iso(H^n)$ is a topological group. In the sequel we mainly consider H^n although much of the material holds in an appropriate way also for the euclidean space E^n resp. the sphere S^n of constant sectional curvature 0 resp. $+1$.

Realise H^n in the upper half space $E_+^n = \{ x = (x_1, \ldots, x_n) \in E^n \,|\, x_n > 0 \}$ equipped with the hyperbolic metric $ds^2 = |dx|^2/x_n^2$ (cf. [23]; see Figure 1). In particular, the distance between two points $x, y \in H^n$ lying on the n-axis is given by $|\log \frac{x}{y}|$. Geodesics are either open semicircles or vertical lines orthogonal to $E^{n-1} = \{ x_n = 0 \}$. A similar picture holds for hyperplanes. The set $\partial H^n = \widehat{E}^{n-1} = E^{n-1} \cup \{\infty\}$ is called the ideal boundary of H^n.

In this model, the group $Iso(H^n)$ of isometries consists of Möbius transformations leaving invariant E_+^n. By Poincaré extension, it is isomorphic to the group of Möbius transformations of $\widehat{E}^{n-1}$.

Consider the real Clifford algebra C_{n-2} with generators $i_1, \ldots, i_{n-2}$ different from 1 together with its well-known involutions such as $I = i_{k_1} \cdots i_{k_r} \mapsto I^* = i_{k_r} \cdots i_{k_1}$. The Clifford group G_{n-2} denotes the multiplicative group of products of non-zero vectors in C_{n-2}. Identify the subset of all Clifford vectors with the euclidean $(n-1)$-space and H^n with $\{ x = a_0 + a_1 i_1 + \cdots + a_{n-2} i_{n-2} + t i_{n-1} \,|\, t > 0 \}$. Then, by a result of Ahlfors (for references, cf. [19]), the group of direct Möbius

---

2000 *Mathematics Subject Classification.* Primary 51M20, 51M16; Secondary 57S30.
Partially supported by the Swiss National Science Foundation 20-67619.02.

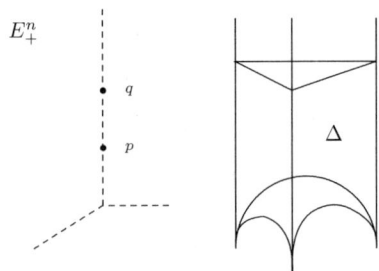

FIGURE 1. Upper half space and an ideal tetrahedron Δ

transformations $Iso^+(H^n)$ is isomorphic to $PSL(2, C_{n-2})$, the projective linear group of Clifford matrices with entries $a, b, c, d \in G_{n-2} \cup \{0\}$ such that $a^*b, c^*d, c^*a, d^*b \in E^{n-1}$ and with pseudo-determinant $ad^* - bc^* = 1$ acting by fractional linear transformations on the ideal boundary of H^n. In particular, $Iso^+(H^3) \cong PSL(2, \mathbb{C})$.

According to the fixed point behavior a Möbius transformation is either elliptic, parabolic, or loxodromic. For example, if $\gamma \in Iso(H^n)$ has precisely one resp. two fixed points in $\widehat{E}^{n-1}$ and none in E^n_+, then γ is parabolic resp. loxodromic. Suppose that the parabolic element γ fixes the point ∞ and consider the subspaces $S_r(\infty) = \{x \in H^n \,|\, x_n = t = r > 0\}$ of H^n. $S_r(\infty)$ inherits a euclidean metric and is called a horosphere based at ∞. Moreover, γ acts as euclidean isometry on each such horosphere. A subgroup $\Gamma \subset Iso(H^n)$ is elementary if Γ has a finite orbit in $\overline{H^n}$. It is said to be of elliptic, parabolic resp. loxodromic type if Γ has a finite orbit in H^n, fixes precisely one ideal point and has no other finite orbit in $\overline{H^n}$ resp. is neither elliptic nor parabolic. These notions are conjugacy invariant characterizations.

Another important frame for hyperbolic geometry is the Lorentz-Minkowski space $E^{n,1}$ equipped with the bilinear form $<x, y> = x_1 y_1 + \cdots + x_n y_n - x_{n+1} y_{n+1}$ of signature $(n, 1)$ (cf. [23]). Then, H^n can be interpreted as vector subset $\{x \in E^{n,1} \,|\, <x, x> = -1, \; x_{n+1} > 0\}$. In this setting, hyperplanes H in H^n arise as orthogonal complements of normed spacelike vectors e, that is, $H = e^\perp$ with $<e, e> = 1$. Let $P \subset H^n$ denote a convex polytope bounded by finitely many hyperplanes $H_i = e_i^\perp$, $i \in I$, with e_i directed inwards to P. Assume that P is acute-angled (all non-right dihedral angles are strictly less than $\pi/2$) of finite volume. Then, the Gram matrix $G(P) = (<e_i, e_k>)_{i,k \in I}$ of P is an indecomposable symmetric matrix of signature $(n, 1)$ with entries on the diagonal equal to 1 and beside the diagonal equal to the following geometrical entities (cf. [26]).

$$
- <e_i, e_k> = \begin{cases} 0 & \text{if } H_i \perp H_k \\ \cos \alpha_{ik} & \text{if } H_i, H_k \text{ intersect on } P \text{ at the angle } \alpha_{ik} \\ 1 & \text{if } H_i, H_k \text{ are parallel} \\ \cosh l_{ik} & \text{if } H_i, H_k \text{ admit a common perpendicular of length } l_{ik} \end{cases}
$$

Of particular interest are simplices with many right dihedral angles. An orthoscheme $R = R(\alpha_1, \ldots, \alpha_n)$ in S^n, E^n or H^n is an n-simplex bounded by hyperplanes $H_0, \ldots, H_n$ with Gram matrix $G(R)$ consisting of elements

$$< e_i, e_k >= \begin{cases} 1 & \text{if } i = k \\ -\cos \alpha_i & \text{if } i = k + 1 \\ 0 & \text{else} \end{cases}$$

Orthoschemes were first studied by L. Schläfli (cf. [9], for example). They generalise the notion of right triangle to higher dimensions in a natural way. Let p_k denote the vertex of an orthoscheme R opposite to H_k for $0 \leq k \leq n$. Then, $p_0 p_1, \ldots, p_{n-1} p_n$ form an orthogonal edge path joining p_0 with p_n. Therefore, in H^n, only p_0 and p_n may be ideal points. Orthoschemes appear in different contexts and especially in connexion with regular polytopes (cf. [10]). A regular polytope $P = P_{reg}$ is a convex n-polytope such that the symmetry group $Sym(P)$ acts transitively on its j-dimensional faces ($0 \leq j \leq n - 1$). $Sym(P)$ permutes the vertices of P and has a unique fixed point, the centre of P. It is the centre of the insphere and of the circumsphere of P. As a consequence of the definition, all faces of P are identical regular polytopes and all vertex figures (each one arises as intersection of P with a sufficiently small sphere centered at a vertex) are congruent regular polytopes of codimension one. Let $\{r\}$ denote a regular r-gon. One can associate with P a generalised symbol by induction as follows. The Schläfli symbol of a regular polytope P with face $\{r_1, \ldots, r_{n-2}\}$ and vertex figure $\{r_2, \ldots, r_{n-1}\}$ is defined by $\{r_1, \ldots, r_{n-1}\}$. As an example, $\{3, \ldots, 3\}$ is the Schläfli symbol of a regular simplex. The 120-cell, a terminology introduced by Coxeter, is a regular 4-polytope P with Schläfli symbol $\{5, 3, 3\}$. Its 120 faces are regular dodecahedra with Schläfli symbol $\{5, 3\}$.

Consider a regular n-polytope $P = \{r_1, \ldots, r_{n-1}\}$. Draw successively perpendiculars from the centre p_n of P to its faces. This decomposes P into isometric n-orthoschemes as follows. Let F^{j-1} be a face of the j-dimensional face $F^j \subset P$, $1 \leq j \leq n - 1$. The centres $F_0 =: p_0, p_1, \ldots, p_n$ of the flag $\{F^0, F^1, \ldots, F^{n-1}, F^n := P\}$ form the characteristic orthoscheme R of P. In fact, P decomposes into $\operatorname{ord} Sym(P)$ orthoschemes all isometric to R, and R is a fundamental domain for the action of $Sym(P)$ on P.

1.2. Hyperbolic space forms.

Let Γ denote a discrete subgroup of $Iso(H^n)$. Then, by Dirichlet's construction, Γ has a convex fundamental polytope P. The orbit space $Q = H^n/\Gamma$ modelled on P is a complete hyperbolic n-space form. If Γ possesses elliptic elements, Q is a hyperbolic orbifold with singularities. If Γ is without elliptic elements, then the quotient H^n/Γ is a hyperbolic manifold usually denoted by M.

Well-known examples for hyperbolic 2-manifolds are Riemannian surfaces of genus > 1. There are several different constructions of hyperbolic 3-space forms (cf. [20]). Many knot and link complements on the sphere S^3 yield hyperbolic 3-manifolds. There are arithmetic constructions of hyperbolic 3-space forms by embedding matrix groups with coefficients in the ring of integers of a number field into $PSL(2, \mathbb{C})$. By Poincaré's fundamental polyhedron theorem, a method to construct hyperbolic n-space forms consists of identifying suitably the faces of a hyperbolic n-polytope satisfying certain cycle relations. For example, a regular dodecahedron $\{5, 3\}$ with dihedral angle $2\pi/5$ gives rise to the compact hyperbolic 3-manifold of Weber-Seifert. The Gieseking manifold is the non-compact hyperbolic 3-manifold of finite volume related to an ideal regular tetrahedron $\{3, 3\}$ with

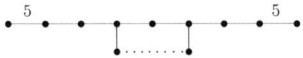

FIGURE 2. Bugaenko's compact hyperbolic Coxeter polytope in H^8

dihedral angle $\pi/3$. By a similar procedure, the 120-cell $\{5,3,3\}$ with dihedral angle $2\pi/5$ gives rise to the compact hyperbolic 4-manifold of Davis [**12**].

2. Hyperbolic Coxeter groups

2.1. Definitions and results. The most basic class of hyperbolic n-space forms arises as quotients by hyperbolic Coxeter groups or by suitable subgroups. An elliptic, parabolic or hyperbolic Coxeter group Γ_C is a discrete group generated by finitely many reflexions R in hyperplanes H of S^n, E^n or H^n subject to relations

$$R^2 = 1 \quad ; \quad (RR')^p = 1 \quad \text{if } \angle(H, H') = \pi/p \quad \text{for } p > 1 .$$

A convex fundamental polytope of Γ_C is called a Coxeter polytope. Its dihedral angles are of the form π/p for integers $p \geq 2$. As an example, the characteristic orthoscheme of a regular n-polytope $P = \{r_1, \ldots, r_{n-1}\}$ is a Coxeter orthoscheme R_C giving rise to a particularly nice Coxeter group Γ_C. This is best illustrated by means of the Coxeter diagram. A Coxeter diagram Σ of a Coxeter group Γ_C or its Coxeter polytope P_C is a labelled graph whose nodes i, k correspond to generators R_i, R_k of Γ_C. Suppose that $(R_i R_k)^p = 1$. If $p = 2$, the nodes i, k are not connected. For $p = 3$ resp. for $p \geq 4$, they are connected by a single edge resp. by an edge marked p. Notice that hyperbolic Coxeter diagrams of finite covolume are connected graphs (cf. §1.1).

Remark. In a similar way, one can associate with any convex polytope in S^n, E^n or H^n bounded by finitely many hyperplanes H_i, $i \in I$, a weighted graph called elliptic, parabolic or hyperbolic. Label the edges connecting the nodes i, k with α_{ik} resp. l_{ik} or mark them ∞ otherwise (cf. §1.1).

As an example, consider the Coxeter diagram in Figure 2. The dotted edge indicates that the corresponding mirror hyperplanes do not intersect and have a common perpendicular whose length is not specified. The associated Gram matrix G is of order 11 and of signature $(8, 1)$. It describes a compact Coxeter polytope P_C in H^8 discovered by V. Bugaenko in 1992 using arithmetic methods. The combinatorial properties of P_C can be read off from the principal submatrices of G (cf. [**26**], part II).

2.2. The classification problem. (For references, compare part II of [**26**] and [**10**]). The elliptic and parabolic Coxeter groups were classified by Coxeter in 1934 [**6**]. Hyperbolic Coxeter groups, however, are only partially classified. Well known are the triangle groups denoted by (p, q, r) which are generated by the reflexions in the sides of a triangle with angles $\pi/p, \pi/q, \pi/r$ satisfying the existence condition $1/p + 1/q + 1/r < 1$. The list of all hyperbolic Coxeter orthoschemes (and regular polytopes) is surprisingly short. They exist only up to dimension five (cf. Figure 3). Hyperbolic Coxeter simplices exist only up to dimension nine.

More generally, there are non-existence bounds due to E. Vinberg resp. M. Prokhorov and A. Khovanskij saying that cocompact resp. finite covolume Coxeter groups do not exist in H^n for $n \geq 30$ resp. $n \geq 996$. The top-dimensional Coxeter

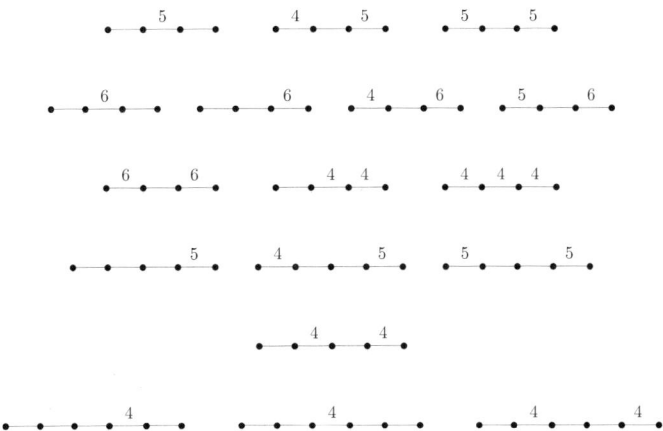

FIGURE 3. Hyperbolic Coxeter n-orthoschemes for $n \geq 3$

group Γ_* known up to the present acts on H^{21} with finite covolume and was constructed by R. Borcherds in 1987 by arithmetic means. Its fundamental polytope has ideal vertices and 210 faces.

3. Volume and sphere packing density

3.1. Volumes of hyperbolic orthoschemes. Consider a hyperbolic 3-orthoscheme $R = R(\alpha_1, \alpha_2, \alpha_3)$ and define the additional parameter $\theta \in [0, \pi/2[$ by

$$\tan \theta = \frac{\sqrt{\cos^2 \alpha_2 - \sin^2 \alpha_1 \sin^2 \alpha_3}}{\cos \alpha_1 \cos \alpha_3} \; .$$

Then, N. Lobachevsky's formula expresses the volume of R in terms of the dilogarithm function

$$\text{Л}(x) = -\int_0^x \log|2 \sin t| \, dt = \frac{1}{2} \sum_{k=1}^{\infty} \frac{\sin(2kx)}{k^2}$$

as follows (cf. [**7**], [**14**], for example).

$$\text{vol}_3(R) = \frac{1}{4} \Big\{ \text{Л}(\alpha_1 + \theta) - \text{Л}(\alpha_1 - \theta) + \text{Л}(\frac{\pi}{2} + \alpha_2 - \theta) + \text{Л}(\frac{\pi}{2} - \alpha_2 - \theta) + $$
$$ + \text{Л}(\alpha_3 + \theta) - \text{Л}(\alpha_3 - \theta) + 2\text{Л}(\frac{\pi}{2} - \theta) \Big\} \; .$$

It follows that $\text{vol}_3(\bullet\!\!-\!\!\bullet\!\!-\!\!\bullet\overset{6}{-}\!\!\bullet) = \frac{1}{8} \text{Л}(\frac{\pi}{3}) \simeq 0.0423$.

Another application is J. Milnor's formula for the volume of an ideal tetrahedron Δ. Its dihedral angles at opposite edges are equal so that Δ is characterised by at most three different angles satisfying the (horospherical) identity $\alpha + \beta + \gamma = \pi$. Then,

$$\text{vol}_3(\Delta) = \text{Л}(\alpha) + \text{Л}(\beta) + \text{Л}(\gamma) \; .$$

Next, consider orthoschemes $R = R(\alpha_1, \ldots, \alpha_{2n})$ in H^{2n} or S^{2n} and associate with R its graph Σ with weights $\alpha_1, \ldots, \alpha_{2n}$. In order to express L. Schläfli's volume reduction formula (cf. [**15**], for example), normalize the volume functional on S^m resp. H^m of curvature $K = +1$ resp. $K = -1$ as follows.

FIGURE 4. The Coxeter group Γ_{fcc}

$$f_m^{\pm} := K^{m/2} \frac{2^{m+1}}{\text{vol}_m(S^m)} \text{vol}_m \quad , \quad f_0 := 1 .$$

Then, the reduction formula expresses the volume of a $2n$-dimensional orthoscheme R with graph Σ in the following way (cf. [15]).

$$f_{2n}^{\pm}(\Sigma) = \sum_{k=0}^{n} \frac{(-1)^k}{k+1} \binom{2k}{k} \sum_{\sigma} f_{2n-(2k+1)}^{+}(\sigma) \quad , \quad \sum f_{-1}^{+} = 1,$$

where σ runs through all elliptic subgroups of order $2(n-k)$ of Σ all of whose components are of even order.

It follows that $\text{vol}_4(\overset{5}{\bullet\!-\!\bullet}\ \bullet\ \overset{5}{\bullet\!-\!\bullet}) = \frac{13\pi^2}{5,400}$ (cf. also [15]). By a simple dissection argument, the above formula allows to identify the covolume of the Coxeter group Γ_{fcc} by $\frac{\pi^2}{1,440}$ (see Figure 4).

Remark. Consider an arbitrary hyperbolic polytope P of even dimension. In [29], T. Zehrt obtained a volume formula for P in terms of the combinatorial data and the dihedral angles. In particular, the covolume of an arbitrary discrete group $\Gamma < Iso(H^4)$ can be expressed by means of a fundamental polytope $P_\Gamma \subset H^4$ as follows. Denote by Ω^d the set of all d-dimensional faces of P_Γ where $0 \le d \le 3$. Γ induces an equivalence relation on each set Ω^d in an obvious way. Let μ^d be the number of equivalence classes in Ω^d. For $i = 1, \ldots, \mu^d$, denote by $\Omega_{(i)}^d$ an equivalence class and let l_i^d be the number of its elements. Denote by $a^0(\Omega_{(i)}^2)$ the number of vertices of an element in $\Omega_{(i)}^2$. Furthermore, let g_i^d be the order of the stabilizer of elements in the class $\Omega_{(i)}^d$ in Γ. Then,

$$\frac{3}{4\pi^2} K^2 \text{vol}_4(P_\Gamma)$$

$$= \sum_{i=1}^{\mu^0} \left(\frac{1}{g_i^0} - \frac{1}{2} l_i^0 \right) + \sum_{i=1}^{\mu^2} \left(1 - \frac{1}{2} a^0(\Omega_{(i)}^2) \right) \cdot \left(\frac{1}{g_i^2} - \frac{1}{2} l_i^2 \right) + 1 - \frac{1}{2} a_{inf}^0(P_\Gamma) ,$$

where $K = -1$ and $a_{inf}^0(P_\Gamma)$ denotes the number of ideal vertices of P_Γ. This formula generalises C. L. Siegel's result [24] about the area of fundamental polygons P_Γ for discrete subgroups Γ of $Iso(H^2)$ which says that

$$\frac{1}{2\pi} K \text{vol}_2(P_\Gamma) = \sum_{i=1}^{\mu^0} \left(\frac{1}{g_i^0} - \frac{1}{2} l_i^0 \right) + 1 - \frac{1}{2} a_{inf}^0(P_\Gamma) .$$

This formula allows to identify the triangle group $(2, 3, 7)$ as the unique minimal covolume discrete group acting on H^2.

3.2. Density of sphere packings. Let $\mathcal{B}$ be a packing of S^n, E^n or H^n with r-balls B (for balls in S^n, we assume that $r < \pi/4$).

Associate to each ball B of the packing $\mathcal{B}$ its Dirichlet-Voronoï cell D consisting of all points closer to B than to any other ball of $\mathcal{B}$. Consider the local density

n	$d_n \simeq$
2	0.90690
3	0.77964
4	0.64782
5	0.52571
6	0.41924
7	0.32999

FIGURE 5. Euclidean simplicial density d_n

of B in D given by $ld(B, D) = \mathrm{vol}_n(B)/\mathrm{vol}_n(D) < 1$. More precisely, the local density can be estimated from above by the simplicial density function $d_n(r)$. For its definition, consider $n+1$ r-balls B mutually touching one another. Their centers give rise to a regular n-simplex S_{reg} of edge length $2r$ and of dihedral angle 2α, say. In the hyperbolic case, for example, these parameters are related according to

$$\frac{1}{\cos(2\alpha)} = n - 1 + \frac{1}{\cosh(2r)} \quad .$$

Now, the simplicial density function is given by

$$d_n(r) = (n+1) \, \frac{\mathrm{vol}_n(B \cap S_{reg})}{\mathrm{vol}_n(S_{reg})} \quad .$$

In the euclidean case, the simplicial density function $d_n(r)$ does not depend on r, and we write $d_n = d_n(r)$. Indeed, one can interpret d_n as limiting density $d_n = \lim_{r \to 0} d_n(r)$ on H^n, for example, by looking at the curvature dependence of the volume element for H^n. As an example, one easily sees that

$$d_2 = \frac{\pi}{2\sqrt{3}} \simeq 0.90690 \ .$$

By a well-known result of A. Thue, this value is the maximal density for disc packings of E^2, and it is attained by the packing associated with the root lattice A_2 (cf. also [8], p. 264).

By results of Coxeter, C. Rogers and K. Böröczky (cf. [2], [8]), the local density can be estimated as follows.

$$ld_n(B, D) \le d_n(r) \quad , \quad \forall \ B \in \mathcal{B} \ .$$

This estimate is sharp if D forms a regular polytope. By means of orthoscheme trigonometry, one can derive an explicit formula for the simplicial density function on E^n, for example. In [16], the following result and table (cf. Figure 5) are elaborated. Let $S_0 \subset S^{n-1}$ denote a regular simplex of dihedral angle $2\alpha_0^n = \arccos \frac{1}{n}$. Then,

$$d_n = \frac{1}{n} \cdot \prod_{k=2}^{n} \left(\frac{k+1}{k-1} \right)^{\frac{n-k+1}{2}} \cdot \mathrm{vol}_{n-1}(S_0) \ .$$

In the hyperbolic case, we allow also arrangements $\mathcal{B}_\infty$ by horoballs B_∞ of infinite radius which can be seen as arrangements by a half space $\{ x_n > c \}$ and by euclidean balls of diameter $< c$ touching the boundary $\partial H^n = \{ x_n = 0 \}$ of $H^n \subset E^n_+$ for some $c > 0$. Consider a horoball B_∞ and a point $p \in H^n$. Then, $\mathrm{dist}(p, B_\infty)$ is defined to be the length of the unique perpendicular from p to the

n	$d_n(\infty) \simeq$
2	0.95493
3	0.85328
4	0.73046
5	0.60695
6	0.49339
7	0.39441
8	0.31114
9	0.24285
10	0.18789

FIGURE 6. Simplicial horoball density $d_n(\infty)$

horosphere S_∞ bounding B_∞, where $\mathrm{dist}(p, B_\infty)$ is taken negative for $p \in B_\infty$. The Dirichlet-Voronoĭ cell of B_∞ is defined to be the convex body

$$D_\infty = D(B_\infty) = \{\, p \in H^n \mid \mathrm{dist}(p, B_\infty) \leq \mathrm{dist}(p, B'_\infty) \,,\ \forall B'_\infty \in \mathcal{B}_\infty \,\} \,.$$

Since both, B_∞ and D_∞ are of infinite volume, the notion of local density has to be modified. Let $q \in \partial H^n$ denote the base point of B_∞, and interpret S_∞ as euclidean $(n-1)$-space. Let $K_{n-1}(R) \subset S_\infty$ be a ball with center $c \in S_\infty$. Then, $q \in \partial H^n$ and $K_{n-1}(R)$ determine a convex cone $C_n(R) := \mathrm{cone}_q(K_{n-1}(R)) \subset \overline{H^n}$ with apex q consisting of all hyperbolic geodesics through $K_{n-1}(R)$ asymptotic to q. With these preparations, the local density of B_∞ with respect to D_∞ is defined by

$$ld(B_\infty, D_\infty) := \limsup_{R \to \infty} \frac{\mathrm{vol}_n(B_\infty \cap C_n(R))}{\mathrm{vol}_n(D_\infty \cap C_n(R))} \quad,$$

and it is independent of the choice of the center c of $K_{n-1}(R)$. By analytical continuation, the simplicial density function can be extended easily to the case $r = \infty$. Consider $n+1$ horoballs B_∞ which are mutually tangent. The convex hull of their base points at infinity is an ideal regular simplex $S_{reg}^\infty \subset \overline{H^n}$ with dihedral angle given by $2\alpha_\infty^n = \arccos \frac{1}{n-1}$. Hence, it is natural to write

$$d_n(\infty) = (n+1) \frac{\mathrm{vol}_n(B_\infty \cap S_{reg}^\infty)}{\mathrm{vol}_n(S_{reg}^\infty)} \,.$$

One easily sees that $d_2(\infty) = 3/\pi$ (cf. [8], p. 169). In [3], one finds the formula

$$d_3(\infty) = \left(\frac{1}{1^2} + \frac{1}{2^2} - \frac{1}{4^2} - \frac{1}{5^2} + \frac{1}{7^2} + \frac{1}{8^2} - + \ldots \right)^{-1} \,.$$

In general, there is the following relation between $d_n(\infty)$ and the volume μ_n of an ideal regular hyperbolic n-simplex (cf. [16]).

$$d_n(\infty) = \frac{n+1}{n-1} \cdot \frac{n}{2^{n-1}} \cdot \prod_{k=2}^{n-1} \left(\frac{k-1}{k+1} \right)^{\frac{n-k}{2}} \cdot \frac{1}{\mu_n} \,.$$

In Figure 6, we present some explicit values (cf. [16]). By means of $d_n(\infty)$, there is the following bound also due to K. Böröczky.

$$ld(B_\infty, D_\infty) \leq d_n(\infty) \quad,\quad \forall \, B_\infty \in \mathcal{B}_\infty \quad,$$

where optimality holds if D_∞ is an ideal regular polytope with in-ball B_∞.

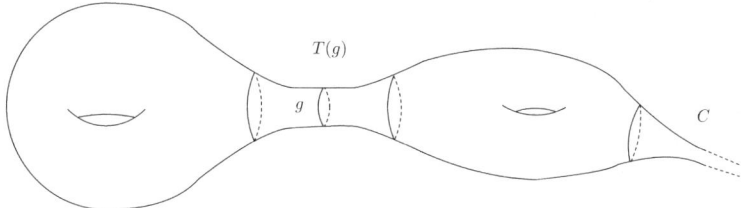

FIGURE 7. A Margulis decomposition

4. Some global results for hyperbolic space forms

4.1. The Lemma of Margulis. Let M be a finite volume hyperbolic manifold of dimension $n \geq 2$. For $\varepsilon > 0$, consider the thick and thin decomposition $M = M_{>\varepsilon} \cup M_{\leq\varepsilon}$ into the part $M_{>\varepsilon}$ where at each point p the injectivity radius $i_p(M)$ is everywhere larger than $\varepsilon/2$ and into the closure $M_{\leq\varepsilon}$ of the complementary thin part $M_{<\varepsilon}$. The famous Margulis Lemma (cf. [**25**], §4.5, for example) implies that there is a universal constant $\varepsilon(n) > 0$ with the following properties. For all $\varepsilon \leq \varepsilon(n)$, the thick part $M_{>\varepsilon}$ is compact, and the components of the thin part $M_{<\varepsilon}$ are either tubular neighborhoods $T(g)$ of simple (that is, without self-intersection) closed geodesics g of length $l(g) < \varepsilon$ homeomorphic to ball bundles over the circle or, in the non-compact case, cusp neighborhoods C homeomorphic to products of compact flat manifolds with a real half line (see Figure 7). For convenience, we call such a decomposition a Margulis decomposition and the constant $\varepsilon(n)$ the n-th Margulis constant.

Suppose for a moment that M is of finite volume with $m \geq 1$ cusps. Each cusp can be written in the form $C = C_q = V_q/\Gamma_q$ with $\Gamma_q < \Gamma$ of parabolic type fixing some point $q \in \partial H^n$ and leaving precisely invariant some horoball $V_q \subset H^n$ based at q. Among these nested cusp neighborhoods C there is a particular one, the canonical cusp C_*. For its description, realise H^n in the upper half space and suppose for simplicity that the parabolic fixed point q equals ∞. Consider the translational lattice Λ in the stabiliser Γ_∞. By Bieberbach theory, Λ is a finite index lattice of rank $n - 1$. Denote by $\mu \in \Lambda$ a shortest non-trivial vector. Then, the horoball $B_\infty(\mu) = \{\, x \in H^n \mid x_n > |\mu| \,\}$ based at ∞ is precisely invariant with respect to Γ and yields the canonical cusp $C_* = B_\infty(\mu)/\Gamma_\infty$ embedded in M. By combining this concept with certain (horo-)packing arguments the following universal volume bound can be derived (for details, cf. [**17**]).

THEOREM 4.1.1. *Let M denote a hyperbolic n-manifold of finite volume with $m \geq 1$ cusps. Let μ_n be the volume of an ideal regular hyperbolic n-simplex and Ω_{n-2} the volume of S^{n-2}. Then,*

$$\mathrm{vol}_n(M) \geq m \cdot \frac{\Omega_{n-2}}{2^{n-2} \cdot (n-1)^2 \cdot d_{n-1} \cdot d_n(\infty)} \geq m \cdot \frac{2^n}{n(n+1)} \cdot \mu_n \quad.$$

For $n > 3$, the inequalities are strict.

Example. A hyperbolic 3-manifold M with $m \geq 1$ cusps satisfies

$$\mathrm{vol}_3(M) \geq m\mu_3 \simeq m \cdot 1.01494 \quad.$$

4.2. Volume spectra of hyperbolic space forms. The Lemma of Margulis implies that there are positive minima for the sets of volumes of hyperbolic manifolds and orbifolds. Consider therefore the spectra $\mathcal{V}_n$ resp. $\mathcal{W}_n$ of all volumes of hyperbolic n-manifolds resp. n-orbifolds (up to isometry). Furthermore, introduce the corresponding subsets $\mathcal{V}_n^\infty$ resp. $\mathcal{W}_n^\infty$ by restricting to cusped space forms. For even n, the Theorem of Gauss-Bonnet-Chern says that $\mathcal{V}_n \subset \frac{1}{2}\Omega_n \mathbb{N}$. For $n > 3$, a result of H. Wang [27] implies that $\mathcal{V}_n$ is a discrete subset of $\mathbb{R}_+$. However, the situation for $n = 3$ is completely different. By a result of T. Jørgensen and W. Thurston, $\mathcal{V}_3$ is not discrete in $\mathbb{R}_+$, but finite-to-one and well-ordered of order type ω^ω (cf. [11], for example).

Since volume is one of the most important and, by Mostow-Prasad rigidity, topological invariants for M, there is some interest in the structure of the different volume spectra and, in particular, in their small parts. Consider the minima $\nu_n, \nu_n^\infty, \omega_n, \omega_n^\infty$ of the volume spectra $\mathcal{V}_n, \mathcal{V}_n^\infty, \mathcal{W}_n, \mathcal{W}_n^\infty$. The case $n = 2$ is well understood. For example, by a result of Siegel [24], $\omega_2 = \pi/42$, uniquely attained by the triangle group $(2, 3, 7)$. Moreover, by a similar argument, $\omega_2^\infty = \pi/6$, uniquely realised by the triangle group $(2, 3, \infty)$.

For $n = 3$, the situation is much more difficult. It was shown by C. Adams that $\nu_3^\infty = \mu_3 = \mathrm{vol}_3(S_{reg}^\infty(\pi/3)) \simeq 1.0149$, uniquely realised by the Gieseking manifold (cf. §4.1 and [17], for example). Moreover, R. Meyerhoff [21] showed that $\omega_3^\infty = \frac{1}{2}\text{JI}(\pi/3)$, and that this value is realised only by the quotient of H^3 modulo the Coxeter group ●—●—●⁶●. In the proof, Meyerhoff used the density bound of §3.2 motivating the result of Theorem 4.1.1. However, it is an unresolved question whether the volume of the compact hyperbolic 3-manifold M_* constructed by Weeks and by Matveev-Fomenko realises ν_3. In any case, $\nu_3 \leq \mathrm{vol}_3(M_*) \simeq 0.9427$. Also ω_3 waits for its determination.

For $n = 4$, we know that $\nu_4^\infty = 4\pi^2/3$ arising as volume of several hundreds of cusped 4-manifolds all obtained from the right-angled ideal 24-cell $\{3, 4, 3\}$ by different face gluings (cf. [16], for example). In the compact case, no example is known at present which minimises the Euler characteristic. The compact Davis manifold (cf. §1.2) has Euler characteristic 26 since it can be dissected into 14,400 Coxeter orthoschemes with diagram ●⁵●—●—●⁵●. In [5], M. Conder and C. Maclachlan construct a compact (non-orientable) hyperbolic 4-manifold of Euler characteristic 8. At the time of writing, this example yields the smallest volume of any known compact hyperbolic 4-manifold. In the orbifold case, we know that $\omega_4^\infty = \pi^2/1,440$, and that it is uniquely realised by the quotient of H^4 modulo the Coxeter group Γ_{fcc} (cf. Figure 4; [13]). Finally, M. Belolipetsky [1] showed that among all arithmetically defined compact oriented hyperbolic 4-orbifolds, the quotient of H^4 by ●—●—●—●⁵● is the orbifold of minimal volume.

4.3. An explicit Margulis decomposition. Let $n \geq 2$. Define $\nu := \left\lceil \frac{n-1}{2} \right\rceil$ and

$$c_\nu := \frac{2^{\nu+1}}{\pi^\nu} \cdot \frac{\Gamma(\frac{\nu+2}{2})^2}{\Gamma(\nu+2)} = \frac{2}{\pi^\nu} \int_0^{\pi/2} \sin^{\nu+1} t \, dt \quad .$$

Let $M = H^n/\Gamma$ denote a hyperbolic n-manifold of finite volume. For simplicity, we assume M always oriented and interpret $\Gamma < PSL(2, C_{n-2})$ (cf. §1.1). Let $\varepsilon > 0$ and consider the thick and thin decomposition $M = M_{>\varepsilon} \cup M_{\leq\varepsilon}$ of M

(cf. §4.1). We present an estimate for the n-th Margulis constant $\varepsilon(n) \geq \varepsilon_n > 0$ yielding a particular Margulis decomposition as follows (for more details, cf. [19]).

THEOREM 4.3.1. *For each* $\varepsilon \leq \varepsilon_n := \frac{c_\nu}{3^{\nu+1}}$, *the connected components of the thin part* $M_{\leq\varepsilon}$ *of* M *is a finite disjoint union of canonical cusps and tubes around simple closed geodesics of length* $l < \varepsilon$ *and radius* r *satisfying*

$$\cosh(2r) = \frac{1 - 3\kappa}{\kappa} \quad , \quad \text{where} \quad \kappa = 2(l/c_\nu)^{\frac{2}{\nu+1}} \quad .$$

A closer investigation shows that $\varepsilon_2 = 2/3 \simeq 0.6666$, $\varepsilon_3 = 1/18 \simeq 0.0555$, $\varepsilon_5 \simeq 0.0050$, ... and that ε_n is strictly decreasing with respect to n.

We describe the different ingredients of the proof (for more details, cf. [19]). Let $\varepsilon \leq \varepsilon_n$. In the first step (1), particular tubular neighborhoods T around sufficiently short simple closed geodesics and cusp neighborhoods C in $M_{\leq\varepsilon}$ are constructed. The second step (2) consists in showing that the constructed neighborhoods are mutually disjoint. The final step (3) is the verification that, for $\varepsilon \leq \varepsilon_n$, each connected component of $M_{\leq\varepsilon}$ is contained either in a tube T or in a cusp C.

(1) Suppose first that $M_{\leq\varepsilon}$ has unbounded components. Therefore, Γ contains parabolic elements. Associate to each non-trivial subgroup of parabolic type in Γ its canonical cusp $C_* \subset M$ as discussed in §4.1.

Next, consider a simple closed geodesic g of length l bounded from above by ε_n. g gives rise to a loxodromic element $\gamma \in \Gamma$ with translational length $l(g)$ along its axis a. An n-dimensional collar theorem due to C. Cao and P. Waterman [4] guarantees the existence of a non-trivial r-cylinder $C(r) = \{\, x \in H^n \mid d(x,a) < r \,\}$ with r as above and which is precisely invariant with respect to Γ. Therefore, it projects to M and yields an embedded tube $T = T_g(r)$ around g in M of radius r as required.

(2) First, one shows that canonical cusps associated to inequivalent parabolic elements are disjoint and that a canonical cusp is disjoint from a tube as constructed in (1). This can be done by using the following analogue for Clifford matrices of Jørgensen's trace inequality for $PSL(2, \mathbb{C})$ (cf. [28]). Consider

$$S = \begin{pmatrix} 1 & \mu \\ 0 & 1 \end{pmatrix}, \mu \in E^{n-1}, \quad \text{and} \quad T = \begin{pmatrix} a & b \\ c & d \end{pmatrix}$$

generating a discrete and non-elementary subgroup in $PSL(2, C_{n-2})$. Then, $|c| \cdot |\mu| \geq 1$.

In addition, the following trigonometrical lemma is needed (cf. [18]; see Figure 8).

LEMMA 4.3.1. *Let* $\gamma \in Iso(H^n)$ *be a loxodromic element with axis* a_γ, *rotational part* R *and translational length* τ. *Let* $p \in H^n$ *be such that* $p \notin a_\gamma$, *and assume that the foot of the perpendicular from* p *to* a_γ *is* $\hat{p}$. *Denote by* $\omega = \omega(p)$ *the angle at* $\hat{p}$ *in the triangle* $(p, \hat{p}, R(p))$. *Let* $d = d(p, \gamma(p))$ *and* $\delta = d(p, a_\gamma)$. *Then,* $\cos\alpha_1 \geq \cos\omega \geq \cos\alpha_\nu$, *and*

$$\cosh d = \cosh\tau + \sinh^2\delta \cdot \left(\cosh\tau - \cos\omega\right) \quad .$$

By means of this lemma, the verification that the tubes T, T' associated with distinct simple closed geodesics g, g' are disjoint can be seen roughly as follows. The geodesics g, g' in $M = H^n/\Gamma$ give rise to loxodromic elements $\gamma, \gamma' \in \Gamma$ with disjoint fixed point pairs and disjoint axes $a_\gamma, a_{\gamma'}$. Denote by $\tau, \tau' \leq \varepsilon_n < c_\nu$ the

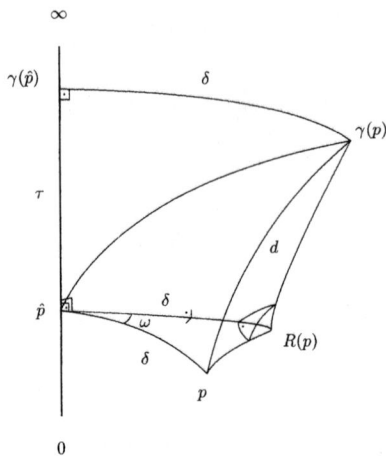

$$\text{FIGURE 8. The trigonometry of a loxodromic element } T$$

translational lengths and by $0 \leq \alpha_1, \ldots, \alpha_\nu, \alpha'_1, \ldots, \alpha'_\nu < 2\pi$ the rotation angles of γ, γ'. In fact, we can normalize $\gamma, \gamma' \in \Gamma$ so that $\kappa' \geq \kappa$, $\kappa \geq \cosh\tau - \cos\alpha_\nu =:$ $k(\gamma) = k$, and similarly for κ' and $k' = k(\gamma')$. Next, let $p \in a_\gamma$ denote the foot point on a_γ of the common perpendicular of a_γ, $a_{\gamma'}$. Then, $\delta := d(g, g') = d(a_\gamma, a_{\gamma'}) = d(p, a_{\gamma'})$. Since by hypothesis the tube T of radius r around g is embedded in M, the axis $\gamma'(a_\gamma) = a_{\gamma'\gamma\gamma'^{-1}}$ of the element $\gamma'\gamma\gamma'^{-1}$ is disjoint from a_γ, and $d := d(p, \gamma'(p)) \geq 2r$. We have to show that $\delta \geq r + r'$.

$$\begin{aligned}
\cosh(2r) \quad &\leq \cosh d \leq \cosh\tau' + \sinh^2\delta \cdot (\cosh\tau' - \cos\alpha'_\nu) = \cosh\tau' + \sinh^2\delta \cdot k' \\
&\leq k' + 1 + \sinh^2\delta \cdot k' = \cosh^2\delta \cdot k' + 1 \quad.
\end{aligned}$$

Since $k' \leq \kappa'$, we deduce that

$$\cosh(2\delta) = 2\cosh^2\delta - 1 \geq 2 \cdot \frac{\cosh(2r) - 1}{\kappa'} - 1 = \frac{1 - 4\kappa}{\kappa\kappa'} + \frac{1 - 4\kappa - \kappa\kappa'}{\kappa\kappa'} \quad.$$

But $\kappa' \geq \kappa$ so that

$$\cosh(2\delta) \geq \frac{\sqrt{1 - 4\kappa}}{\kappa} \cdot \frac{\sqrt{1 - 4\kappa'}}{\kappa'} + \frac{1 - 4\kappa - \kappa\kappa'}{\kappa\kappa'} \quad.$$

Since $l \leq \varepsilon_n = \frac{c_\nu}{3^{\nu+1}}$, $\kappa = 2\,(l(g)/c_\nu)^{\frac{2}{\nu+1}} \leq 2/9$. Therefore, $1 - 3\kappa \leq \sqrt{1 - 4\kappa}$, and

$$\cosh(2r) = \frac{1 - 3\kappa}{\kappa} \leq \frac{\sqrt{1 - 4\kappa}}{\kappa} \quad,$$

and similarly for $\cosh(2r')$. Hence,

$$\frac{\sqrt{1 - 4\kappa}}{\kappa} \cdot \frac{\sqrt{1 - 4\kappa'}}{\kappa'} \geq \cosh(2r) \cdot \cosh(2r') \quad.$$

Finally, it is easy to show that

$$\frac{1 - 4\kappa - \kappa\kappa'}{\kappa\kappa'} \geq \frac{\sqrt{1 - 4\kappa - \kappa^2}}{\kappa} \cdot \frac{\sqrt{1 - 4\kappa' - \kappa'^2}}{\kappa'} \geq \sinh(2r) \cdot \sinh(2r') \quad.$$

This means that $\cosh(2\delta) \geq \cosh(2r + 2r')$ which concludes the proof.

(3) We only prove here that each unbounded component $M_{\leq \varepsilon}$ is contained in a canonical cusp C_*. By means of conjugation, we may suppose that ∞ is a parabolic fixed point of Γ and that the translation $T(x) = x + 1$ is the element of the translation subgroup $\Lambda \subset \Gamma_\infty$ of shortest positive length. Hence, the horoball $B_\infty = \{ x \in H^n \mid x_n > 1 \}$ projects onto the canonical cusp C_* in M. Now, let C be an arbitrary cusp neighborhood in M corresponding to the fixed point ∞. C is covered by a horoball V based at ∞ which is precisely invariant with respect to Γ. This implies that $S(V) \cap V = \emptyset$ for all $S \in \Gamma - \Gamma_\infty$. Let

$$S = \begin{pmatrix} a & b \\ c & d \end{pmatrix} \in \Gamma - \Gamma_\infty$$

so that $c \neq 0$. Then, $S^{-1}(\infty) = -c^{-1}d \in E^{n-1}$. Suppose that $V \supset B_\infty$. Consider a point $x = (-c^{-1}d, x_n) \in V \setminus B_\infty$ above $S^{-1}(\infty)$ so that $0 < x_n \leq 1$. Since the subgroup $< S, T >$ of Γ generated by S and T is discrete and non-elementary, $|c| \geq 1$. This implies that x is a solution of the inequality

$$[S(x)]_n = \frac{x_n}{|cx + d|^2} \geq x_n$$

relating the n-th coordinate of $S(x)$ in terms of x_n. In fact, this inequality is equivalent to $|cx + d| \leq 1$ which, by means of $|c| \geq 1$, leads to

$$1 \geq |cx + d| = |c| \cdot |x + c^{-1}d| = |c| \cdot x_n \geq x_n \quad .$$

Therefore, we obtain a contradiction to the fact that V is precisely invariant with respect to Γ and conclude that $C \subset C_*$.

4.4. Discussion. There are several applications of the Margulis decomposition as presented above (for more results, cf. [19]). We discuss those involving injectivity radii, only. Let M be a hyperbolic manifold of dimension $n \geq 2$. From the tube construction, we see that there is a point $p \in M$ such that the injectivity radius $i_p(M)$ of M at p satisfies

$$i_p(M) > \varepsilon_n/2 = \frac{c_\nu}{2 \cdot 3^{\nu+1}} \quad , \quad \text{where} \quad c_\nu = \frac{2}{\pi^\nu} \int_0^{\pi/2} \sin^{\nu+1} t \, dt \quad .$$

Since $\sin x \geq \frac{2}{\pi} x$ for $x \in [0, \pi/2]$, one obtains

$$\frac{c_\nu}{2 \cdot 3^{\nu+1}} \geq \frac{1}{2(\nu+2)3^{\nu+1}\pi^{\nu-1}} > \frac{1}{2(\nu+2)\pi^{2\nu}} \geq \frac{1}{(n+3)\pi^{n-1}} \quad ,$$

that is,

$$i_p(M) > \frac{1}{(n+3)\,\pi^{n-1}} \quad \text{implying} \quad \mathrm{vol}_n(M) > \frac{\Omega_{n-1}}{n} \, \frac{1}{[(n+3)\pi^{n-1}]^n} \quad .$$

The first inequality can be brought into a form independent of the dimension n of M as follows.

PROPOSITION 4.4.1. *If M has a simple closed geodesic of length $\leq \varepsilon_n$, then there is a point $p \in M$ such that*

$$i_p(M) > 0.2217 \quad .$$

For the proof, consider a simple closed geodesic in M of length $l \leq \varepsilon_n$. Then, there is a tube T embedded in M of radius $r = r(l)$ satisfying $\cosh(2r) = (1-3\kappa)/\kappa$. In particular, $\cosh(2r) \geq \cosh(2r(\varepsilon_n)) = 3/2$ since $\kappa \leq \kappa(\varepsilon_n) = 2/9$. By a result of A. Przeworski [22], there is an embedded ball $B_p(\rho)$ centered at some point $p \in M$ which is of radius $\rho = \operatorname{arcsinh}(\tanh(r)/2)$. Hence, $\rho \geq \rho(\varepsilon_n) > 0.2217$ which yields the conclusion.

THEOREM 4.4.1. *Let M denote a compact hyperbolic n-manifold with diameter* $\operatorname{diam}(M) = \max\{d(p,q) \mid p,q \in M\}$ *and injectivity radius* $i(M) = \min\{i_p(M) \mid p \in M\}$. *Then,*

$$i(M) \geq \frac{c(n)}{(\sinh(\operatorname{diam}(M)))^{[\frac{n+1}{2}]}} \quad,$$

where $c(n)$ is a constant depending on the dimension n, only.

Here are the ingredients of the proof. Suppose that M has a short simple closed geodesic g of length l. Again, there is a tube T around g of radius r satisfying

$$\sinh^2 r = \frac{1}{\kappa} - 2 \quad, \quad \text{where} \quad \kappa = 2\,(l/c_\nu)^{\frac{2}{\nu+1}} \quad.$$

In particular, $\operatorname{vol}_n(M) \geq \operatorname{vol}_n(T) = \frac{2\pi}{n-1} \cdot l \cdot \sinh^{n-1} r$. Since $\sinh^{n-1} r \sim \text{const} \cdot l^{-(n-1)/(\nu+1)}$, for small l, it follows that

$$\operatorname{vol}_n(M) \geq \text{const} \cdot l^{-[\frac{n-2}{2}]/[\frac{n+1}{2}]} \quad,$$

which yields the inequality

$$i(M) \geq \frac{c'(n)}{\operatorname{vol}_n(M)^{[\frac{n+1}{2}]/[\frac{n-2}{2}]}} \quad.$$

Next, consider any simple closed geodesic g of length l in M. By a result of E. Heintze and H. Karcher (cf. [19], for example),

$$l \geq \frac{\operatorname{vol}_n(M)}{\Omega_n} \cdot \frac{2\pi}{(\sinh(\operatorname{diam}(M)))^{n-1}} \quad.$$

This implies that

$$i(M) \geq \text{const} \cdot \frac{1}{i(M)^{[\frac{n-2}{2}]/[\frac{n+1}{2}]} \cdot (\sinh(\operatorname{diam}(M)))^{n-1}} \quad,$$

that is,

$$i(M)^{1+[\frac{n+1}{2}]/[\frac{n-2}{2}]} \geq \text{const} \cdot \frac{1}{(\sinh(\operatorname{diam}(M)))^{n-1}} \quad.$$

The last step of the verification is easy.

References

[1] M. Belolipetsky, *On volumes of arithmetic quotients of $SO(1,n)$*. Preprint, 2003.

[2] K. Böröczky, *Packing of spheres in spaces of constant curvature*. Acta Math. Acad. Sci. Hung. 32 (1978), 243–261.

[3] K. Böröczky, A. Florian, *Über die dichteste Kugelpackung im hyperbolischen Raum*. Acta Math. Acad. Sci. Hung. 15 (1964), 237–245.

[4] C. Cao, P. L. Waterman, *Conjugacy invariants of Möbius groups*. In: Quasiconformal Mappings and Analysis, A Collection of Papers Honoring F. W. Gehring, P. L. Duren et al, Editors, Springer-Verlag, 1998.

[5] M. Conder, C. Maclachlan, *Compact hyperbolic 4-manifolds of small volume.* Preprint, to appear in Proc. A.M.S.

[6] H. S. M. Coxeter, *Discrete groups generated by reflections.* Ann. Math. **35** (1934), 588–621.

[7] H. S. M. Coxeter, *The functions of Schläfli and Lobatschefsky.* Q. J. Math., Oxf. Ser. **6** (1935), 13–29.

[8] H. S. M. Coxeter, *Arrangements of equal spheres in non-Euclidean spaces.* Acta Math. Acad. Sci. Hungar. **5** (1954), 263–274.

[9] H. S. M. Coxeter, *Regular honeycombs in hyperbolic space.* Proc. internat. Congr. Math. 1954 Amsterdam vol. 3 (1956), 155–169.

[10] H. S. M. Coxeter, *Regular polytopes.* Dover, 1973.

[11] M. Gromov, *Hyperbolic manifolds according to Thurston and Jørgensen.* Séminaire Bourbaki, 32e année, 1979/80, no. 546, 1−14.

[12] M. W. Davis, *A hyperbolic 4-manifold.* Proc. Amer. Math. Soc. **93** (1985), 325−328.

[13] T. Hild, R. Kellerhals, *The fcc lattice and the cusped hyperbolic 4-orbifold of minimal volume.* Preprint, 2004.

[14] R. Kellerhals, *On the volume of hyperbolic polyhedra.* Math. Ann. **285** (1989), 541−569.

[15] R. Kellerhals, *On Schläfli's reduction formula.* Math. Z. **206** (1991), 193−210.

[16] R. Kellerhals, *Ball packings in spaces of constant curvature and the simplicial density function.* J. Reine Angew. Math. **494** (1998), 189–203.

[17] R. Kellerhals, *Volumes of cusped hyperbolic manifolds.* Topology **37** (1998), 719−734.

[18] R. Kellerhals, *Quaternions and some global properties of hyperbolic 5-manifolds.* Canad. J. Math. **55** (2003), 1080–1099.

[19] R. Kellerhals, *On the structure of hyperbolic manifolds.* Israel J. Math. **143** (2004), 361–379.

[20] C. Maclachlan, A. W. Reid, *The arithmetic of hyperbolic 3-manifolds.* Graduate Texts in Mathematics. 219. Springer-Verlag, New York, 2003.

[21] R. Meyerhoff, *The cusped hyperbolic 3-orbifold of minimum volume.* Bull. Amer. Math. Soc. **13** (1985), 154−156.

[22] A. Przeworski, *Cones embedded in hyperbolic manifolds.* J. Diff. Geom. **58** (2001), 219−232.

[23] J. G. Ratcliffe, *Foundations of hyperbolic manifolds.* Graduate Texts in Mathematics, 149. Springer-Verlag, New York, 1994.

[24] C. L. Siegel, *Some remarks on discontinuous groups.* Ann. Math. **46** (1945), 708−718.

[25] W. P. Thurston, *The geometry and topology of three-manifolds.* electronic manuscript, march 2002.

[26] E. B. Vinberg (ed.), *Geometry II: spaces of constant curvature.* In: Encyclopaedia of Mathematical Sciences, vol. 29, Springer-Verlag, 1993.

[27] H.-C. Wang, *Topics in totally discontinuous groups.* In: Symmetric Spaces, W. M. Boothby and G. L. Weiss, Editors, Pure Appl. Math., vol. 8, Marcel Dekker, 1972.

[28] P. L. Waterman, *Möbius transformations in several dimensions.* Adv. Math. **101** (1993), 87−113.

[29] T. Zehrt, *Polytopal complexes in spaces of constant curvature.* Ph.D. thesis, University of Basel, 2003.

UNIVERSITY OF FRIBOURG, DEPARTMENT OF MATHEMATICS, CH–1700 FRIBOURG, SWITZERLAND

E-mail address: `Ruth.Kellerhals@unifr.ch`

Regular and Chiral Polytopes in Low Dimensions

Peter McMullen and Egon Schulte

ABSTRACT. There are two main thrusts in the theory of regular and chiral polytopes: the abstract, purely combinatorial aspect, and the geometric one of realizations. This brief survey concentrates on the latter. The dimension of a faithful realization of a finite abstract regular polytope in some euclidean space is no smaller than its rank, while that of a chiral polytope must strictly exceed the rank. There are similar restrictions on the dimensions of realizations of regular and chiral apeirotopes. From the viewpoint of realizations in a fixed dimension, the problems are now completely solved in up to three dimensions, while considerable progress has been made on the classification in four dimensions, the finite regular case again having been solved. This article reports on what has been done already, and what might be expected in the near future.

1. Introduction

Donald Coxeter's work on regular polytopes and groups of reflexions is often viewed as his most important contribution. At its heart lies a dialogue between geometry and algebra which was so characteristic for his mathematics (see, for example, [4, 5, 7]). This paper is yet more evidence for his lasting influence on generations of geometers.

In [16] (see also [17, Sections 7E, 7F]), we classified completely all the faithfully realized regular polytopes and discrete regular apeirotopes in dimensions up to three. Further, in [14], the first author classified the regular polytopes and apeirotopes of maximal rank in each higher dimension, and showed that chiral polytopes could not have full rank. Last, in [19, 20], the second author has found all the chiral apeirohedra in three dimensions.

The present paper surveys the developments on realizations of regular or chiral polytopes, which have occurred since the publication of our book [17]. There are two quite different ways to approach realizations. The first, for which a fairly complete theory exists (at least, in the finite case), asks for a description of the space of all realizations (a kind of "moduli space") of a given abstract regular polytope or apeirotope, with rank playing only a minor rôle (see [17, Sections 5B, 5C] for further details). The second, about which much less is known in general terms, asks

---

2000 *Mathematics Subject Classification.* Primary 51M20; Secondary 52B15.

The second author was supported in part by NSA-grant H98230-04-1-0116.

for a classification of the realizations of all these polytopes and apeirotopes in a euclidean space of given dimension (in this case, it is usual to impose conditions such
as faithfulness and discreteness). This problem is solved in three dimensions. The
finite regular polyhedra have long been known; adding to the Petrie-Coxeter apeirohedra of [3], Grünbaum [11] found all but one of the remaining regular apeirohedra,
while Dress [8, 9] found the missing example, and proved that the classification was
then complete. We refer the reader to [16] for a quick method of arriving at the full
characterization, including a discussion of the geometry of the regular apeirohedra
and presentations of their symmetry groups, as well as for the enumeration of the
regular 4-apeirotopes in three dimensions.

In four dimensions, the currently open problems are those of classifying the
finite regular polyhedra, and the regular apeirohedra and 4-apeirotopes; [14] solves
the problems of the regular 4-polytopes and 5-apeirotopes. The paper [15] in preparation actually settles the first of these problems (the polytopes with planar faces
were classified in [1, 2]); however, the other two, together with the corresponding
classification problems for chiral polytopes, are still open, although some progress
has been made on them.

2. Regular and chiral polytopes

For the general background on abstract regular polytopes, we refer the reader to
the recently published monograph [17]; for the most part, we shall not cite original
papers directly. In this paper, we largely concentrate on the geometric aspects of
the theory, that is, on realizations of regular polytopes.

However, we begin with the more combinatorial picture. An *abstract polytope*
of *rank* n, or simply an (*abstract*) *n-polytope*, is a partially ordered set $\mathcal{P}$ with a
strictly monotone rank function, taking values in $\{-1, 0, \ldots, n\}$. The elements of
rank j are the *j-faces* of $\mathcal{P}$, or *vertices*, *edges* and *facets* of $\mathcal{P}$ if $j = 0, 1$ or $n - 1$,
respectively. The maximal chains are the *flags* of $\mathcal{P}$ and contain exactly $n + 2$ faces,
including a unique minimal face and a unique maximal face (usually omitted from
the notation). Two flags are called *adjacent* if they differ by one element; then $\mathcal{P}$
is *strongly flag-connected*, meaning that, if Φ and Ψ are two flags, then they can be
joined by a sequence of successively adjacent flags $\Phi = \Phi_0, \Phi_1, \ldots, \Phi_k = \Psi$, each of
which contains $\Phi \cap \Psi$. Finally, if F and G are a $(j-1)$-face and a $(j+1)$-face with
$F < G$, then there are exactly *two* j-faces H such that $F < H < G$. An n-polytope
$\mathcal{P}$ is then called *regular* if its combinatorial automorphism group $\Gamma(\mathcal{P})$ (preserving
the partial ordering) is (simply) transitive on its flags; in this case, if Φ is a (fixed)
base flag and, for $j = 0, \ldots, n-1$, ρ_j is the automorphism which maps Φ to the
adjacent flag Φ^j with a different j-face, then $\Gamma(\mathcal{P})$ is generated by $\rho_0, \ldots, \rho_{n-1}$.

We can adopt (see [17, Theorem 2E11]) the viewpoint that an abstract regular
polytope is to be identified with its group. The latter is precisely what is called
a *string C-group*; here, the "C" stands for "Coxeter", though not every C-group
is a Coxeter group. A string C-group Γ is a group generated by n involutions ρ_j
(the *distinguished generators*) with $j \in \mathsf{N} := \{0, \ldots, n-1\}$, such that ρ_j and ρ_k
commute if $0 \leq j \leq k - 2 \leq n - 3$, and

$$(2.1) \qquad \langle \rho_i \mid i \in \mathsf{J} \rangle \cap \langle \rho_i \mid i \in \mathsf{K} \rangle = \langle \rho_i \mid i \in \mathsf{J} \cap \mathsf{K} \rangle$$

for each $\mathsf{J}, \mathsf{K} \subseteq \mathsf{N}$; the last is the *intersection property*. Each string C-group Γ then
determines (uniquely) a regular n-polytope $\mathcal{P}$ with $\Gamma = \Gamma(\mathcal{P})$. The *j-faces* of $\mathcal{P}$

are the right cosets $\Gamma_j \sigma$ of the *distinguished subgroup*

$$\Gamma_j := \langle \rho_i \mid i \neq j \rangle$$

for each $j \in \mathsf{N}$, and two faces are incident just when they intersect (as cosets). In fact, incidence actually induces an order relation:

$$\Gamma_j \sigma \leq \Gamma_k \tau \iff \Gamma_j \sigma \cap \Gamma_k \tau \neq \emptyset \text{ and } j \leq k.$$

Formally, we also adjoin two copies of Γ itself, as the (unique) (-1)- and n-faces of $\mathcal{P}$. The maximal chains (with respect to this ordering) are the *flags* of $\mathcal{P}$; the group Γ is then simply transitive on the flags of $\mathcal{P}$. In particular, for $j = 0, \ldots, n-1$ the distinguished generator ρ_j of Γ takes the *base flag* $\Phi := \{\Gamma_{-1}, \Gamma_0, \Gamma_1, \ldots, \Gamma_{n-1}, \Gamma_n\}$ into the adjacent flag Φ^j which differs from it in Γ_j. Note that the distinguished subgroups $\Gamma_{n-1} = \langle \rho_0, \ldots, \rho_{n-2} \rangle$ and $\Gamma_0 = \langle \rho_1, \ldots, \rho_{n-1} \rangle$ are themselves string C-groups; the corresponding polytopes are the *facet* and *vertex-figure* of $\mathcal{P}$, respectively (the latter consists of the faces of $\mathcal{P}$ with vertex Γ_0). As we said earlier, [**17**, Theorem 2E11] shows that this description of a regular polytope $\mathcal{P}$ in terms of (its C-group) $\Gamma(\mathcal{P})$ and the previous one in terms of the face poset are equivalent.

The distinguished generators of $\Gamma = \Gamma(\mathcal{P})$ satisfy relations

$$(2.2) \qquad (\rho_i \rho_j)^{p_{ij}} = \varepsilon \quad (i, j = 0, \ldots, n-1),$$

with $p_{ii} = 1$, $p_{ij} = p_{ji} \geq 2$ if $i \neq j$, and $p_{ij} = 2$ if $|i - j| \geq 2$ (hence the term "string" C-group). The numbers $p_j := p_{j-1,j}$ $(j = 1, \ldots, n-1)$ determine the *Schläfli type* $\{p_1, \ldots, p_{n-1}\}$ of $\mathcal{P}$. To avoid cases which, in our context, turn out to be trivial, we always assume that adjacent generators ρ_{j-1} and ρ_j of Γ do not commute (this is justified in Section 3); in other words, $p_j > 2$ (possibly, $p_j = \infty$). If the polytope is determined just by the p_j, then we have the *universal* regular polytope (of that Schläfli type), for which we use the same symbol $\{p_1, \ldots, p_{n-1}\}$ (but without qualification); we write $[p_1, \ldots, p_{n-1}]$ for the corresponding *Coxeter* group. Generally, however, the group Γ will satisfy additional relations as well, for some of which we introduce special notation later.

The underlying face-set of a polytope $\mathcal{P}$ can be finite or infinite. An infinite n-polytope is also called an (*abstract*) n-apeirotope; when $n = 2$, we also refer to it as an *apeirogon*, and when $n = 3$ as an *apeirohedron*.

A central question in the abstract theory is that of the amalgamation of polytopes of lower rank. If a regular $(n+1)$-polytope has facets (of type) the n-polytope $\mathcal{P}$ and vertex-figures the n-polytope $\mathcal{Q}$, then the facets of $\mathcal{Q}$ must be isomorphic to the vertex-figures of $\mathcal{P}$. Conversely, if $\mathcal{P}$ and $\mathcal{Q}$ satisfy this latter criterion, then we write $\langle \mathcal{P}, \mathcal{Q} \rangle$ for the class of all regular $(n+1)$-polytopes with facet $\mathcal{P}$ and vertex-figure $\mathcal{Q}$. The question has two parts. First, is $\langle \mathcal{P}, \mathcal{Q} \rangle \neq \emptyset$; in other words, does there exist any such regular $(n+1)$-polytope at all? If so, then there is a *universal* member $\{\mathcal{P}, \mathcal{Q}\}$ in the family $\langle \mathcal{P}, \mathcal{Q} \rangle$, of which every other one is a quotient (in the sense that its group is an appropriate quotient). Second, given that it exists, we ask what $\{\mathcal{P}, \mathcal{Q}\}$ is. (See [**17**, Section 4B] for further details.) In the present context, we often pose this question in the form: is a given regular polytope, whose facet and vertex-figure are known, actually universal of its kind?

There are several general techniques for constructing new regular polytopes from old ones. In particular, two different regular polytopes may be related by what is called a *mixing operation*; the distinguished generators of the second group are certain products of those of the first (see [**17**, Chapter 7]). Apart from the *duality* operation δ, which just reverses the order of the distinguished generators (and the order relation on the faces), there are two others we mention here; one further operation (for chiral polyhedra) will occur in Section 6. Let $\Gamma = \langle \rho_i \mid i \in \mathsf{N} \rangle$ be a string C-group, let $j \neq k$, and consider the operation

$$(\rho_0, \ldots, \rho_{n-1}) \mapsto (\rho_0, \ldots, \rho_{j-1}, \rho_j \rho_k, \rho_{j+1}, \ldots, \rho_{n-1}) =: (\sigma_0, \ldots, \sigma_{n-1}).$$

Since adjacent generators of Γ do not commute, we easily see that the group $\Delta :=$ $\langle \sigma_0, \ldots, \sigma_{n-1} \rangle$ cannot possibly be a string C-group unless $(j, k) = (2, 0)$ or $(n - 3, n - 1)$. The former will rule itself out later for geometric reasons (see Section 3); the latter, namely,

$$(2.3) \quad \pi: (\rho_0, \ldots, \rho_{n-1}) \mapsto (\rho_0, \ldots, \rho_{n-4}, \rho_{n-3}\rho_{n-1}, \rho_{n-2}, \rho_{n-1}) =: (\sigma_0, \ldots, \sigma_{n-1}),$$

which we denote by $\Gamma \mapsto \Gamma^\pi$, is called the *Petrie operation*, since it generalizes the operation with the same name when $n = 3$. Even when $n = 3$, the Petrie operation π does not always yield a C-group (though such cases are rather exceptional), but, for higher rank, each application has to be checked directly. However, if in fact Γ^π is a C-group, then we write $\mathcal{P} \mapsto \mathcal{P}^\pi$ to indicate the effect of the operation on the corresponding polytope $\mathcal{P}$; the new polytope $\mathcal{P}^\pi$ is called the *Petrial* of $\mathcal{P}$. One general case (see [**14**]) can be settled easily.

PROPOSITION 2.1. *If $\Gamma = \langle \rho_0, \ldots, \rho_{n-1} \rangle$ is a string C-group with $n \geq 4$ for which p_{n-3} is odd, then the Petrial Γ^π is not a C-group.*

Mixing operations are particularly powerful when applied to regular polyhedra or apeirohedra $\mathcal{P}$. For example, the Petrial $\mathcal{P}^\pi$ can be obtained from $\mathcal{P}$ by replacing the 2-faces by the *Petrie polygons* of $\mathcal{P}$ (while keeping the vertices and edges); the geometric picture of a Petrie polygon here is one which shares two successive edges of each 2-face which it meets, but not a third. An important class of regular polyhedra or apeirohedra consists of those which are completely determined by their Schläfli type and the length of their Petrie polygons. We write $\{p, q\}_r$ for the polyhedron (possibly infinite) of Schläfli type $\{p, q\}$, whose Petrie polygons of length r determine it. Its group is the Coxeter group $\langle \rho_0, \rho_1, \rho_2 \rangle = [p, q]$, with the imposition of the single extra relation

$$(2.4) \qquad\qquad (\rho_0 \rho_1 \rho_2)^r = \varepsilon.$$

We note that, if it is a genuine polyhedron, then the Petrial of $\{p, q\}_r$ is $\{r, q\}_p$.

In the context of polyhedra, another operation is also of great importance. The (*second*) *facetting operation* φ_2 is given by

$$(2.5) \qquad\qquad \varphi_2: (\rho_0, \rho_1, \rho_2) \mapsto (\rho_0, \rho_1 \rho_2 \rho_1, \rho_2),$$

and replaces the 2-faces of a polyhedron $\mathcal{P}$ by the *holes* (while keeping the vertices and edges); a hole of $\mathcal{P}$ is an edge-circuit which exits from the *second* edge (in some local orientation) emanating from a vertex from the edge by which it entered. The designation of a (possibly infinite) regular polyhedron of Schläfli type $\{p, q\}$, which is determined by its holes of length h, is $\{p, q \mid h\}$. The corresponding relation to be imposed on the Coxeter group $\langle \rho_0, \rho_1, \rho_2 \rangle = [p, q]$ is

$$(2.6) \qquad\qquad (\rho_0 \rho_1 \rho_2 \rho_1)^h = \varepsilon.$$

Various examples of such polyhedra occur later; for now, let us observe that the three Petrie-Coxeter apeirohedra are, as abstract regular polyhedra, $\{4, 6 \mid 4\}$, $\{6, 4 \mid 4\}$ and $\{6, 6 \mid 3\}$.

In [**17**, Section 7A] we also introduced the notion of a mix of two regular polytopes (or corresponding C-groups). The following abstract construction is a special case of this mix and occurs when one polytope is 1-dimensional, that is, a segment. Again, suppose that $\Gamma = \langle \rho_i \mid i \in \mathsf{N} \rangle$ is a string C-group. Let τ be an involution which commutes with all ρ_j, and consider the operation

$$(2.7) \qquad (\rho_0, \ldots, \rho_{n-1}, \tau) \mapsto (\rho_0\tau, \rho_1 \ldots, \rho_{n-1}) =: (\sigma_0, \ldots, \sigma_{n-1}).$$

This is called *mixing with a segment*, because τ can be regarded as the generating involution of the group of the segment { } (see Section 3 for the notation). We have (see [**17**, Theorem 7A8])

THEOREM 2.2. *Mixing a string C-group Γ with the group of a segment always yields another C-group. This is isomorphic to Γ if all edge-circuits in the associated regular polytope $\mathcal{P}$ have even length; otherwise, it is isomorphic to the direct product $\Gamma \times \mathcal{C}_2$ of Γ with a cyclic group $\mathcal{C}_2$ of order 2.*

The resulting regular polytope (which we again say is obtained from $\mathcal{P}$ by mixing with a segment) is denoted by $\mathcal{P} \Diamond$ { }. This has twice as many vertices as $\mathcal{P}$ precisely when some edge-circuit of $\mathcal{P}$ has odd length.

We also require another basic technique for constructing regular polytopes from certain groups by what are called *twisting operations* (see [**17**, Chapter 8]). In this, a given group (usually itself a C-group) is augmented by means of one or more group automorphisms. This technique has been extremely successful in various classification problems for regular polytopes. In the present context, it assumes great importance in the enumeration of the regular polyhedra in $\mathbb{E}^4$; see Section 7 below.

Roughly speaking, chiral polytopes have half as many possible automorphisms as have regular polytopes. More technically, the n-polytope $\mathcal{P}$ is *chiral* if it has two orbits of flags under its group $\Gamma(\mathcal{P})$, with adjacent flags in different orbits. A chiral n-polytope $\mathcal{P}$ is then identified with a group of the form $\Gamma = \langle \sigma_1, \ldots, \sigma_{n-1} \rangle$, on which there are relations

$$(2.8) \qquad \begin{cases} \sigma_j^{p_j} = \varepsilon, & j = 1, \ldots, n-1, \\ (\sigma_j\sigma_{j+1}\cdots\sigma_k)^2 = \varepsilon, & 1 \leq j < k \leq n-1. \end{cases}$$

We again refer to $\{p_1, \ldots, p_{n-1}\}$ as the *Schläfli type* of $\mathcal{P}$.

The relationship between the group and the corresponding (abstract) polytope is a little less obvious than is the case for regular polytopes (see [**21**] for more details). The distinguished generator σ_j permutes the $(j-1)$- and j-faces cyclically in the appropriate section of the base flag $\Phi = \{F_0, F_1, \ldots, F_{n-1}\}$; if F_j' replaces F_j in the adjacent flag Φ^j, then $F_{j-1}'\sigma_j = F_{j-1}$ and $F_j\sigma_j = F_j'$. The vertices of $\mathcal{P}$ are (identified with) the right cosets of the subgroup $\Gamma_0 := \langle \sigma_2, \ldots, \sigma_{n-1} \rangle$, with $F_0 = \Gamma_0$ itself the base vertex. The involutory element $\tau := \sigma_1\sigma_2$ interchanges the two vertices of the base edge, taking Φ into $(\Phi^0)^2 = (\Phi^2)^0$; it is often useful to replace σ_1 as a generator by τ (compare [**19**]).

In a chiral polytope, adjacent flags are not equivalent under the group. If Φ is replaced by an adjacent flag, Φ^0 (say), then the respective generators are $\sigma_1^{-1}, \sigma_1^2\sigma_2, \sigma_3, \ldots, \sigma_{n-1}$. Thus a chiral polytope occurs in two (*combinatorially*) *enantiomorphic forms*, each specified by the choice of an orbit of base flags (Φ or Φ^0), or, equivalently, a conjugacy class of sets of generators (represented by

$\sigma_1, \ldots, \sigma_{n-1}$ or $\sigma_1^{-1}, \sigma_1^2\sigma_2, \sigma_3, \ldots, \sigma_{n-1}$, respectively). For a regular polytope, these two enantiomorphic forms can be identified (under the generator ρ_0 of Γ).

3. Realizations

There are many candidates for spaces in which regular polytopes $\mathcal{P}$ might be realized geometrically. The usual (and generally most useful) context of realizations is of those in euclidean spaces, because it is in these that we obtain the richest structure. However, initially at least, it is appropriate for us to broaden the definition. Thus, for the time being, E is a k-dimensional spherical space $\mathbb{S}^k$, euclidean space $\mathbb{E}^k$ or hyperbolic space $\mathbb{H}^k$, for some k. If $\mathcal{P}$ is a finite polytope, then E will be spherical; if $\mathcal{P}$ is an apeirotope, then, since we are generally interested only in discrete realizations, E will be euclidean or hyperbolic.

We begin with a brief review of some definitions (see [**17**, Chapter 5] for the general background here). Let $\mathcal{P}$ be an abstract regular polytope (or apeirotope – for the moment, we use the generic term, not distinguishing between the finite and infinite cases), and let $\Gamma := \Gamma(\mathcal{P})$. For a *faithful realization* of $\mathcal{P}$ we have two ingredients. First, we need a suitable space E which admits a group $\mathcal{G}$ of isometries isomorphic to Γ; this is the *symmetry group* of the realization of $\mathcal{P}$. It is convenient to identify the *reflexion* R_j in $\mathcal{G}$ corresponding to the involution ρ_j in Γ with its *mirror*

$$\{x \in E \mid xR_j = x\}$$

of fixed points; we thus use the same symbol E for the ambient space to denote the identity mapping. The intersection

$$W := R_1 \cap \cdots \cap R_{n-1}$$

is called the *Wythoff space* of the realization. The realization of $\mathcal{P}$ associated with $\mathcal{G}$ and its generators R_j then arises from some choice of *initial vertex* $v \in W$. The vertex-set of the realization is $V := v\mathcal{G}$, the orbit of v under $\mathcal{G}$, and we always assume that E is spanned by V (as a subspace of the appropriate kind), so that E is thought of as the *ambient space* of the realization, namely, the space (of one of the three kinds) of smallest dimension which contains it.

Note that, if $\mathcal{G}$ were to be such that $R_j = E$, the identity mapping, then $R_k = E$ for all $k > j$ as well and the realization would not be faithful. In particular, this will happen if $p_j = 2$, which is why we excluded this possibility in Section 2.

The induced geometric structure, the actual *realization* P of $\mathcal{P}$, is defined as follows. Write $F_0 := v$, and, for $j \geq 1$, let

$$F_j := F_{j-1}\langle R_0, \ldots, R_{j-1}\rangle;$$

these are the basic faces. Then the j-faces of the realization are the $F_j G$ with $G \in \mathcal{G}$, with the order relation given by iterated membership. Thus *edges* are composed of the two vertices which belong to them (we also think of an edge as the line-segment between its vertices – there will be no ambiguity, even in the spherical case, because antipodal points of the sphere will never determine an edge), 2-faces of the edges which belong to them, and so on up to the *ridges* or $(n-2)$-faces and *facets* or $(n-1)$-faces. We sometimes refer to the realization P as a *geometric polytope*. Its *dimension* is defined by $\dim P := \dim E$, and its vertex-set is denoted by $V(P) := V$. Finally, for the realization to be *faithful*, we demand that, for each $j = 1, \ldots, n-1$, a j-face be uniquely determined by the $(j-1)$-faces which belong to it. Recall here our initial assumption that $\mathcal{G}$ and Γ be isomorphic, so for a faithful realization we then have natural bijections between the sets of j-faces of $\mathcal{P}$ and P for each j. Some regular polytopes do not admit faithful realizations, because this latter condition implies a corresponding purely combinatorial condition on $\mathcal{P}$.

A realization of an abstract regular n-polytope $\mathcal{P}$ determines a realization of each of its faces or co-faces (iterated vertex-figures). In particular, F_{n-1} (and its induced structure, with the same initial vertex v) gives a realization of the facet of $\mathcal{P}$; its symmetry group is the image $\mathcal{G}_{n-1}$ of Γ_{n-1}. If we write w for the midpoint of the edge between v and vR_0, then w is the initial vertex of a realization of the vertex-figure of $\mathcal{P}$, with symmetry group the image $\mathcal{G}_0$ of Γ_0. (This suffices for our purposes. However, in the hyperbolic case of a polytope with vertices on the absolute, then the initial vertex w is well-defined as the intersection of the mirror R_0 with the line between v and vR_0 – in any event, w will always lie in this intersection.) Faithfulness is hereditary; that is, if the original realization of $\mathcal{P}$ is faithful, then the realizations of the facet and vertex-figure of $\mathcal{P}$ are also faithful. In a similar way, $\langle R_0, \ldots, R_{j-1} \rangle$ is the symmetry group of the basic j-face F_j of P, while $\langle R_{j+1}, \ldots, R_{n-1} \rangle$ is that of the basic co-j-face P/F_j, which is the $(j+1)$-fold iterated vertex-figure. Thus the vertex-figure itself is P/F_0. Even more generally, $\langle R_{j+1}, \ldots, R_{k-1} \rangle$ is the symmetry group of the section F_k/F_j (for $j \leq k - 2$), the $(j+1)$-fold iterated vertex-figure of the basic k-face F_k.

We often find it more convenient to use vR_0 rather than w as the initial vertex of the vertex-figure; for most purposes, this makes little difference, since the combinatorics are not altered.

For regular polytopes of rank at most 2 we have the following spherical or euclidean realizations. In $\mathbb{E}^0$ we just have the point (realizing the 0-polytope), the finite regular 1-polytopes are segments { }, which are naturally realized in the 0-sphere $\mathbb{S}^0$, while the regular apeirogon $\{\infty\}$ is naturally realized discretely in $\mathbb{E}^1 = \mathbb{R}$. In the unit circle $\mathbb{S}^1$, there is an infinite family of (finite) regular polygons. Their mirrors R_0 and R_1 are lines through its centre at a *rational* angle π/p, meaning that $p > 2$ is a rational number (always in its lowest terms); the resulting regular polygon is denoted $\{p\}$. In addition, $\{\infty\}$ has non-discrete faithful realizations in $\mathbb{S}^1$. As we mentioned before, we shall not address here the question of finding all possible realizations of a given abstract regular polytope; a fairly complete theory has been described in [**17**, Sections 5B, 5C]. Suffice it to remark that the realization space has been determined for several interesting classes of polytopes; see, for example, [**18**].

There are important restrictions on faithful realizations; we refer to [**17**, Sections 5B, 5C] for proofs.

THEOREM 3.1. *Let P be a faithful realization of an abstract regular polytope $\mathcal{P}$, whose ambient space E is a spherical, euclidean or hyperbolic space. Then* $\dim P \geq \operatorname{rank} \mathcal{P} - 1$.

THEOREM 3.2. *Let P be a faithful realization of an abstract regular n-polytope in E, with group $\mathcal{G} = \langle R_0, \ldots, R_{n-1} \rangle$. Then* $\dim R_j \geq j$ *for* $j = 0, \ldots, n - 2$, *and* $\dim R_{n-1} \geq n - 2$.

In both theorems, if the polytope is finite, so that the ambient space is spherical, then, regarded as euclidean realizations, each of the dimensions must be increased by 1.

If we have (not necessarily faithful) realizations of the abstract regular polytope (or apeirotope) $\mathcal{P}$ in two euclidean spaces, say P with mirrors $S_0, \ldots, S_{n-1}$ in L and Q with mirrors $T_0, \ldots, T_{n-1}$ in M (possibly some $S_j = L$ or $T_j = M$), then their *blend* has mirrors $S_j \times T_j$ in $L \times M$ for $j = 0, \ldots, n-1$. Indeed, if $v \in S_1 \cap \cdots \cap S_{n-1}$ and $w \in T_1 \cap \cdots \cap T_{n-1}$ are the initial vertices of the two realizations, then (v, w) can be chosen as the initial vertex of the blend, which we then write $P \# Q$. A realization which cannot be expressed as a blend in a non-trivial way is called *pure*.

One main tool for classifying regular polytopes of a fixed rank n in a fixed dimension is the *dimension vector* $(\dim R_0, \dim R_1, \ldots, \dim R_{n-1})$ of the possible realizations; the first step in any enumeration is to determine which dimension vectors can occur.

It is worth noting that, in general, duals of faithfully realizable regular polytopes are not necessarily faithfully realizable at all (Petrials are particular examples), let alone in the same space.

There is a similar realization theory for chiral polytopes. Indeed, let us call a realization P of an abstract polytope $\mathcal{P}$ *chiral* if P has two orbits of flags under its symmetry group $\mathcal{G}(P)$, with adjacent flags lying in different orbits. It is clear that the original polytope $\mathcal{P}$ must be regular or chiral. Note that there exist (already in $\mathbb{E}^3$) faithful realizations of polytopes with two flag orbits under $\mathcal{G}(P)$ which are not chiral (see [22] for examples).

It is helpful to remark that, if $\mathcal{P}$ is a regular n-polytope with group $\Gamma = \langle \rho_0, \ldots, \rho_{n-1} \rangle$, then its combinatorial rotation subgroup $\Gamma^+(\mathcal{P})$ has generators

$$\sigma_j := \rho_{j-1} \rho_j, \qquad j = 1, \ldots, n-1.$$

Thus a chiral realization of a polytope may be thought of as having only rotational symmetries. Moreover, if the abstract polytope $\mathcal{P}$ is at least chiral, in that its group Γ contains the automorphisms $\sigma_1, \ldots, \sigma_{n-1}$ in the definition of chirality, then $\mathcal{P}$ is actually regular if we can adjoin any one of the involutions ρ_j for $j = 0, \ldots, n-1$. (We then have $\rho_i = \sigma_{i+1} \rho_{i+1}$ for $i = 0, \ldots, j-1$, or $\rho_i = \rho_{i-1} \sigma_i$ for $i = j+1, \ldots, n-1$.)

Chiral realizations are derived by a variant of Wythoff's construction, applied to a suitable representation $\mathcal{G} = \langle S_1, \ldots, S_{n-1} \rangle$ of the underlying combinatorial group $\Gamma := \langle \sigma_1, \ldots, \sigma_{n-1} \rangle$; the latter is $\Gamma(\mathcal{P})$ or $\Gamma^+(\mathcal{P})$ according as the abstract polytope $\mathcal{P}$ is chiral or regular. The Wythoff space now is the fixed set of the subgroup $\mathcal{G}_0 := \langle S_2, \ldots, S_{n-1} \rangle$. We describe the 3-dimensional case in more detail in Section 6.

It is clear that an abstract regular polytope may have chiral realizations, though not necessarily faithful ones; it is an interesting open question whether it could actually have faithful chiral realizations. It is an elementary observation that a realized polygon with full rotational symmetry group is actually regular. Similar arguments to those used in the proof of Theorem 3.1 then yield

PROPOSITION 3.3. *If P is a faithful chiral realization of an abstract polytope, whose ambient space is a spherical, euclidean or hyperbolic space E, then $\dim P \geq \operatorname{rank} \mathcal{P} - 1$.*

When the abstract polytope $\mathcal{P}$ is finite, we usually assume that the centroid of the vertex-set V of its (chiral or regular) realization P is the origin o of E, so that $\mathcal{G}$ is an orthogonal group. If $\mathcal{P}$ is infinite, in which case we again call P a *(geometric) apeirotope*, we will additionally demand of P that it be discrete, so that the group $\mathcal{G}$ acts discretely on the ambient space E. Moreover, in order to avoid constant repetition of various fixed phrases subsequently, we adopt the conventions that, in the geometric context of realizations, *regular polytope* will mean "faithfully realized finite abstract regular polytope", while *regular apeirotope* will mean "discrete faithfully realized abstract regular apeirotope"; we also adopt the corresponding terminology for chiral polytopes and chiral apeirotopes.

We end the section with two general remarks. Let S and T be linear reflexions. First, since $ST = (-S)(-T) = S^{\perp}T^{\perp}$ (thus identifying $-S$ with its mirror $S^{\perp}$, and so on), then $S \cap T$ and $S^{\perp} \cap T^{\perp}$ are both pointwise fixed by the product. That is, the axis (fixed set) of ST is

$$(3.1) \qquad (S \cap T) + (S^{\perp} \cap T^{\perp}) \ (= (S \cap T) + (S + T)^{\perp}).$$

In particular, if S and T commute, then (3.1) is the mirror of their product $ST = TS$, which is again a reflexion.

Second, we have a general construction from [**14**], of which special cases already occur in [**16**]. Let X be a point-set in a euclidean space E. We call X *rational* if the points of X can be chosen to have rational coordinates with respect to some (linear or affine) coordinate system in E. The following remark is obvious.

LEMMA 3.4. *Let E be a euclidean space, and let X be a finite point-set in E. Let $\mathcal{R}(X)$ be the group generated by the point-reflexions (inversions) in the points of X. Then $\mathcal{R}(X)$ is discrete if and only if X is rational.*

If P is a regular polytope with ambient space E, then we similarly call P *rational* if its vertex-set is rational. We have the following.

THEOREM 3.5. *Let P be a rational regular n-polytope in the euclidean space E, with symmetry group $\mathcal{G}_0 = \langle R_1, \ldots, R_n \rangle$ and initial vertex w, and suppose that $v \in R_1 \cap \cdots \cap R_n$. Let $R_0 = \{w\}$ be the point-reflexion in the point w. Then $\mathcal{G} := \langle R_0, \ldots, R_n \rangle$ is the group of a discrete regular $(n+1)$-apeirotope* apeir P, *with 2-faces apeirogons, and vertex-figure P at the initial vertex v.*

We call apeir P the *free abelian apeirotope on P*, or *with vertex-figure P*, and *base vertex v*. When we apply this construction, it will usually be the case that P itself is finite and full-dimensional in E, so that v is the centre of P.

4. Regular polytopes of full rank

If P is a realization of a regular polytope $\mathcal{P}$ for which equality holds in Theorem 3.1, then we say that P is *of full rank*. The emphasis is placed this way round, because our aim (as explained in Section 1) is to classify regular (and chiral) polytopes by dimension. In this case, we can go further than Theorem 3.2, and place further restrictions on the dimensions of the mirrors of the generating reflexions of the realizations. We refer to [**14**] for a proof.

THEOREM 4.1. *Let P be a faithful realization of full rank of a regular n-polytope $\mathcal{P}$ in the ambient space E, with symmetry group $\mathcal{G} = \langle R_0, \ldots, R_{n-1} \rangle$. Then $\dim R_j = j$ or $n - 2$ for $j = 0, \ldots, n - 3$, and $\dim R_{n-2} = \dim R_{n-1} = n - 2$.*

For finite polytopes, we now find it convenient to revert to the former definition of realization in euclidean spaces. In other words, henceforth we regard a sphere which carries the vertices of a realization P of a finite regular polytope as sitting in the euclidean space of one larger dimension with centre the origin o. The mirrors

R_j of its euclidean group $\mathcal{G}$ are then thought of as linear subspaces, also of one larger dimension than before; in particular, in the minimal case, R_0 is either a line or a hyperplane. Finally, we shall use the more familiar I for the identity (in a sense, E is no longer quite appropriate).

REMARK 4.2. If R is a linear reflexion in a euclidean space E, then $-R = (-I)R$, the product of R with the central inversion $-I$, is the reflexion in the orthogonal complement $R^{\perp}$ of R. Replacing a mirror by its orthogonal complement is often a useful tool in studying realizations. In particular, in the case of a faithful realization of full rank of a finite regular n-polytope with centre o, if the mirror R_0 is a line, then $-R_0$ is a hyperplane reflexion. If we replace R_0 by $-R_0$, then at worst we have replaced the symmetry group $\mathcal{G}$ by $\mathcal{G} \times \mathcal{C}_2$, with $\mathcal{C}_2 = \{\pm I\}$; in any event, we always have another finite group. Thus the mirror replacement often produces groups closely related to finite groups generated by hyperplane reflexions.

Remark 4.2 enables us to introduce some important geometric operations on finite polytopes of full rank, which are the key to their enumeration. For such polytopes P, since o is the sole fixed point of the ambient space E under the group $\mathcal{G}$, it follows that

$$K_0 := R_0 \cap \cdots \cap R_{n-1} = \{o\}.$$

Thus the central reflexion $-I$, identified with its mirror $\{o\}$, is K_0, so the mirror replacement of Remark 4.2 is $R_0 \mapsto R_0 K_0$. Moreover, it is extremely useful to have variant operations, which act on the co-$(j-1)$-face P/F_{j-1} for some j and also apply to apeirotopes when their co-$(j-1)$-faces are finite. With

$$K_k := R_k \cap \cdots \cap R_{n-1} \quad (0 \le k \le n-1),$$

we see that (the reflexion in) K_k induces the central inversion on the affine hull of P/F_{k-1}; recall here our assumption of full rank. For $0 \le j \le k \le n-1$, we then define the operation κ_{jk} on $\mathcal{G}$ by
(4.1)
$$\kappa_{jk} \colon (R_0, \ldots, R_{n-1}) \mapsto (R_0, \ldots, R_{j-1}, R_j K_k, R_{j+1}, \ldots, R_{n-1}) =: (S_0, \ldots, S_{n-1}).$$

This produces a new group with generators $S_0, \ldots, S_{n-1}$. We abbreviate κ_{jj} to κ_j, because this is the most important case (and here usually only with $j = 0, 1$), but κ_{02} is also useful. Thus κ_j interchanges the two possibilities for R_j which can occur in Theorem 4.1. Just as with the Petrie operation, though, it must be emphasized that it is by no means generally the case that κ_{jk} will yield a C-group when it is applied to another; for example, for S_j to be an involution, we need $j = k$ or $j \le k - 2$. Observe also that $K_{n-1} = R_{n-1}$, so that the Petrie operation of (2.3) can be written as $\pi = \kappa_{n-3,n-1}$.

One result, for which we only have a case-by-case (but not a general) proof, is the following.

THEOREM 4.3. *If P is a finite regular polytope of full rank, then P^{κ_0} is also a finite regular polytope of full rank.*

It is instructive to see how the operation κ_0 acts geometrically on simple examples. In fact, κ_0 may do one of three things, even when the original group $\mathcal{G}$ is a hyperplane reflexion group: it may double the order, leave it the same, or even halve it. To illustrate this, in $\mathbb{E}^3$ take, respectively, the (group of the) tetrahedron, octahedron and cube; note that, in each case, whereas the old facets were of full rank, the new ones (of the polyhedron associated with the new group) are skew polygons, and so are not. In the planar case, we have $\{p\}^{\kappa_0} = \{q\}$, where $\frac{1}{p} + \frac{1}{q} = \frac{1}{2}$.

REMARK 4.4. If $K_k \in \langle R_j, \ldots, R_{n-1} \rangle$, then κ_{jk} results in a mixing operation.

It would be inappropriate to reproduce all the details of [14] here, even in outline form. However, let us note a few of the salient facts. We shall say more about three and four dimensions in later sections; from five dimensions on, things settle in a common pattern. Recall our conventions that "regular (or chiral) polytope" will mean "faithfully realized finite abstract regular (or chiral) polytope", while "regular (or chiral) apeirotope" will mean "discrete faithfully realized abstract regular (or chiral) apeirotope".

For the regular n-polytopes in $\mathbb{E}^n$, we add to the simplex, cross-polytope and cube the results of applying κ_0 to each. From the n-simplex $\{3^{n-1}\}$ we obtain a polytope $\{3^{n-1}\}^{\kappa_0}$ with $2(n+1)$ vertices, those of the simplex and its dual; its group is $S_{n+1} \times C_2$. For the n-cross-polytope, $\{3^{n-2}, 4\}^{\kappa_0}$ has the same vertices and symmetry group as $\{3^{n-2}, 4\}$. With the n-cube $\{4, 3^{n-2}\}$, there is a distinction between even and odd dimensions n. When n is even, $\{4, 3^{n-2}\}^{\kappa_0}$ has the same vertices and symmetry group; however, when n is odd, $\{4, 3^{n-2}\}^{\kappa_0}$ is isomorphic to the *half-cube* $\{4, 3^{n-2}\}/2 \cong \{4, 3^{n-2}\}_n$, obtained by identifying opposite vertices of the cube.

For the regular $(n+1)$-apeirotopes in $\mathbb{E}^n$, we can apply the "apeir" construction to each of the six n-polytopes of the last paragraph. We also have $\{4, 3^{n-2}, 4\}$, the tiling of space by n-cubes, and, finally, $\{4, 3^{n-2}, 4\}^{\kappa_1}$, which is obtained from it by replacing its vertex-figure $\{3^{n-2}, 4\}$ with $\{3^{n-2}, 4\}^{\kappa_0}$. This last is very interesting; its 3-face is the Petrie-Coxeter apeirohedron $\{4, 6 | 4\}$, and, more generally, its facet is the n-face of $\{4, 3^{m-2}, 4\}^{\kappa_1}$ for each $m \geq n$.

The following table lists the numbers of regular polytopes and apeirotopes of full rank, according to dimension.

dimension	polytopes	apeirotopes
0	1	-
1	1	1
2	∞	6
3	18	8
4	34	18
≥ 5	6	8

We end the section by quoting another result from [14]. If equality occurs in Proposition 3.3, then (as before) we say that P is *of full rank*. This result shows that including chiral polytopes does not add any new examples to the previous classification.

THEOREM 4.5. *There are no chiral realizations of polytopes of full rank.*

5. Regular polytopes in three dimensions

The paper [16] was devoted to the complete classification of the regular polytopes and apeirotopes in $\mathbb{E}^3$, and so we confine ourselves here to the briefest mention of the techniques employed.

With rank at most 2, we have the segment in rank 1, and the polygons (planar and zigzag) and apeirogons (linear, zigzag and helical) in rank 2. We say no more about them.

With rank 3, we first note that the three regular planar tessellations and their Petrials are planar. There are nine "classical" regular polyhedra (the so-called Platonic solids and the Kepler-Poinsot polyhedra – see [17, Section 1A] for discussion of truer attributions), and nine others, which (as a family) can be regarded either as their Petrials, or as the result of applying κ_0 to them. There are twelve apeirohedra which are blends of the six planar ones with a segment or apeirogon, and twelve others which are pure (unblended); of these, except for the Petrie-Coxeter apeirohedra of [3], all but one were found by Grünbaum [11], and the last was discovered by Dress [8, 9].

The last case of the twelve pure apeirohedra is possibly the most interesting, at least for the methods employed. A geometric discussion shows that the possible dimension vectors (of the mirrors of the generating reflexions) are given by $(2, 1, 2)$, $(1, 1, 2)$, $(1, 2, 1)$ and $(1, 1, 1)$. If these mirrors are R_0, R_1, R_2 (we assume that the initial vertex is o, so that R_1, R_2 are linear mirrors), define S_0' to be the translate of R_0 through o, $S_j' := R_j$ for $j = 1, 2$, and finally $S_j := S_j'$ or $-S_j'$, according as R_j is a plane or line. This relates the original symmetry group to one of the crystallographic Coxeter groups $[3, 3]$, $[3, 4]$ or $[4, 3]$ (we need both the latter forms) or the corresponding regular polyhedra; then the three groups, each with four dimension vectors, result in the twelve apeirohedra.

These apeirohedra are listed in the following table; for any notation not introduced hitherto, we refer to [16] or [17, Section 7E].

	$\{3, 3\}$	$\{3, 4\}$	$\{4, 3\}$
(2,1,2)	$\{6, 6 \mid 3\}$	$\{6, 4 \mid 4\}$	$\{4, 6 \mid 4\}$
(1,1,2)	$\{\infty, 6\}_{4,4}$	$\{\infty, 4\}_{6,4}$	$\{\infty, 6\}_{6,3}$
(1,2,1)	$\{6, 6\}_4$	$\{6, 4\}_6$	$\{4, 6\}_6$
(1,1,1)	$\{\infty, 3\}^{(a)}$	$\{\infty, 4\}_{\cdot, *3}$	$\{\infty, 3\}^{(b)}$

The entries in the left column are the dimension vectors $(\dim R_0, \dim R_1, \dim R_2)$, and the remaining columns are indexed by the corresponding finite regular polyhedra. Of these twelve apeirohedra, nine occur naturally as distinguished members in large families of polyhedra (generally apeirohedra), in which all but two polyhedra are chiral (the two exceptional polyhedra are regular); we elaborate on this in Section 6.

Finally, there are eight regular 4-apeirotopes in $\mathbb{E}^3$ (see [17, Section 7F]). There is the regular tiling $\{4, 3, 4\}$ of space by cubes, the result $\{\{4, 6 \mid 4\}, \{6, 4\}_3\}$ of applying κ_1 (or π) to it, and six more obtained by applying the "apeir" operation to the six rational regular polyhedra, namely, the tetrahedron, octahedron and cube and their Petrials.

6. Chiral polytopes in three dimensions

We now proceed with the enumeration of the (discrete and faithful) chiral polyhedra in $\mathbb{E}^3$, following [19, 20]. Again, we shall not go into details and therefore only briefly summarize the results.

The symmetry group $\mathcal{G} := \mathcal{G}(P)$ of a chiral polyhedron P has two orbits on the flags, such that adjacent flags are in distinct orbits. If $\mathcal{P}$ is the underlying abstract polyhedron, then $\mathcal{G}$ is isomorphic to $\Gamma(\mathcal{P})$ or $\Gamma^+(\mathcal{P})$ according as $\mathcal{P}$ is chiral or regular. In either case, $\mathcal{G} = \langle S_1, S_2 \rangle$, where S_1, S_2 are the distinguished generators of $\mathcal{G}$ associated with a base flag Φ of P and corresponding to the generators σ_1, σ_2 of $\Gamma(\mathcal{P})$ or $\Gamma^+(\mathcal{P})$, respectively. If P is of type $\{p, q\}$, then

$$S_1^p = S_2^q = (S_1 S_2)^2 = I,$$

but in general there are also other independent relations. If Φ is replaced by Φ^2 (the adjacent flag with a different 2-face), then the new pair of generators of $\mathcal{G}$ are $S_1 S_2^2, S_2^{-1}$. Thus S_1, S_2 and $S_1 S_2^2, S_2^{-1}$ are the pairs of generators representing the two enantiomorphic forms of P.

As we remarked in Section 3, a chiral polyhedron P can be obtained from a variant of Wythoff's construction, applied to a group $\mathcal{G} = \langle S_1, S_2 \rangle$ with initial vertex a point v fixed by S_2 (but not S_1). If we set $T := S_1 S_2$, which must be a reflexion in a line or plane, then the base vertex, edge and facet of P are v, $v \langle T \rangle$ and $(v \langle T \rangle) \langle S_1 \rangle$, respectively; as usual, the other vertices, edges and facets are their images under $\mathcal{G}$.

The first step is to determine the possible special groups and their generators. Recall that, if $R \colon x \mapsto xR' + t$ is a general element of $\mathcal{G}$, with $R' \in O_3$, the orthogonal group, and $t \in \mathbb{E}^3$ a translation vector, then the linear mappings R' form the *special group* $\mathcal{G}_0$ of $\mathcal{G}$. In the present context, $\mathcal{G}$ must be a crystallographic group in $\mathbb{E}^3$ and $\mathcal{G}_0 = \langle S_1', S_2' \rangle$ a finite subgroup of O_3. If $T(\mathcal{G})$ denotes the subgroup of all translations in $\mathcal{G}$, then $\mathcal{G}_0 \cong \mathcal{G}/T(\mathcal{G})$. It turns out that the only possible special groups are $[3, 3]$ and $[3, 4]$ (possibly as $[4, 3]$), the full tetrahedral and octahedral group, respectively, and their rotation subgroups $[3, 3]^+$ and $[3, 4]^+$ (possibly as $[4, 3]^+$), as well as the group $[3, 3]^*$ obtained from $[3, 3]^+$ by adjoining the central inversion in the invariant point of $[3, 3]^+$. In particular, this limits the possible Schläfli types to $\{4, 6\}$, $\{6, 4\}$, $\{6, 6\}$, $\{\infty, 3\}$ and $\{\infty, 4\}$.

A chiral polyhedron in $\mathbb{E}^3$ cannot be finite (by Theorem 4.5) or be a blend (its group must be affinely irreducible). Thus each chiral polyhedron is infinite and pure.

The possible apeirohedra fall into six infinite 2-parameter families (up to congruence). In each family, all but two polyhedra are chiral; the two exceptional polyhedra are regular. The following table lists the families of polyhedra by the structure of their special group, along with the two regular polyhedra occurring in each family; in three families, one exceptional polyhedron is finite.

$[3, 3]^*$	$[4, 3]$	$[3, 4]$	$[3, 3]^+$	$[4, 3]^+$	$[3, 4]^+$
$P(a, b)$	$Q(c, d)$	$Q(c, d)^*$	$P_1(a, b)$	$P_2(c, d)$	$P_3(c, d)$
$\{6, 6\}_4$	$\{4, 6\}_6$	$\{6, 4\}_6$	$\{\infty, 3\}^{(a)}$	$\{\infty, 3\}^{(b)}$	$\{\infty, 4\}_{\cdot, *3}$
$\{6, 6 \mid 3\}$	$\{4, 6 \mid 4\}$	$\{6, 4 \mid 4\}$	$\{3, 3\}$	$\{4, 3\}$	$\{3, 4\}$

The columns are indexed by the special groups to which the respective polyhedra correspond; some groups occur twice but with different pairs of generators. The second row contains the six families; as we said before, possibly with one exception, all members of a family are apeirohedra. For the first three families, discreteness forces the parameter pairs a, b and c, d, respectively, to be relatively prime integers; however, for the last three families, the parameters can be reals. (Thus, when the polyhedra are considered up to similarity, there is a single rational or real parameter, as appropriate.)

In particular, the chiral polyhedra $P(a,b)$, $Q(c,d)$ and $Q(c,d)^*$ (the dual of $Q(c,d)$) have finite skew faces and skew vertex-figures, and are of types $\{6,6\}$, $\{4,6\}$ or $\{6,4\}$, respectively; remarkably, in each family, the two regular polyhedra have planar faces or vertex-figures. Recall that no regular polyhedron has finite skew faces and skew vertex-figures (see [**17**, Section 7E]). On the other hand, the polyhedra $P_1(a,b)$, $P_2(c,d)$ and $P_3(c,d)$ have infinite faces consisting of helices over triangles, squares or triangles, respectively, and are of types $\{\infty,3\}$, $\{\infty,3\}$ or $\{\infty,4\}$.

The last two rows of the table comprise nine of the twelve pure regular apeirohedra in $\mathbb{E}^3$, namely those listed in the table of Section 5 with dimension vectors $(1,2,1)$, $(1,1,1)$ or $(2,1,2)$, as well as the three (finite) "crystallographic" Platonic polyhedra. The three remaining pure regular apeirohedra $\{\infty,6\}_{4,4}$, $\{\infty,4\}_{6,4}$ and $\{\infty,6\}_{6,3}$ all have dimension vector $(1,1,2)$ and do not occur in families alongside chiral polyhedra.

We now display the families of polyhedra, with the various known relationships among them. These complement the known relationships between regular polyhedra (see [**17**, Section 7E]). Four operations on (chiral or regular) polyhedra and their groups $\mathcal{G}$ are involved: the duality operation δ, the second facetting operation φ_2, the halving operation η, and one further optation κ (see Section 2 or [**19**]). In terms of the generators of $\mathcal{G}$ they are defined as follows:

$$
\begin{aligned}
\delta: &\quad (S_1, S_2) &\mapsto&\quad (S_2^{-1}, S_1^{-1}), \\
\varphi_2: &\quad (S_1, S_2) &\mapsto&\quad (S_1 S_2^{-1}, S_2^2), \\
\eta: &\quad (S_1, S_2) &\mapsto&\quad (S_1^2 S_2, S_2^{-1}), \\
\kappa: &\quad (S_1, S_2) &\mapsto&\quad (S_1(-I), (-I)S_2).
\end{aligned}
$$

In each case, the pair of elements on the right are the generators for the group of a new polyhedron, namely the image of the given polyhedron under δ, φ_2 or η, respectively.

The following diagram emphasizes operations relating families rather than individual polyhedra. In particular, we drop the parameters from the notation; for example, P_1 denotes the family of polyhedra $P_1(a,b)$.

$$(6.1)$$

In the diagram we may replace φ_2 by κ. The circular arrow in the diagram indicates the self-duality of the family (in fact, of each of its polyhedra). The operations δ and φ_2 map a polyhedron to one with the same parameter pair, either a,b or c,d. However, η replaces c,d by the new pair $c-d, c+d$, and κ applied to $Q^*(c,d)$ replaces c,d by $\frac{1}{2}d, -c$. Moreover, note that φ_2, when applicable, and κ map a polyhedron to one in the same row of the table.

For a discussion of other classes of highly symmetric polyhedra in $\mathbb{E}^3$ we refer the reader to, for example, [12].

7. Regular polytopes in four dimensions

Just as is the case with the classical regular polytopes and apeirotopes, the richest family of full rank occurs in $\mathbb{E}^4$. Again, we do not wish to go into the results of [14] in great detail; instead, we shall concentrate on a few plums.

We have already accounted for the effects of κ_0; we merely note that the sixteen classical regular (convex and star-) 4-polytopes give rise to another sixteen in this way. However, we can also apply π to the 4-cube $\{4, 3, 3\}$, to obtain

$$\{4, 3, 3\}^\pi = \{\{4, 4 \mid 4\}, \{4, 3\}_3\}.$$

That is, the facets are toroids, and the vertex-figure is the half-3-cube; moreover, the polytope is universal of this kind. The final instance (of the 34 in the table of Section 4) is obtained by applying κ_0 to this last.

These two finite polytopes just mentioned contribute two regular 5-apeirotopes via the "apeir" construction. Leaving aside the examples already discussed in Section 4, then for the 5-apeirotopes there remain those obtained from $\{3, 3, 4, 3\}$ and its dual $\{3, 4, 3, 3\}$. For the first, we can apply κ_1 (that is, apply κ_0 to its vertex-figure $\{3, 4, 3\}$); we get an apeirotope whose 3-faces are Petrie-Coxeter apeirohedra $\{6, 6 \mid 3\}$. For the other, we can first apply κ_1; the resulting apeirotope has 3-faces the last Petrie-Coxeter apeirohedron $\{6, 4 \mid 4\}$. To both of these (that is, $\{3, 4, 3, 3\}$ and $\{3, 4, 3, 3\}^{\kappa_1}$), we can now apply π as well; the 3-faces remain as they were (that is, octahedra $\{3, 4\}$ or $\{6, 4 \mid 4\}$, respectively); the facet of the first is the universal apeirotope $\{\{3, 4\}, \{4, 4 \mid 4\}\}$ (we comment on this further in Section 8).

It is a striking fact that all three Petrie-Coxeter apeirohedra in $\mathbb{E}^3$ occur as 3-faces of regular 5-apeirotopes in $\mathbb{E}^4$ (one of them twice).

We next discuss the recent (as yet unpublished) classification of the four-dimensional (finite) regular polyhedra. Those polyhedra with planar faces were all found in [1, 2]; the methods we employ in [15] are akin to those used in [14, 16], and are, we feel, much simpler.

As we have already pointed out in Section 3, our strategy is to determine what possible dimension vectors can occur, and then to enumerate every polytope in the corresponding subclasses. Theorem 3.2 provides a starting point; in the current case, the dimension vector must satisfy

$$\dim R_0 \geq 1, \quad \dim R_1 \geq 2, \quad \dim R_2 \geq 2.$$

We now proceed as follows. As essentially the same trick we perform in $\mathbb{E}^3$, if the mirror R_0 satisfies $\dim R_0 = 1$, then we can replace it by

$$-R_0 = R_0^\perp,$$

its orthogonal complement, which (as an isometry) is its product with the central inversion $-I$; we refer to this more general operation as κ_0 as well. We always obtain another finite group $\mathcal{G}'$; in fact,

$$|\mathcal{G}'| = \tfrac{1}{2}|\mathcal{G}|, \ |\mathcal{G}|, \ \text{or } 2|\mathcal{G}|.$$

Next, if $\dim R_0 = 2$ and $\dim R_2 = 3$ (or vice versa, but this case will have to be excluded), then we can replace R_0 by $R_0 R_2$, that is, apply (or reverse) the Petrie operation π; bearing in mind (3.1), the new R_0 has dim $R_0 = 1$ or 3, and in the former case we can proceed as previously.

Finally, as long as our (possibly new) group contains a hyperplane reflexion (that is, dim $R_j = 3$ for some j), we can regard $\mathcal{G}$ as a reflexion (Coxeter) group, on which certain involutions with 2-dimensional mirrors act as automorphisms (more precisely, $\mathcal{G}$ is the corresponding semi-direct product). When we have carried out the foregoing procedures, only the dimension vectors $(3, 2, 3)$ and $(2, 3, 2)$ need to be considered. For classification purposes, we then reverse the procedure: the starting point is a Coxeter group, not necessarily with standard generators, which can be represented by a diagram that permits permutation of its nodes.

We give a couple of simple examples of what happens in the cases $(3, 2, 3)$ and $(2, 3, 2)$ in a little detail, and then comment on the remaining cases (with the exception of $(2, 2, 2)$) more briefly. We list them according to their dimension vectors.

- $(3, 2, 3)$: from the group $[3, 4, 3]$ of the regular 24-cell, we derive the diagrams

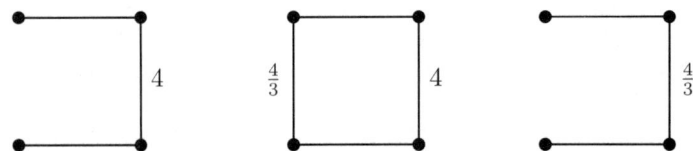

each of which permits a top-to-bottom flip, and thereby gives two dual regular polyhedra with dimension vectors $(3, 2, 3)$. (From the first diagram, we obtain the polyhedra $\{4, 8 | 3\}$ and $\{8, 4 | 3\}$ of [**3**].) Similar examples derive from the diagram

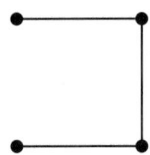

- $(2, 3, 2)$: the general case is derived from a diagram

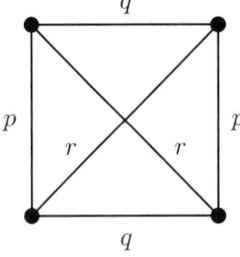

with horizontal and vertical flips. This gives rise to a polyhedron of type $\{2p, 2q\}_{2r}$, from which are obtained up to five others by duality and Petriality. (There is a restriction on q: it must not be a fraction with even denominator.) As a specific instance, the full family of six is obtained when $\{p, q, r\} = \{3, 4, \frac{4}{3}\}$.

- $(3, 3, 3)$: this corresponds to three-dimensional polyhedra, and so is excluded (but only on these grounds).
- $(1, 3, 3)$: this is allowed; κ_0 can be applied to the case $(3, 3, 3)$.
- $(2, 3, 3)$: this is obtained from $(3, 3, 3)$ or $(1, 3, 3)$ by Petriality; therefore, the first possibility must be excluded.
- $(3, 3, 2)$: this would be obtained from $(2, 3, 3)$ by duality; however, in the allowed case, the faces of the original are centred at o, and so the dual must be excluded.
- $(1, 3, 2)$: this would be obtained from $(3, 3, 2)$ by applying κ_0, and so is also disallowed.
- $(1, 2, 3)$: this arises from $(3, 2, 3)$ by applying κ_0.
- $(2, 2, 3)$: this is obtained from $(3, 2, 3)$ or $(1, 2, 3)$ by Petriality.
- $(3, 2, 2)$: this would arise from $(2, 2, 3)$ by duality. However, it may be seen that (with either possibility) the product $R_0 R_1$ of the corresponding reflexions R_0 and R_1 in the original is a double rotation (in two orthogonal planes), since $R_0 \cap R_1 = \{o\}$; it follows that the class cannot occur.
- $(1, 2, 2)$: this would be obtained from $(3, 2, 2)$ by applying κ_0, and so it too must be excluded.

It is notable that only the groups $[3, 3, 3]$ and $[3, 4, 3]$ give rise to polyhedra in the classes $(3, 2, 3)$ and $(2, 3, 2)$ and those derived from them. Even though other finite reflexion groups in $\mathbb{E}^4$ permit diagram automorphisms (for suitably chosen generators), these are inner, and then the corresponding "polyhedra" degenerate.

The anomalous case is dimension vector $(2, 2, 2)$, to which the notion of a Coxeter group with outer automorphisms is inapplicable. Indeed, some examples of this kind cannot be related to Coxeter groups in any meaningful way. The approach here is through quaternions. Each isometry which occurs in such a group is a rotation (that is, lies in SO_4), and so can be represented by a quaternionic transformation of the form

$$(7.1) \qquad\qquad\qquad x \mapsto \bar{a} x b,$$

where a, b are unit quaternions (recall that $a^{-1} = \bar{a}$). In keeping with our usual conventions, mappings are thought of as acting on the right; thus it must be the inverse of a quaternion which provides an appropriate mapping when acting on the left. For the mapping (7.1) to be a reflexion, both a and b must be pure imaginary. Our symmetry group $\mathcal{G}$ gives rise to two groups $\mathcal{G}_L$ and $\mathcal{G}_R$ of the left-acting quaternions a and right-acting quaternions b; then $\mathcal{G}$ is a certain quotient of $\mathcal{G}_L \times \mathcal{G}_R$ (for further details at this stage, we refer the reader to [10]). Further, there are then quotients G_L, G_R of $\mathcal{G}_L$, $\mathcal{G}_R$ in SO_3, each by normal subgroups of index 2, and these are generated by half-turns about lines in $\mathbb{E}^3$. If $a = \cos\vartheta + u\sin\vartheta$, with u pure imaginary, then the image of a under the homomorphism from $\mathcal{G}_L$ to G_L is a rotation through 2ϑ about the axis in $\mathbb{E}^3$ through u, when the latter is regarded as a unit vector in $\mathbb{E}^3$. Thus, for example, if a is pure imaginary, then its image is the half-turn about the axis in $\mathbb{E}^3$ through a; it is important to note that this half-turn lifts to two pure imaginary quaternions $\pm a$. The only groups which can occur as such groups G_L or G_R are dihedral, octahedral or icosahedral; the cyclic and tetrahedral groups do not contain enough half-turns. Finally, if the generating reflexions are

$$x R_j := \bar{a}_j x b_j = -a_j x b_j \quad (j = 0, 1, 2),$$

then (as scalar products of vectors in $\mathbb{E}^3$),

$$\langle a_1, a_2 \rangle = \pm \langle b_1, b_2 \rangle,$$

because the product $R_1 R_2$ must have a 2-dimensional axis. However, the opposite must be true for the product $R_0 R_1$, because this has to be a double rotation.

In summary, the following ingredients go into the enumeration. First, two groups in $\mathbb{E}^3$ generated by half-turns: these are a dihedral group D_{2k} (k can only take the values 2, 3 or 5), the octahedral group $S_4 = [3,3] = [3,4]_3$ or the isosahedral group $A_5 = [3,5]_5$. Second, for $j = 1, 2$, two regular polyhedra of type $\{r_j, q\}$ (with the same q); here, we must allow $r_j > 1$, rather than the usual $r_j \geq 2$, to account for two possible liftings of the half-turns contributing to R_0. We then obtain a polyhedron of type $\{p, q\}$, where the face $\{p\}$ is of the form $\{p_1\} \# \{p_2\}$, with

$$\frac{1}{p_j} = \frac{1}{2} \left(\pm \frac{1}{r_1} \pm \frac{1}{r_2} \right),$$

where the signs are chosen so that $p_j > 2$ for $j = 1, 2$. It is convenient to write the face, instead, as

$$\left\{ \frac{p}{d_1, d_2} \right\}, \qquad \text{with} \quad p_j = \frac{p}{d_j}$$

(in lowest terms) for $j = 1, 2$.

As a specific example, if $r_1 = 3$, $r_2 = \frac{5}{2}$ and $q = 5$, then we obtain a polyhedron of type

$$\{ \tfrac{30}{1,11}, 5 \}.$$

However, if we replace $\frac{5}{2}$ by $\frac{5}{3}$ (or 3 by $\frac{3}{2}$), indicating a different choice of lifting for R_0, then we obtain type

$$\{ \tfrac{15}{2,7}, 5 \}.$$

REMARK 7.1. A further comment is in order here. An opposite orthogonal transformation of $\mathbb{E}^4$ is of the form

$$x \mapsto \overline{a} \, \overline{x} b,$$

with a, b as before. In a group $\mathcal{G}$ containing such transformations, the corresponding left and right groups $\mathcal{G}_L$ and $\mathcal{G}_R$ must be conjugate in the whole group of unit quaternions. Thus one could also use quaternions to investigate the classes other than $(2, 2, 2)$; however, the methods which we have already described are more efficacious.

8. Open problems

As the dimension increases, so there are more possibilities for the ranks of faithfully realized regular or chiral polytopes or apeirotopes. In full rank, the regular cases are classified, and chirality does not occur. In $\mathbb{E}^4$, therefore, the open cases are the (finite) chiral polytopes of rank 3, and the regular or chiral apeirotopes of ranks 3 and 4.

We look at the regular cases first; we begin with rank 4. Each of the eight regular 4-apeirotopes in $\mathbb{E}^3$ can be blended (mixed) with a segment or an apeirogon; this gives 16 blended examples. Next, the "apeir" construction described at the end of Section 3 can be applied to any of the four-dimensional rational regular polyhedra. Finally, certain of the facets of the regular apeirotopes of full rank in

$\mathbb{E}^4$ are 4-apeirotopes. It is possible that there are not too many more examples which do not fall under one of these three categories, and maybe even none at all.

Incidentally, there is only one four-dimensional 4-apeirotope whose facets are finite regular polyhedra. This is the universal $\{\{3,4\},\{4,4\,|\,4\}\}$, with facet the octahedron $\{3,4\}$ and vertex-figure the toroid $\{4,4\,|\,4\}$, which, as noted in Section 7, is the facet of the 5-apeirotope $\{3,4,3,3\}^\pi$. (Compare [17, Theorem 10B3] with $s = 4$ in the dual form, and the preceding discussion.) To see that this is the only example, observe that there are no four-dimensional (finite) regular polyhedra with triangular faces (nor with pentagons or pentagrams either, but these must be excluded on crystallographic grounds). Hence, the only possible vertex-figure has square faces, which means that the facet must be an octahedron or its Petrial $\{6,4\}_3$. In turn, the vertex-figure must be a regular polyhedron with square faces, and circumradius equal to its edge-length; this forces it to be $\{4,4\,|\,4\}$. Finally, direct calculation shows that, in fact, $\{6,4\}_3$ cannot actually be a facet in such a way.

As for the four-dimensional regular apeirohedra, a mere glance at some of the possibilities shows that the enumeration problem is likely to be rather hard. For example, in $\mathbb{E}^2$ the apeirohedron $\{\frac{5}{2},10\}$ is non-discrete; however, when it is blended with its isomorphic copy $\{5,\frac{10}{3}\}$ in $\mathbb{E}^4$, a discrete regular apeirohedron of type $\{5,10\}$ is obtained. Several similar examples also occur.

There are also examples derived from complex reflexion groups in $\mathbb{C}^2$, which we regard as real groups in $\mathbb{E}^4$ generated by reflexions with 2-dimensional mirrors. A curiosity is the following. We can twist the first of the two diagrams below by the dihedral group D_3 (or symmetric group S_3), and the second by C_2. We then actually obtain the same geometric group; however, the outer automorphisms of one correspond to the generating reflexions of the other (and vice versa). We refer to [17, Section 9D] for the background here.

We now turn to chiral polytopes and apeirotopes. For the latter, various infinite families of chiral apeirohedra were described in [19, 20] (see Section 6); each such apeirohedron can be blended with a segment or an apeirogon to give a four-dimensional chiral apeirohedron. Finally, there are plenty of finite chiral polyhedra in $\mathbb{E}^4$; for example, each chiral toroid $\{4,4\}_{(s,t)}$ is realizable. Whether there exist non-toroidal finite chiral polyhedra in $\mathbb{E}^4$ is a nice open question.

Finally, presentations for the symmetry groups have only been fully worked out for the 3-dimensional regular polyhedra and apeirotopes (see [17, Sections 7E, 7F]). For higher dimensions, presentations are known for certain classes of polytopes, for example, the regular star-polytopes (see [17, Section 7D] or [13]). In this context, the main tool is the so-called "circuit criterion", which states that the automorphism group of an abstract polytope $\mathcal{P}$ (and thus the symmetry group of a faithful realization) is determined by the group of its vertex-figure and the circuit structure of the edge-graph of $\mathcal{P}$ (see [17, Section 2F] for more details). A variant of this method should also succeed in the chiral case. In particular, there is an interest in presentations for the symmetry groups of the 3-dimensional chiral apeirohedra. Here we do not know if the corresponding abstract apeirohedra are also chiral or if they are regular. Settling this question may have to be the first step in arriving at presentations for their symmetry groups.

References

[1] Arocha, J.L., Bracho, J. and Montejano, L. *Regular projective polyhedra with planar faces, I.* Aequationes Math. **59** (2000), 55–73.

[2] Bracho, J. *Regular projective polyhedra with planar faces, II.* Aequationes Math. **59** (2000), 160–176.

[3] Coxeter, H.S.M. *Regular skew polyhedra in 3 and 4 dimensions and their topological analogues.* Proc. London Math. Soc. (2) **43** (1937), 33–62. (Reprinted with amendments in *Twelve Geometric Essays*, Southern Illinois University Press, Carbondale, 1968, 76–105.)

[4] Coxeter, H.S.M. *Regular Polytopes* (3rd edition), Dover, New York, 1973.

[5] Coxeter, H.S.M. *Regular Complex Polytopes* (2nd edition), Cambridge University Press, Cambridge, 1991.

[6] Coxeter, H.S.M. *Regular and semi-regular polytopes, III.* Math. Z. **200** (1988), 3–45. (Reprinted in *Kaleidoscopes: Selected Writings of H.S.M. Coxeter*, eds. Sherk, F.A., McMullen, P., Thompson, A.C. and Weiss, A.I., Wiley-Interscience, New York, etc., 1995, 313–355.)

[7] Coxeter, H.S.M. and Moser, W.O.J. *Generators and Relations for Discrete Groups* (4th edition), Springer, New York, etc., 1980.

[8] Dress, A.W.M. *A combinatorial theory of Grünbaum's new regular polyhedra, I: Grünbaum's new regular polyhedra and their automorphism group.* Aequationes Math. **23** (1981), 252–265.

[9] Dress, A.W.M. *A combinatorial theory of Grünbaum's new regular polyhedra, II: complete enumeration.* Aequationes Math. **29** (1985), 222–243.

[10] Du Val, P. *Homographies, Quaternions and Rotations*, Oxford University Press, Oxford, 1964.

[11] Grünbaum, B. *Regular polyhedra – old and new.* Aequationes Math. **16** (1977), 1–20.

[12] Grünbaum, B. *Acoptic polyhedra.* In *Advances in Discrete and Computational Geometry*, Contemp. Math., vol. 223, eds. Chazelle, B. et al., American Mathematical Society, Providence, RI, 1999, 163–199.

[13] McMullen, P. *The groups of the regular star-polytopes.* Canad. J. Math. (2) **50** (1998), 426–448.

[14] McMullen, P. *Regular polytopes of full rank.* Discrete Comput. Geom. **32** (2004), 1–35.

[15] McMullen, P. *Four-dimensional regular polyhedra* (in preparation).

[16] McMullen, P. and Schulte, E. *Regular polytopes in ordinary space.* Discrete Comput. Geom. **17** (1997), 449–478.

[17] McMullen, P. and Schulte, E. *Abstract Regular Polytopes*, Encyclopedia of Mathematics and its Applications, vol. 92, Cambridge University Press, Cambridge, 2002.

[18] Monson, B.R. and Weiss, A.I. *Realizations of regular toroidal maps.* Canad. J. Math. (6) **51** (1999), 1240–1257.

[19] Schulte, E. *Chiral polytopes in ordinary space, I.* Discrete Comput. Geom. **32** (2004), 55–99.

[20] Schulte, E. *Chiral polytopes in ordinary space, II* (in preparation).

[21] Schulte, E. and Weiss, A.I. *Chiral polytopes.* In *Applied Geometry and Discrete Mathematics (The Victor Klee Festschrift)*, DIMACS Series in Discrete Mathematics and Theoretical Computer Science, vol. 4, eds. Gritzmann, P. and Sturmfels, B., Amer. Math. Soc. and Assoc. Computing Machinery, 1991, 493–516.

[22] Wills, J.M. *Combinatorially regular polyhedra of index 2.* Aequationes Math. **34** (1987), 206–220.

DEPARTMENT OF MATHEMATICS, UNIVERSITY COLLEGE LONDON, GOWER STREET, LONDON WC1E 6BT, ENGLAND
E-mail address: `p.mcmullen@ucl.ac.uk`

DEPARTMENT OF MATHEMATICS, NORTHEASTERN UNIVERSITY, BOSTON, MA 02115 USA
E-mail address: `schulte@neu.edu`

Polytopes, Honeycombs, Groups and Graphs

Barry Monson and Asia Ivić Weiss

ABSTRACT. H. S. M. Coxeter had a deep understanding of several branches of geometry; but it was to regular and chiral structures that he returned most often during his long, productive and influential life. Here we take a tour through some of his many contributions to the theories of regular polytopes, maps, and symmetric graphs. We end with a mention of recent results of our own concerning regular self-dual 4-polytopes and 3-transitive graphs.

Dedicated to the memory of Donald Coxeter, our friend, teacher and supervisor (PhD's 1979 and 1981, respectively)

1. Introduction: from maps to honeycombs and polytopes

The study of abstract polytopes may be said to have begun with the study of regular maps, which are essentially rank 3 abstract polytopes with a high degree of symmetry. The theory of regular maps is beautifully conveyed in the frequently cited book [24], written jointly by Coxeter and his student Willy Moser (PhD 1957). There they credit the first notion of a map to Kepler, who in 1619 stellated the regular dodecahedron to obtain the star polyhedron $\{\frac{5}{2}, 5\}$, which can be considered a map of type $\{5, 5\}$, that is, a map on a surface of genus 4, consisting of twelve pentagons, with five meeting at each vertex.

Following Brahana [4], Coxeter [10, 12] defined a *regular map* to be a map whose symmetry group contains two particular automorphisms: σ_1 which cyclically permutes the edges of a face and σ_2 which cyclically permutes the edges meeting a vertex of this face. (See Figure 1; we note that the term 'regular' is commonly used in the narrower sense explained in Section 4 below.) If p and q are the periods of σ_1 and σ_2 respectively, the map is said to have type, or Schläfli symbol, $\{p, q\}$; all faces are then p-gons, with q of them surrounding each vertex.

In 1948 Coxeter enumerated all regular maps on a torus. (The corresponding enumeration for genus 2 was done earlier by Threlfall in 1932 [45].) Indeed, any regular map of type $\{p, q\}$ can be obtained by suitable identifications from its universal cover, namely from the corresponding two-dimensional honeycomb or tessellation

2000 *Mathematics Subject Classification.* Primary 51M20; Secondary 05C25, 51F15.
The first author acknowledges support from the NSERC of Canada Grant #4818.
The second author acknowledges support from the NSERC of Canada Grant #8857.

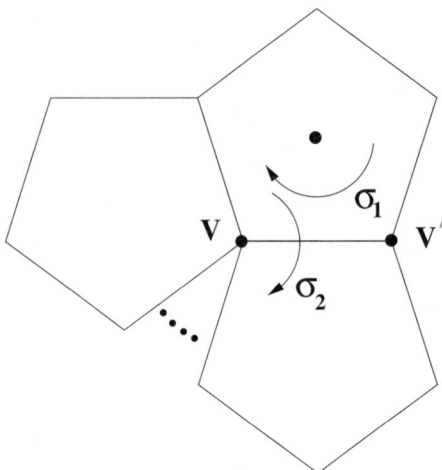

FIGURE 1. Generating symmetries for a regular map.

with Schläfli symbol $\{p, q\}$ on the sphere, the Euclidean plane or the hyperbolic plane. As Coxeter proved in [**12**], the Euclidean tessellations $\{4, 4\}, \{3, 6\}$, and $\{6, 3\}$ cover all possible regular maps on a torus.

The product $\sigma_1 \sigma_2$ reverses an edge by interchanging its two vertices (such as V and V' in Figure 1), as well as the two faces containing that edge. Thus $(\sigma_1 \sigma_2)^2 = 1$. The symmetry group of the map is in fact generated by σ_1 and σ_2, which satisfy at least the relations

$$(1.1) \qquad\qquad \sigma_1^p = \sigma_2^q = (\sigma_1 \sigma_2)^2 = 1.$$

The various commutators of σ_1, σ_2 move a vertex of the map two steps along a *Petrie polygon*, in which each two consecutive sides, but no three, belong to a face of the regular map. (J. F. Petrie was the first to recognize the importance of this device, which is actually a skew polygon in the familiar setting of the regular polyhedra.) When a regular map can actually be obtained from its universal cover by simply identifying vertices separated by r steps along Petrie polygons, it is customary to denote the map by $\{p, q\}_r$.

Petrie's mathematical ideas were mainly transmitted to us by his friend, Donald Coxeter [**10, 18**]. In 1926, Petrie was first to generalize the concept of a regular skew polygon to that of a *regular skew polyhedron*. He discovered two infinite skew polyhedra: one consisting of squares, six at each vertex, and the other of hexagons, four at each vertex. Donald, after being told of this marvel, at once found the third and final possibility in ordinary Euclidean space: hexagons, six at each vertex. These polyhedra can be described by an extension of the Schläfli symbol, namely $\{p, q \,|\, n\}$, in which holes are specified as having length n. (A *hole* is a polygonal chain of edges in the map, which at each vertex leaves two faces of the map on say the right side, and so $q - 2$ on the left (see Figure 2).) Then the three possible skew polyhedra are: $\{4, 6 \,|\, 4\}, \{6, 4 \,|\, 4\}$ and $\{6, 6 \,|\, 3\}$. The group of $\{p, q \,|\, n\}$ is easily seen to be defined by (1.1), along with the relation

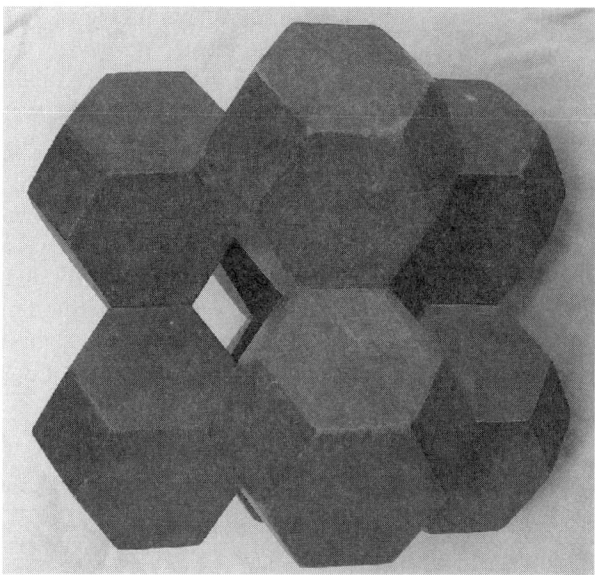

FIGURE 2. Photo of a model of $\{6, 4|4\}$ from Coxeter's collection.

$$(\sigma_1 \sigma_2^{-1})^n = 1.$$

These ideas of Petrie inspired Donald to work on the topological extension of the theory and write in 1937 the paper [10], which can be considered a cornerstone of the modern theory of abstract polytopes. This work deals with symmetry groups of polyhedra of type $\{p, q \,|\, n\}$ and is followed by the closely related paper [11], published in 1939 and dealing with the symmetry group $G^{p,q,r}$ of a map $\{p, q\}_r$. (The finite maps of this type known to 1980 or so are listed in Table 8 of [24].)

Coxeter calls a regular map *reflexible* if it possesses an automorphism that interchanges the two vertices V, V' without interchanging the faces (Figure 1). The group of automorphisms of such a map must be transitive on flags in the poset of vertices, edges and faces of the map, as ordered by rank and incidence. If there is no such automorphism, the flags of a regular map will fall into two orbits. In the first edition of *The Fifty-nine Icosahedra* [21], the term 'unselfreflexible' was used for polyhedra with this feature. In later editions of the book and in other more recent publications (see for example [23]), the different term *chiral* is used instead, to indicate that there are two versions of a polyhedron, each of which can be reflected into the other (like left to right hand). However, 'chiral' still applied to polyhedra whose group was not necessarily transitive on the faces, like the snub cube or snub dodecahedron. Today, the term chiral (cf. Section 4) is mainly reserved for abstract polytopes whose group of symmetries has exactly two flag orbits, with adjacent flags in distinct orbits.

Quite early on Coxeter was intrigued with the idea of moving from maps and two-dimensional honeycombs to dimension three [14, 15]; but it wasn't until 1970 that he surveyed his efforts in the monograph *Twisted Honeycombs* [17]. There Coxeter followed Sommerville [44] in defining a *regular honeycomb* to be a symmetrical division of a 3-dimensional manifold into polyhedral cells, all alike, such that each rotational symmetry of a cell is also a symmetry of the whole configuration. Several examples of such honeycombs, mostly derived from tessellations of hyperbolic 3-space, are listed in Tables 2 and 3 of [17].

Regular tessellations of hyperbolic 3-space were studied by Schlegel in 1883 [39], who found that there are just four regular honeycombs

$$\{3,5,3\}, \ \{4,3,5\}, \ \{5,3,4\}, \ \text{and} \ \{5,3,5\},$$

each of whose cells is a finite polyhedron (that is, with each vertex an ordinary point in hyperbolic space). In 1950, Coxeter and Whitrow extended the list to non-compact type, thereby allowing infinite cells, or vertices where infinitely many cells meet [27]. In an invited address at the International Congress of Mathematicians in 1954 in Amsterdam, Coxeter gave a complete classification of regular hyperbolic honeycombs (see Table 1).

TABLE 1. Tessellations of $\mathbb{H}^n$, $n \geq 2$.

Rank	Schläfli symbol
$n = 2$	$\{p,q\}$ with $1/p + 1/q < 1/2$
$n = 3$	$\{3,5,3\}\{4,3,5\}\{5,3,4\}\{5,3,5\}$
	$\{3,3,6\}\{6,3,3\}\{3,4,4\}\{4,4,3\}$
	$\{6,3,5\}\{5,3,6\}\{4,3,6\}\{6,3,4\}$
	$\{3,6,3\}\{6,3,6\}\{4,4,4\}$
$n = 4$	$\{3,3,3,5\}\{5,3,3,3\}\{4,3,3,5\}\{5,3,3,4\}\{5,3,3,5\}$
	$\{3,4,3,4\}\{4,3,4,3\}$
$n = 4$	$\{3,3,3,4,3\}\{3,4,3,3,3\}\{3,3,4,3,3\}\{3,4,3,3,4\}\{4,3,3,4,3\}$

The direct symmetry group $[p,q,r]^+$ of a 3-dimensional regular spherical, Euclidean or hyperbolic honeycomb $\{p,q,r\}$ is generated by three rotations σ_1, σ_2, and σ_3 satisfying the relations

$$(1.2) \qquad \begin{aligned} \sigma_1^p = \sigma_2^q = \sigma_3^r &= 1, \\ (\sigma_1\sigma_2)^2 = (\sigma_2\sigma_3)^2 = (\sigma_1\sigma_2\sigma_3)^2 &= 1 \end{aligned}$$

[46, p. 217], [17, p. 28]. In 1970, Coxeter gave the first survey of rank 4 chiral polytopes [17]. (However, there is considerable earlier work on 3-manifolds whose fundamental groups are say binary polyhedral groups; some of these spaces are essentially chiral 4-polytopes; see for example [42, §61–62].) Coxeter's examples are derived from the Euclidean and hyperbolic tessellations by identifications along Petrie polygons (which for 3-dimensional honeycombs are defined to be polygons in which every three, but no four, consecutive edges belong to a Petrie polygon of a cell). The so-called 'twisted honeycombs' then arise by 'forcing' right-handed and left-handed Petrie polygons of a tessellation to be of different lengths. The length of one of these Petrie polygons is the order of $\sigma_1\sigma_3$ and the length of the other is the order of $\sigma_1\sigma_3^{-1}$. If the orders of these two elements are distinct after identifications, then the lengths of the two Petrie polygons are different; the symmetry group of the honeycomb contains no reflections; and the honeycomb must be chiral.

Many of these honeycombs were obtained through their groups, often by enumerating cosets of suitable subgroups of groups of type $((p,r,t;q))$, in Coxeter's notation. The well-known algorithm used here for coset enumeration was devised

FIGURE 3. W.A. Wythoff's projection of the 120-cell $\{5,3,3\}$, drawn about 1920, presented to Coxeter by S.L. van Oss about 1930. This manuscript is now at York University.

by Coxeter and Todd [**47**] in 1936 (and implemented on a computer by John Leech and by Abraham Sinkov). The Todd-Coxeter algorithm is a systematic procedure for enumerating the cosets of a subgroup of finite index in a group given by generators and relations. The algorithm has been an important component in most computer programs to date dealing with symbolic calculation in algebra.

Although Petrie's discoveries were an important incentive in Coxeter's study of reflexible maps and their topological analogues, Coxeter's interest in the subject without doubt arose naturally from his earlier work on reflection groups.

At Cambridge, both Coxeter (PhD 1931) and Todd (PhD ca 1930) were students of H. F. Baker. Inspired by the work of his fellow student, Coxeter began to investigate the properties of groups generated by reflections. Todd [**46**] had studied the groups of symmetries of the classical regular polytopes. For each such group he also computed the period of the product (in a certain order) of all the generating reflections. Later, Coxeter noted the connection with Petrie polygons in the polytope [**18**, p. 92] and derived simple formulae for its length. In 1933, while at Princeton, Coxeter then proved that all continued products of the specified generators of a finite group generated by reflections are conjugate. Such an element of the group is now commonly called a *Coxeter element* and its order the *Coxeter number* for the group.

The earliest accounts of Coxeter's research on reflection groups are found in [**7**] and [**6**]. He there investigated the properties of a group W generated by reflections in the facets $R_0, \ldots, R_{n-1}$ of a polytope P, in which the dihedral angle between the j-th and k-th facets is π/p_{jk}, where each p_{jk} is an integer ≥ 2 (or equals ∞ when the facets are parallel). Most importantly, W is an example of what is now called a *Coxeter group*, that is an abstract group generated by elements ρ_j, $j \in \{0, \ldots, n-1\}$, subject to defining relations

$$(\rho_j \rho_k)^{p_{jk}} = 1,$$

where $p_{jj} = 1$ and $p_{jk} = p_{kj} \in \{2, 3, \ldots, \infty\}$, for all $j \neq k \in \{0, \ldots, n-1\}$. The spherical and Euclidean reflection groups were first studied systematically and finally completely classified by Coxeter in a sequence of papers [**7, 8, 9**] appearing in the early 1930s. To facilitate the exposition, he made use of a graph, now known as a *Coxeter* (or *Dynkin*) *diagram*, whose nodes represent the facets of the polytope P. Two such nodes are connected by a branch labeled by the integer p_{jk}, whenever the angle between the corresponding facets is π/p_{jk} with $p_{jk} \geq 3$. If instead the facets are perpendicular, so that $p_{jk} = 2$, the corresponding nodes are not connected.

2. Coxeter groups

Another fellow student at Cambridge (1927-1931) was Gilbert de B. Robinson, with whom Donald had a very long friendship and who was instrumental in bringing Donald to Toronto. As a student of A. Young, Robinson had worked mostly in group theory; but he had as well an interest in geometry and so attended Baker's tea parties, where in 1928 he met Coxeter. Robinson pointed out to Coxeter how the orbit of a certain point in a fundamental region of a reflection group yields the vertices of a symmetric polytope. This procedure, first carried out for the polytope $\{3, 3, 5\}$ by Wythoff [**50**], had been generalized by G. de B. Robinson [**38**]. The construction, used by Coxeter in his dissertation, in an attempt to classify uniform polytopes, is now a standard procedure in the construction of regular polytopes [**18**, pp. 196-204], [**34**, ch. 5.4].

3. Symmetric graphs

Certain graphs arise naturally in the study of maps and polytopes. It is therefore not surprising that Coxeter had considerable interest in graphs. We recall an incident in 1980, when Donald encountered a very beautiful graph with 28 vertices. When Asia, then Donald's PhD student, arrived at his office for their regular Friday meeting, Donald excitedly asked her if she had ever seen such a graph. Surely, being so symmetric and having so few vertices, it must be known? Asia could not help, so Donald proceeded to phone Bill Tutte. After carefully listening to the description, Tutte informed Donald that the graph was indeed known. To Coxeter's great surprise and delight, the graph was commonly called the *Coxeter graph*, it having been invented some thirty years earlier by Coxeter himself. This graph is

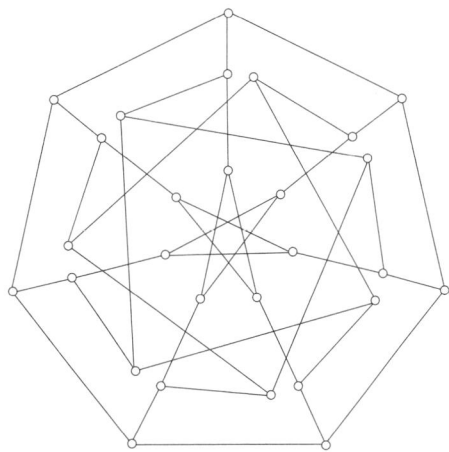

FIGURE 4. The Coxeter graph.

easily described as consisting of the four heptagons: $\{\frac{7}{0}\}$ (that is a heptagon with no edges), $\{\frac{7}{1}\}$, $\{\frac{7}{2}\}$, and $\{\frac{7}{3}\}$, with corresponding vertices joined as shown in the Figure 4, where the $\{\frac{7}{0}\}$ and $\{\frac{7}{2}\}$ are concyclic. (The graph was used as the logo for the Coxeter Legacy Conference.)

Several weeks later, having discovered some new and interesting connections between the graph and the finite projective plane $PG(2,7)$, Donald wrote the manuscript for *My graph* [20]. It was about then that Asia reminded Donald that they might have to postpone their next weekly meeting – she was expecting and soon due to give birth to her first child. When Donald realized that she might well have to spend several hours in the labour room, he gave her the fresh manuscript commenting, "This is something you can read while waiting", with a reassuring "Don't worry. It is easy to read, mostly a survey".

In late 1940s, Coxeter began the first of many investigations into configurations and graphs. He justified the study using the general principle that interesting configurations are represented by interesting graphs [13, p. 416]. A deeper mathematical study of the symmetric graphs was really initiated by Tutte in his PhD dissertation (Cambridge 1948). Immediately following his studies, and at Coxeter's urging, Tutte joined the faculty of the University of Toronto, where he remained until 1962, when he moved to the University of Waterloo, remaining there until retirement. Although they had no joint papers, Tutte and Coxeter had extensive discussions and were each very much appreciative and aware of the other's work, which was frequently cited.

In this general area, Coxeter was mainly interested in self-dual configurations in the projective plane, such as the Pappus 9_3, the Desargues 10_3 and the Cremona-Richmond 15_3 (whose Levi graph is Tutte's 8-cage) [13]. Central to the investigation are several elegant and general results of Tutte [48]. Because of their intrinsic interest, and because we need them later, we summarize here some key ideas (see [1]).

A finite simply connected graph G, whose automorphism group Aut(G) acts transitively on vertices, is said to be *vertex transitive*. Clearly G is then *k-valent* for some integer $k \geq 0$, so that k edges are incident with each vertex. Of particular interest to us are *3-valent* (or *cubic*) graphs (such as the Coxeter graph).

A vertex transitive graph is *symmetric* if the stabilizer of each vertex is transitive on the vertices adjacent to it. Hence, a symmetric graph is transitive on edges, even on 1-arcs. By a *t-arc* we mean an ordered list of vertices $[v] = [v_0, v_1, \ldots, v_t]$ such that $\{v_{i-1}, v_i\}$ is an edge for $1 \leq i \leq t$, but no $v_{i-1} = v_{i+1}$. The graph G is *t-transitive* if $\text{Aut}(G)$ is transitive on t-arcs, but not on $(t+1)$-arcs. (Coxeter and some others have used the equivalent but here confusing term 't-regular'.) One remarkable property of a trivalent t-transitive graph is that $\text{Aut}(G)$ must be sharply transitive on the t-arcs in G. Tutte also proved that if $\text{Aut}(G)$ is transitive on the t-arcs in the trivalent graph G, then $t \leq 5$. For example, the Coxeter graph is 3-transitive, and its automorphism group is $PGL_2(7)$ of order 336.

The first systematic survey of symmetric trivalent graphs was undertaken in the 1920s by two electrical engineers, R. M. Foster and G. M. Campbell, who worked on telephone substation and repeater circuits for AT&T. In 1932, Foster published drawings of nine symmetric cubic graphs [**29**]. He continued working on such graphs and in 1966, at the Conference on Graph Theory and Combinatorial Analysis held at the University of Waterloo, distributed a list of symmetric cubic graphs with up to 400 vertices. In 1988, when Foster was 92, the *Foster Census* was assembled by Izak Bouwer and his colleagues (including Barry). The *Census* includes a foreword by Coxeter, as well as a biographical preface by another Coxeter student, Sy Schuster (PhD 1953). The *Census* contains graphs with up to 512 vertices, and, remarkably, only five of the 198 graphs are known to have been missed by Foster [**1**, p. 147]. Coxeter was aware of Foster's work already in 1950. In [**13**] he credits Foster with the discovery of the three graphs which arise as the Levi graphs of the configurations 8_3, 10_3, and 12_3.

Soon after the Waterloo Conference, Foster suggested, in a letter to Coxeter, further study of *0-symmetric graphs*, namely finite vertex transitive trivalent graphs which are not symmetric. (See the preface to [**22**].) Foster compiled and distributed a list of such graphs in 1975. This was the impetus for the book [**22**], dedicated to Foster by Coxeter and his coauthors.

4. Abstract polytopes: regular and chiral

We indicated in Section 1 that Coxeter's earliest work on skew polyhedra anticipated more recent interest in generalizing the idea of regular polytope as a combinatorial object. For example, in [**33**] P. McMullen defined the combinatorial regularity of a convex polytope by using flag-transitivity, a property which is intrinsic to the face lattice. Then, in 1976, a real push toward the development of a comprehensive theory was made by Grünbaum, in a short but very influential paper [**31**]. There Grünbaum proposed that structures as diverse as polytopes, graphs and projective planes be viewed as posets of a very general sort, which he called *polystromata*. Even with the imposition of further conditions, such an object might behave locally like a classical polytope, yet globally have far more general topological type. In fact, at about this time, Coxeter and G. Shephard [**25**] produced a rank 4 example with toroidal cells. In combinatorial terms, this object is an abstract regular polytope, the general theory of which was really initiated by L. Danzer and E. Schulte in 1980 [**28**].

An (*abstract*) *n-polytope* is a partially ordered set $\mathcal{P}$, equipped with a strictly monotone rank function having range $\{-1, 0, 1, \ldots, n\}$. An element of rank j is called a *j-face* and is typically denoted by F_j; faces of rank $0, 1$ and $n-1$ are called *vertices*, *edges* and *facets*, respectively. We further require that $\mathcal{P}$ have two (improper) faces: a unique least face F_{-1} and a unique greatest face F_n; that each flag of $\mathcal{P}$ have precisely $n+2$ faces; that $\mathcal{P}$ be strongly flag-connected; and that $\mathcal{P}$ have a homogeneity property: whenever $F < G$ and $\text{rank}(G) - \text{rank}(F) = 2$, there are precisely two faces H such that $F < H < G$. We refer to [**34**, ch. 2] for further details.

The *automorphism group* $\text{Aut}(\mathcal{P})$ for $\mathcal{P}$ consists of all order preserving bijections of $\mathcal{P}$. If $\mathcal{P}$ also admits an order reversing bijection, or *duality*, we say that $\mathcal{P}$ is *self-dual*. Clearly, in this case $\text{Aut}(\mathcal{P})$ is a subgroup of index 2 in the extended group $D(\mathcal{P})$ of all automorphisms and dualities. An involutory duality is called a *polarity*.

It follows at once from the definitions that a 0-polytope can be seen as a single point and a 1-polytope as an ordinary line segment. Each 2-polytope will have an equal number, say p, of vertices and edges arranged in familiar cyclic fashion. Naturally then we call this rank 2, self-dual polytope a *p-gon* and denote it by $\{p\}$, where $p \in \{2, 3, \ldots, \infty\}$.

Suppose for $1 \leq j \leq n-1$ that G and H are incident faces of $\mathcal{P}$ with ranks $j-2$, $j+1$, respectively. Then the rank 2 section $\{F \in \mathcal{P} \,|\, G \leq F \leq H\}$ is a 2-polytope in its own right. If, for each fixed $j \in \{1, \ldots, n-1\}$, all such sections are isomorphic to some p_j-gon, we say that $\mathcal{P}$ is *equivelar*, with type (or Schläfli symbol) $\{p_1, p_2, \ldots, p_{n-1}\}$. Each facet of an equivelar polytope is equivelar of type $\{p_1, \ldots, p_{n-2}\}$; and each face of rank $n-3$ of $\mathcal{P}$ is shared by p_{n-1} facets. If $\mathcal{P}$ is a self-dual equivelar polytope of type $\{p_1, \ldots, p_{n-1}\}$, then $p_j = p_{n-j}$ for $j = 1, \ldots, n-1$.

An n-polytope $\mathcal{P}$ is said to be *regular* if $\text{Aut}(\mathcal{P})$ is transitive on flags. Regular polytopes are necessarily equivelar, say of type $\{p_1, \ldots, p_{n-1}\}$. Then one can prove that $\text{Aut}(\mathcal{P})$ is generated by n involutions $\rho_0, \rho_1, \ldots, \rho_{n-1}$ satisfying at least the relations

$$(4.1) \qquad \begin{array}{rcll} \rho_j^2 & = & 1, & j = 0, \ldots, n-1; \\ (\rho_{j-1}\rho_j)^{p_j} & = & 1, & j = 1, \ldots, n-1; \\ (\rho_j\rho_k)^2 & = & 1, & |k-j| \geq 2. \end{array}$$

Thus $\text{Aut}(\mathcal{P})$ is a quotient of the Coxeter group with the string diagram

$$\bullet \!\!-\!\!\!-\!\!\!-\!\!\! \bullet \!\!-\!\! \quad \cdots \quad -\!\! \bullet \!\!-\!\!\!-\!\!\!-\!\!\! \bullet \ .$$
$$\qquad p_1 \qquad\qquad\qquad\quad p_{n-1}$$

Having fixed a flag Φ in $\mathcal{P}$, then ρ_j can be uniquely defined as the automorphism which maps Φ to the adjacent flag Φ^j (differing from Φ in the j-face only). In analogy with the classical theory, we call such involutions reflections.

In addition to (4.1), the generators of $\text{Aut}(\mathcal{P})$ also satisfy an intersection condition

$$(4.2) \qquad \langle \rho_j \,|\, j \in I \rangle \cap \langle \rho_j \,|\, j \in J \rangle = \langle \rho_j \,|\, j \in I \cap J \rangle \,,$$

for all $I, J \subseteq \{0, 1, \ldots, n-1\}$. Conversely (see [34]), given a group Γ generated by involutions satisfying (4.1) and (4.2), there is regular n-polytope with Schläfli symbol $\{p_1, \ldots, p_{n-1}\}$ and having Γ as its group of automorphisms.

We also note that a regular polytope $\mathcal{P}$ is self-dual if and only if $\mathrm{Aut}(\mathcal{P})$ admits an involutory group automorphism interchanging ρ_j with ρ_{n-j} for $j = 0, \ldots, n-1$ [32, 34]. Hence each regular self-dual polytope possesses a polarity.

Next we consider the 'rotations' $\sigma_j := \rho_{j-1}\rho_j$, $j = 1 \ldots n-1$, which generate a subgroup $\mathrm{Aut}^+(\mathcal{P})$ having index 1 or 2 in $\mathrm{Aut}(\mathcal{P})$. In the latter case, $\mathcal{P}$ is said to be *directly regular*; and then certain natural properties of $\mathrm{Aut}^+(\mathcal{P})$ suggest an abstract setting for the notion of chirality [40, 41].

Assuming $n \geq 3$, we define a *chiral* polytope to be an n-polytope $\mathcal{P}$ for which $\mathrm{Aut}(\mathcal{P})$ has precisely two flag orbits, with adjacent flags in distinct orbits. Hence, chiral polytopes come in two enantiomorphic forms, one associated with a base flag and the other with any adjacent flag. As defined, $\mathcal{P}$ cannot be regular; and now $\mathrm{Aut}(\mathcal{P})$ is generated by $n-1$ automorphisms $\sigma_1, \ldots, \sigma_{n-1}$, which we call *rotations* and which satisfy at least the relations

$$(4.3) \qquad \begin{aligned} \sigma_1^{p_1} &= \ldots = \sigma_{n-1}^{p_{n-1}} = 1, \\ (\sigma_j \sigma_{j+1} \ldots \sigma_k)^2 &= 1 \qquad , \text{ for } 1 \leq j < k \leq n-1, \end{aligned}$$

where each $p_j \in \{2, 3, \ldots, \infty\}$. The group of a chiral polytope also satisfies an intersection condition, which is quite intricate to state for arbitrary rank n. For instance, in rank 4 the condition (on generators $\sigma_1, \sigma_2, \sigma_3$) is simply:

$$(4.4) \qquad \begin{aligned} \langle \sigma_1 \rangle \cap \langle \sigma_2 \rangle &= \{1\} = \langle \sigma_2 \rangle \cap \langle \sigma_3 \rangle, \\ \langle \sigma_1, \sigma_2 \rangle \cap \langle \sigma_2, \sigma_3 \rangle &= \langle \sigma_2 \rangle. \end{aligned}$$

The groups of chiral polytopes and (directly) regular polytopes have been characterized in [40, Th. 1]. Given a group Γ with generators $\sigma_1, \ldots, \sigma_{n-1}$ satisfying the relations (4.3) and an intersection condition (stated only for $n = 4$ in (4.4)), there is an associated chiral or directly regular polytope $\mathcal{P}$ with rotation group Γ. The polytope $\mathcal{P}$ is now regular if and only if there exists an involutory group automorphism ρ of Γ such that $\sigma_1^\rho = \sigma_1^{-1}$, $\sigma_2^\rho = \sigma_1^2 \sigma_2$, and $\sigma_j^\rho = \sigma_j$ for $j \geq 3$.

If $\mathcal{P}$ is a self-dual chiral polytope and δ a duality, then δ either preserves the flag orbits or interchanges them. If δ preserves the orbits, then all dualities must do so, and we say that the polytope is *properly self-dual*. Otherwise, $\mathcal{P}$ is *improperly self-dual*. Furthermore in these circumstances, it is proved in [32] that $\mathcal{P}$ is

(1) properly self-dual if and only if there exists an involutory group automorphism ω of $\mathrm{Aut}(\mathcal{P})$ such that $\sigma_j^\omega = \sigma_{n-j}^{-1}$ for $j = 1, \ldots, n-1$;

(2) improperly self-dual if and only if there exists a group automorphism ω of $\mathrm{Aut}(\mathcal{P})$ such that $\sigma_j^\omega = \sigma_{n-j}^{-1}$ if $j < n-2$; $\sigma_{n-2}^\omega = \sigma_1 \sigma_2 \sigma_1^{-1}$; and $\sigma_{n-1}^\omega = \sigma_1$.

The basic theory of chiral polytopes with rank $n \geq 3$ was established by Schulte and Weiss in 1991 [40] (see also [41]). Of considerable interest are chiral rank 3 polytopes, each of which provides a map on some orientable surface. We mentioned above Coxeter's early work in genus 1 [12]; this gives three infinite families of toroidal chiral maps.

For larger genus the occurrence of chiral maps is rather sporadic, and for some time there was little progress. J. R. Edmonds did find a chiral map of type $\{7, 7\}$ on a surface of genus 7 (unpublished work described in [16, p. 388]). Then in 1962, F. A. Sherk, another of Coxeter's students (PhD 1957), constructed a family of chiral maps of type $\{6, 6\}$ [43], with the smallest map in the family having genus 8. In 1969 Garbe [30] proved that there are no chiral maps on surfaces of genus 2, 3, 4, 5 or 6. Chiral maps of low genus were also studied by Wilson [49]. More recently, all chiral maps of genus up to 15 have been enumerated by Conder and Dobcsányi [5]. Up to enantiomorphism and duality, for genera 7 to 15 there are only thirteen chiral rank 3 polytopes.

5. Symmetric trivalent graphs embedded in polytopes

In 1980, Coxeter observed an interesting connection between the four-dimensional regular convex polytopes and certain symmetric graphs [19]. He wrote, "In 4 dimensions, the convex regular polytope $\{p, q, r\}$ has p-gonal faces; every edge belongs to r of them ... Joining the center of each face to the midpoints of its q edges, we obtain a bipartite graph with alternate p-valent and r-valent vertices". In fact, the same construction works for any abstract equivelar polytope $\mathcal{P}$ of that Schläfli type. Of particular interest are the polytopes of type $\{3, q, 3\}$, which yield trivalent graphs (see [19, 26]). As we shall see below, if $\mathcal{P}$ is self-dual, and either regular or chiral, the associated graph is symmetric.

So henceforth, let us assume that $\mathcal{P}$ is a rank 4 equivelar polytope with Schläfli symbol $\{3, q, 3\}$, where $q \geq 2$. From each such $\mathcal{P}$ we can construct a simple graph $G(\mathcal{P})$, whose vertex set is comprised of all rank 1 and rank 2 faces of $\mathcal{P}$; two such graph vertices are connected by an edge in the graph whenever they are incident as faces in $\mathcal{P}$. It is easy to check that $G(\mathcal{P})$ is connected, bipartite, and trivalent.

Now suppose that $\mathcal{P}$ is regular with fixed base flag $\Phi = \{F_{-1}, F_0, \ldots, F_4\}$ and associated involutory generators ρ_j. These satisfy at least the relations

$$\rho_0^2 = \rho_1^2 = \rho_2^2 = \rho_3^2 = 1,$$
$$(\rho_0\rho_2)^2 = (\rho_0\rho_3)^2 = (\rho_1\rho_3)^2 = 1,$$
(5.1) $$(\rho_0\rho_2)^3 = (\rho_1\rho_2)^q = (\rho_2\rho_3)^3 = 1.$$

If also $\mathcal{P}$ is self-dual, then the extended group $D(P)$ of automorphisms and dualities contains a polarity δ such that

(5.2) $$\delta\rho_0\delta = \rho_3 \text{ and } \delta\rho_1\delta = \rho_2.$$

We now label the vertices of $G(\mathcal{P})$, so that $v_1 := F_1$, $v_2 := F_2$, and $v_0 := v_2\rho_2$, $v_3 := v_1\rho_1$, $v_4 := v_0\rho_1$ (see Figure 5). Thus δ swaps v_1 and v_2, as well as v_0 and v_3; and it is easy to check that $\tau_1 := \rho_2\delta$ and $\tau_2 = \rho_0\rho_2\delta$ map the 3-arc $[v_0, v_1, v_2, v_3]$ to its two 'successor' arcs $[v_1, v_2, v_3, z]$. (In Figure 5, $z = v_4$ and w_3, respectively.) It follows from [48, 7.54] that $\text{Aut}(G(\mathcal{P}))$ is transitive on 3-arcs; furthermore the 'shunts' τ_1, τ_2 actually generate the full automorphism group if the graph is precisely 3-transitive [1, Prop. 18.2].

It is easy to see that the regularity of $\mathcal{P}$ implies that $G(\mathcal{P})$ is transitive on t-arcs, where t is at least 3; it is a little harder to prove that $G(\mathcal{P})$ is precisely 3-transitive [36]:

THEOREM 5.1. *If $\mathcal{P}$ is a finite, regular self-dual polytope of type $\{3, q, 3\}$, then the associated graph $G(\mathcal{P})$ is 3-transitive and $\text{Aut}(G(\mathcal{P})) = D(\mathcal{P})$.*

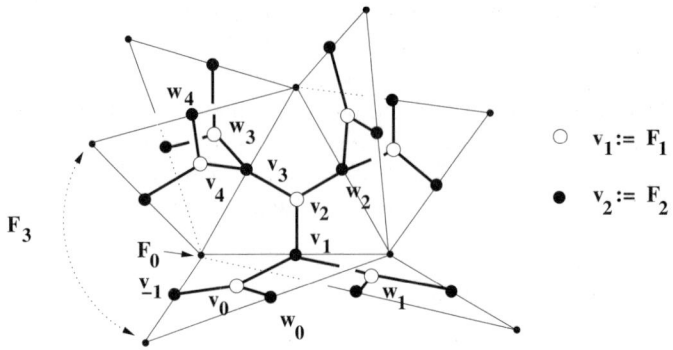

$$\bigcirc \quad v_1 := F_1$$
$$\bullet \quad v_2 := F_2$$

FIGURE 5. The medial layer graph of a polytope of type $\{3, q, 3\}$.

We note that some condition such as self-duality is required to guarantee that the graph be symmetric. For example, there is a regular, but not self-dual polytope $\mathcal{P}$ of type $\{3, 6, 3\}$, whose facets are toroidal maps of type $\{3, 6\}_{(3,0)}$ and whose vertex-figures are other toroidal maps of type $\{6, 3\}_{(1,1)}$ [35, Th. 6.1]. There are $54 = 27 + 27$ faces of ranks 1 and 2, and in fact $G(\mathcal{P})$ is the Gray graph, which is edge-transitive but not vertex-transitive [2, 37].

A little surprisingly, there is a sort of converse to Theorem 5.1. That is, from a graph G with certain natural properties, one can reconstruct an abstract symmetric structure resembling a polytope, in fact one of Grünbaum's polystromata [31]. For our purposes, we can define a polystroma as a partially ordered set with a strictly monotone rank function $\{-1, 0, \ldots, n\}$, and with a unique least face F_{-1} and unique greatest face F_n. (No homogeneity property or flag-connectedness is assumed; but duality and self-duality make sense.)

Assume therefore that the graph G is finite, connected, trivalent, bipartite, and 3-transitive; and let $v = [v_0, v_1, v_2, v_3]$ be a fixed 3-arc of G. Let τ_1 and τ_2 be the shunts mapping v to its successors; thus $\mathrm{Aut}(G) = \langle \tau_1, \tau_2 \rangle$. Let δ be the unique automorphism of G which reverses v; and define $\rho_0 := \tau_1 \tau_2^{-1}$, $\rho_1 := \delta \tau_1$, $\rho_2 := \tau_1 \delta$, $\rho_3 := \tau_1^{-1} \tau_2$. In [36] we show that these ρ_j satisfy the relations (5.1), though not likely the intersection condition (4.2) (for $n = 4$). Guided by [34, ch. 2E], we can then construct a polystroma which is symmetric under the action of the group Γ generated by the ρ_j's. Let's take stock:

THEOREM 5.2. *Suppose that G is a finite, connected, trivalent, bipartite, and 3-transitive graph. The involutions $\rho_0, \rho_1, \rho_2,$ and ρ_3 defined above satisfy the relations implicit in the Coxeter diagram*

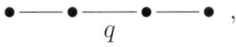

$$q$$

where $2q$ is the order of the shunt τ_1. Furthermore, the group $\Gamma := \langle \rho_0, \rho_1, \rho_2, \rho_3 \rangle$ is a subgroup of index 2 in $\mathrm{Aut}(G)$, and the arc-reversing involution δ conjugates each ρ_j to ρ_{3-j}. If also Γ is a C-group, that is, if it satisfies the intersection condition (4.2), then $\mathrm{Aut}(G)$ is the extended group of a regular self-dual polytope with Schläfli symbol $\{3, q, 3\}$.

In fact, the intersection condition can only fail in a rather subtle way, since the 'facet' group and 'vertex figure' group of the polystroma are always the automorphism groups of regular maps.

For example, $K_{3,3}$ (the Thomsen graph, or Tutte's 4-cage) induces a spherical honeycomb $\{3, 2, 3\}$ [**19**, p. 309], which in fact is a polytope. On the other hand, Foster's graph **56C** induces a polystroma of type $\{3, 7, 3\}$ which is not a polytope.

The situation with chiral polytopes seems to be more complex, perhaps because of subtleties concerning proper and improper self-duality. However, it is easy to show, by way of analogy to Theorem 5.1, that when $\mathcal{P}$ is chiral and self-dual, the graph $G(\mathcal{P})$ is transitive on t-arcs, for some t satisfying $2 \leq t \leq 5$.

References

[1] Biggs, N. *Algebraic Graph Theory* (2nd Ed.), Cambridge Mathematical Library, Cambridge University Press, Cambridge, 1993.

[2] Bouwer, I. Z. *An Edge but not Vertex Transitive Cubic Graph* Canad. Math. Bull. **11** (1968), 533–535.

[3] Bouwer, I. Z. , et al (eds) *The Foster Census: R.M. Foster's Census of Connected Symmetric Trivalent Graphs*, The Charles Babbage Research Centre, Winnipeg, Manitoba, 1988.

[4] Brahana, H. R. *Regular maps and their groups* Amer. J. Math. **49** (1927), 268–284.

[5] Conder, M. and Dobcsányi, P. *Determination of all regular maps of small genus* J. Combin. Theory Ser. B **81** (2001), 224–242.

[6] Coxeter, H. S. M. *The polytopes with regular-prismatic vertex figures, Part I* Philos. Trans. Roy. Soc. London Ser. A **229** (1930), 329–425.

[7] Coxeter, H. S. M. *Groups whose fundamental regions are simplexes* J. London Math. Soc. **6** (1931), 132–136.

[8] Coxeter, H. S. M. *Discrete groups generated by reflections* Ann. of Math. **35** (1934), 588–621.

[9] Coxeter, H. S. M. *The complete enumeration of finite groups $R_i^2 = (R_i R_j)^{k_{ij}}$* J. London Math. Soc. **10** (1935), 21–25.

[10] Coxeter, H. S. M. *Regular skew polyhedra in three and four dimensions, and their topological analogues* Proc. London Math. Soc. (2) **43** (1937), 33–62.

[11] Coxeter, H. S. M. *The abstract groups $G^{m,n,p}$* Trans. Amer. Math. Soc. **45** (1939), 73–150.

[12] Coxeter, H. S. M. *Configurations and maps* Rep. Math. Colloq.(2) **8** (1948), 18–38.

[13] Coxeter, H. S. M. *Self-dual configurations and regular graphs* Bull. Amer. Math. Soc. **56** (1950), 413–455.

[14] Coxeter, H. S. M. *Regular honeycombs in elliptic space* Proc. London Math. Soc. (3) **4** (1954), 471–501.

[15] Coxeter, H. S. M. *Regular honeycombs in hyperbolic space* Proc. Internat. Congress Math. Amsterdam (1954), Vol. 3, North-Holland, Amsterdam, 1956, 155–169.

[16] Coxeter, H. S. M. *Introduction to Geometry* (2nd Ed.), Wiley, New York, 1969.

[17] Coxeter, H. S. M. *Twisted Honeycombs*, Conf. Board Math. Sci. Regional Conference Series in Mathematics No. 4, Amer. Math. Soc., Providence, RI, 1970.

[18] Coxeter, H. S. M. *Regular Polytopes* (3rd Ed.), Dover, New York, 1973.

[19] Coxeter, H.S.M. *The edges and faces of a 4-dimensional polytope* Congr. Numer. **28** (1980), 309–334.

[20] Coxeter, H. S. M. *My graph* Proc. London Math. Soc. (3) **46** (1983), 117–136.

[21] Coxeter, H. S. M., Du Val, P., Flather, H. T. and Petrie, J. F. *The Fifty-nine Icosahedra*, Math. Series No. 6, The University of Toronto Press, Toronto, 1938.

[22] Coxeter, H. S. M., Frucht, R. and Powers, D. L. *Zero-symmetric Graphs: trivalent graphical regular presentations of groups*, Academic Press, New York (1981).

[23] Coxeter, H. S. M. and Huybers, P. *A new approach to the chiral Archimedean solids* C. R. Math. Rep. Acad. Sci. Canada **1** (1979), 269–274.

[24] Coxeter, H.S.M. and Moser, W.O.J. *Generators and Relations for Discrete Groups* (4th Ed.), Springer, New York, 1980.

[25] Coxeter, H. S. M. and Shephard, G. C. *Regular 3-complexes with toroidal cells* J. Combin. Theory B **22** (1977), 131–138.

[26] Coxeter, H.S.M. and Weiss, A.I. *Twisted honeycombs* $\{3,5,3\}_t$ *and their groups* Geom. Dedicata **17** (1984), 169–179.

[27] Coxeter, H. S. M. and Whitrow, G. J. *World structure and non-Euclidean honeycombs* Proc. Roy. Soc. London Ser. A **201** (1950), 417–437.

[28] Danzer, L. and Schulte, E. *Reguläre Inzidenzkomplexe I* Geom. Dedicata **13** (1982), 295–308.

[29] Foster, R. M. *Geometrical circuits of electrical networks* Trans. A.I.E.E. **51** (1932), 309–317.

[30] Garbe, D. *Über die regulären Zerlegungen geschlossener orientierbarer Flächen* J. Reine Angew. Math. **237** (1969), 39–55.

[31] Grünbaum, B. *Regularity of graphs, complexes and designs* Problèmes combinatoires et théorie des graphes, Colloq. Internat. C.N.R.S. No. 260, Orsay (1977), 191–197.

[32] Hubard, I. and Weiss, A. I. *Self-duality of Chiral Polytopes*, J. Combinatorial Theory Ser. B, **111** (2005), 128–136.

[33] McMullen, P. *Combinatorially regular polytopes* Mathematika **14** (1967), 142-150.

[34] McMullen, P. and Schulte, E. *Abstract Regular Polytopes*, Encyclopedia of Mathematics and its Applications vol. 92, Cambridge University Press, Cambridge, 2002.

[35] Monson, B. and Weiss, A. I. *Eisenstein Integers and Related C-groups* Geometriae Dedicata, **66** (1997), 99–117.

[36] Monson, B. and Weiss, A. I. *Medial Graphs of Equivelar 4-Polytopes*, to appear in European J. Combin.

[37] Monson, B., Pisanski, T., Schulte, E. and Weiss, A. I. *Semi-Symmetric Graphs from Polytopes*, in preparation.

[38] Robinson, G. de B. *On the fundamental region of a group, and the family of configurations which arise therefrom* J. London Math. Soc. **6** (1931), 70–75.

[39] Schlegel, V. *Theorie der homogenen zusammengesetzten Raumgebilde* Nova Acta Leop. Carol. **44** (1883), 343–459.

[40] Schulte, E. and Weiss, A. I. *Chiral polytopes* Applied Geometry and Discrete Mathematics. The Victor Klee Festschrift (Gritzmann, P. and Sturmfels, B. eds.), DIMACS Series in Discrete Mathematics and Theoretical Computer Science **4** (1991), 493–516.

[41] Schulte, E. and Weiss, A. I. *Chirality and projective linear groups* Discrete Mathematics **131** (1994), 221–261.

[42] Seifert, H. and Threlfall, W. *Lehrbuch der Topologie*, Chelsea, New York, 1947.

[43] Sherk, F. A. *A family of regular maps of type* $\{6,6\}$ Canad. Math. Bull. **5** (1962), 13–20.

[44] Sommerville, D. M. Y. *An Introduction to the Geometry of n Dimensions*, Methuen, London, 1929.

[45] Threlfall, W. *Gruppenbilder* Abh. Sächs. Akad. Wiss. Math.-phys. Kl. **41** (1932), 1–59.

[46] Todd, J. A. *The groups of symmetries of the regular polytopes* Proc. Camb. Phil. Soc. **27** (1931), 212-231.

[47] Todd, J. A. and Coxeter, H. S. M. *A practical method for enumerating cosets of a finite abstract group* Proc. Edinburgh Math. Soc. (2) **5** (1936), 25–34.

[48] Tutte, W. T. *Connectivity in Graphs*, University of Toronto Press, Toronto, 1966.

[49] Wilson, S. *The smallest non-toroidal chiral maps* J. Graph Theory **2** (1978), 315-318.

[50] Wythoff, W. A. *A relation between the polytopes of the* C_{600}*-family* Konink. Akad. Weten. Amsterdam, Proc. Sect. Sciences **20** (1918), 966–970.

DEPARTMENT OF MATHEMATICS & STATISTICS, UNIVERSITY OF NEW BRUNSWICK, FREDERICTON, NB, CANADA E3B 5A3

E-mail address: bmonson@unb.ca

DEPARTMENT OF MATHEMATICS & STATISTICS, YORK UNIVERSITY, TORONTO, ON, CANADA M3J 1P3

E-mail address: weiss@yorku.ca

Equivelar Polyhedra

Jörg M. Wills

ABSTRACT. We give a survey on polyhedral 2-manifolds with planar faces, embedded in Euclidean 3-space and with local regularity properties: i.e. all faces are p-gons and all vertices are q-valent. We give the basic definitions, questions, results and the three (perhaps) most striking open problems.

In Section 6 we construct a heptagon–dodecahedron, which has the minimal number of faces of any equivelar polyhedron of genus $g \geq 2$.

1. Basic definitions and facts

The purpose of this survey is to investigate polyhedral manifolds with local regularity properties.

A polyhedral 2-manifold (or: a polyhedral embedding, or briefly: a polyhedron) is a compact 2-manifold, built up of planar convex or nonconvex Jordan polygons without self-intersections in Euclidean 3-space E^3, so necessarily oriented. The well-known Möbius–Czaszar torus is a typical example [C49], and its dual, the Szilassi torus is an example where all faces are nonconvex [S86]. If we want to distinguish between these two, we speak of c–polyhedra or nc–polyhedra.

The embeddability in E^3 is the crucial restriction, so we do not consider polyhedra with self-intersections as e.g. the four Kepler–Poinsot star polyhedra. A 2-manifold consists of 0-cells (vertices), 1-cells (edges) and 2-cells (faces). The intersection of any two faces is either an edge or a vertex or it is empty. The number of vertices, edges and faces is denoted by v, e and f (or: f_0, f_1, f_2), respectively. The genus of the manifold is as usual denoted by g, and the Euler–Characteristic is $\chi = 2 - 2g$, because we only consider oriented polyhedral manifolds. For more details see [BW93] or [MSW82] or [MS02]. The basic relation is Euler's equation

$$(1.1) \qquad v - e + f = \chi \qquad (= 2 - 2g).$$

A polyhedron is said to be of type $\{p, q\}$ or equivelar, if all its faces are convex or nonconvex p-gons and all its vertices are q-valent. The name equivelar was introduced in [MSW82] and means "equal flags", because all its flags, i.e. the triplets of incident vertices, edges and faces are combinatorially equivalent.

This is a local regularity property; e.g. the five Platonic solids are equivelar of type $\{3,3\}, \{3,4\}, \{4,3\}, \{3,5\}$ and $\{5,3\}$. For the Platonic solids this local

2000 *Mathematics Subject Classification.* Primary 52B70; Secondary 52B10, 52B15.

regularity already implies global regularity, which we will explain later. For $g > 0$ this is by no means the case, but also here equivelarity is a strong restriction.

As each edge of a polyhedron is incident with exactly two vertices and two faces, equivelarity yields the two elementary equations

$$(1.2) \qquad\qquad pf = 2e = qv.$$

Here the positive integers p, q and v, e, f are not arbitrary. The fact that polyhedra and not arbitrary combinatorial patterns are investigated, leads to the simple inequalities

$$(1.3) \qquad p \geq 3, \ q \geq 3, \ f \geq 4, \ v \geq 4, \ e \leq \min\left(\binom{v}{2}, \binom{f}{2}\right).$$

The equations (1.1) and (1.2) together with the inequalities (1.3) govern the combinatorial structure of equivelar manifolds.

In particular for any $g \geq 2$ there are only finitely many equivelar manifolds (up to isomorphism) as for $g = 0$.

2. Equivelarity and embeddability

A simple consequence of (1.1) and (1.2) is

$$\text{sign}\left(\frac{1}{p} + \frac{1}{q} - \frac{1}{2}\right) = \text{sign } \chi$$

and hence the distinction between $\chi > 0(g = 0)$ and $\chi = 0(g = 1)$ and $\chi < 0(g \geq 2)$. For $g = 0$ the five Platonic solids are the only equivelar polyhedra. For $g = 1$, i.e., the torus, there are the three infinite series $\{3, 6\}$, $\{6, 3\}$ and $\{4, 4\}$ (cf. [**CM80**]). The best known equivelar tori are Czaszar's torus of type $\{3, 6\}$ with 7 vertices and Szilassi's torus [**S86**] of type $\{6, 3\}$ with 7 (nonconvex) faces, which is minimal in both cases.

For $g \geq 2$ one has the greatest variety, because for any integer pair $\{p, q\}$ with

$$(2.1) \qquad\qquad \frac{1}{p} + \frac{1}{q} < \frac{1}{2}$$

the equations (1.1), (1.2) and the inequalities (1.3) are fulfilled for appropriate v, e, f and g. If one does not consider any further geometric or topological conditions, this leads to general and elegant results, e.g. in [**EEK82**], where even the restrictions (1.3) and orientability is skipped.

Also for abstract polyhedra, i.e., with (1.3) but without embeddability in any Euclidean space E^d, one obtains remarkably general results, as was shown in some recent papers [**DN01, BDN02, D04**].

We do not consider these abstract patterns and polyhedra here.

The restriction to embeddability with planar faces into E^3 is the crucial condition from geometry and topology, which makes the results less general, less elegant and more complicated.

It turns out that many (or most) of the abstract equivelar patterns or polyhedra are not embeddable in E^3. But sufficiently many, or more precisely: infinitely many, are embeddable, and some of them are of particular interest, as we will point out later.

As already mentioned, the conditions (1.1)–(2.1) guarantee, that for any $g \geq 2$ there are only finitely many equivelar polyhedra as in the classical case $g = 0$ (and in sharp contrast to the toroidal case $g = 1$). This restriction makes equivelarity attractive. On the other hand, the notation $\{p, q; g\}$ as well as (v, e, f) does not necessarily define an equivelar polyhedron uniquely. There may be two or more non-isomorphic polyhedra with the same $\{p, q; g\}$. In Figure 1, e.g., there is a $\{3, 7; 3\}$ which is not combinatorially regular whereas F. Klein's $\{3, 7\}_8$ is combinatorially regular (cf. Section 4). So one can formulate the general problem for equivelar polyhedra:

Given a triplet $\{p, q; g\}$ with (1.1), (1.2), (1.3) and (2.1).

(a) Is there a polyhedron $\{p, q; g\}$?
(b) Is this polyhedron $\{p, q; g\}$ unique up to isomorphism?

We will mainly deal with part (a) of this general problem. Consequently, several more specific problems follow from this general problem. In the following sections we point out three of them which seem to be of particular interest.

3. General results

The following results are for polyhedra with convex faces (c–polyhedra). But there is no reason to exclude polyhedra with nonconvex faces (nc–polyhedra).

The restriction to c–polyhedra comes from the methods and proofs, where convexity is an essential tool. Here we do not give proofs.

Perhaps the first equivelar polyhedron with $g > 0$ is the Möbius–Czaszar torus [**C49**]. Interesting other individuals, in particular for small g were found by Coxeter [**C73**], Grünbaum and Shephard [**GS84, GS88**], Barnette [**B82**], Betke and Gritzmann [**BG84**], Brehm [**B87**], Bokowski [**B89**], [**BB89**] et al.

The first general result, without the special and simpler case $g = 1$ was given in [**MSW82**] for c-polyhedra.

THEOREM 3.1. [**MSW82**]
The following equivelar c–polyhedra exist

$\{3, 7; g\}$ for $g \geq 2$
$\{3, 8; g\}$ for $g \geq 3$
$\{4, 5; g\}$ for $g \geq 4$
$\{5, 4; g\}$ for $g \geq 5$ $(\neq 6, 8)$
$\{3, 9; g\}, \{4, 6; g\}$ and $\{6, 4; g\}$ for $g \geq 6$ $(\neq 7, 8, 11)$.

Here the small genera are the more difficult cases, which usually require an extra proof, whereas for the larger genera one general construction suffices. Obviously there is no symmetry between p and q, i.e., there is no general duality as for convex polytopes or abstract patterns.

The main reason here is the restriction to c–polyhedra, but in general duality can not be expected for polyhedral manifolds, as was pointed out in [**GS88**]. In Figure 1 we show an nc–polyhedron $\{7, 3; 3\}$, which is not combinatorially regular. An essential difference is the planarity of the faces, whereas vertices do not have a corresponding restriction. Vertices can be arbitrarily "wild" and nonconvex , as was shown in [**BG84**].

Theorem 3.1 shows for seven classes $\{p, q\}$ that for all but finitely many g there is at least one c–polyhedron $\{p, q; g\}$.

Theorem 3.2 shows for infinitely many classes $\{p, q\}$ that there are infinitely many c–polyhedra $\{p, q; g\}$:

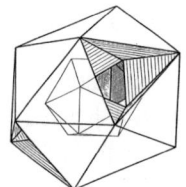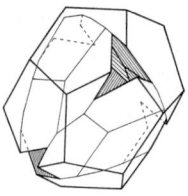

FIGURE 1. $A\{3,7;3\}$ and $\{7,3;3\}$

THEOREM 3.2. [**MSW83**] *For each of the following classes $\{p,q\}$ there are infinitely many g such that an equivelar c–polyhedron $\{p,q;g\}$ exists:*

(a) $\{3,q;g\}$ *for $q \geq 7$*
(b) $\{4,q;g\}$ *for $q \geq 5$*
(c) $\{p,4;g\}$ *for $p \geq 5$.*

Theorem 3.2 is much harder to prove than Theorem 3.1. And it has an interesting consequence, as we will see in the last section.

Again the asymmetry in $\{p,q\}$ comes mainly from the restriction to c–polyhedra. It is remarkable that the p–gons in (c) are all convex, even if p is large. A brief look at Theorems 3.1 and 3.2 shows that there are no equivelar polyhedra with $p \geq 5$ and $q \geq 5$.

This is surprising as there is no such restriction for equivelar maps or abstract polyhedra (cf. [**EEK82, DN01**]).

In particular in [**D04**] there are infinitely many abstract polyhedra $\{k,k;g\}$ with $k \geq 5$. Moreover two of the four Kepler–Poinsot polyhedra are of type $\{5,5\}$, but of course with self-intersections ([**C73**]).

And there are infinite "equivelar" polyhedra, e.g. Coxeter's apeirohedron (cf. [**C73**]) $\{6,6\}_\infty$, which is even regular and J. Gott's $\{5,5\}_\infty$ (cf. [**G67**]).

So this leads to

PROBLEM 3.3. Find an equivelar polyhedron $\{p,q;g\}$ with $p \geq 5$ and $q \geq 5$, or prove that there is no such polyhedron.

4. Combinatorially regular polyhedra for $g \geq 2$

A polyhedron is combinatorially regular if the group $\mathcal{A}$ of its incidence–preserving permutations acts transitively on its flags (cf. e.g. [**C73, C49**], or [**MS02**]).

So, combinatorially regular polyhedra are geometric realizations of regular maps (cf. e.g. [**CM80**]).

As the automorphism group acts transitively on the flags and as each edge is incident with exactly four flags, one has for the order of the automorphism group: $|\mathcal{A}| = 4e$. With (1.1), (1.2), (1.3), (2.1) one obtains for $g \geq 2$:

$$|\mathcal{A}| = 2\chi \left(\frac{1}{p} + \frac{1}{q} - \frac{1}{2}\right)^{-1} \leq 84\,|\chi|$$

which is the famous Riemann–Hurwitz relation [**CM80**]. Equality holds for $\{3,7\}$ and $\{7,3\}$, so these classes deserve particular attention, e.g. F. Klein's regular map of genus 3.

Combinatorial regularity is a global property and clearly implies equivelarity. Regular maps are the natural combinatorial analogues of the Platonic solids. They have been intensively studied (e.g. in [**CM80, W76**]).

There is a natural interest in their realization as polyhedral embeddings. The combinatorially regular polyhedra are much closer related to the Platonic solids than ordinary equivelar polyhedra and they can be considered as the jewels among them. Usually it is much harder to find a combinatorially regular polyhedron than an ordinary equivelar polyhedron, because many more conditions have to be checked.

The following theorem collects the known results for $g \geq 2$. We underline that the first notation (from [**CM80**]) characterizes the polyhedron uniquely up to isomorphism, whereas the second does not, as already mentioned in the previous section.

THEOREM 4.1. *For $g \geq 2$ the following combinatorially regular polyhedra exist:*
$g=3$:

F. Klein's	$\{3,7\}_8$	$(=\{3,7;\ 3\})$	[**SW85**]
W. Dyck's	$\{3,8\}_6$	$(=\{3,8;\ 3\})$	[**B89**]
			and [**B87**]

$g=6$:

Coxeter's	$\{4,6 \mid 3\}$	$(=\{4,6;\ 6\})$	[**SW88**]
and	$\{6,4 \mid 3\}$	$(=\{6,4;\ 6\})$	

$g=73$:

Coxeter's	$\{4,8 \mid 3\}$	$(=\{4,8;\ 73\})$	[**SW88**]
and	$\{8,4 \mid 3\}$	$(=\{8,4;\ 73\})$	

Coxeter's [**C73**] and Ringel's [**RY68**] series

$$\{p,4;\ g\}, \quad p = 5,6,7,... \quad g = (p-4)2^{p-3} + 1$$
$$\text{and}\quad \{4,q;\ g\}, \quad q = 5,6,7,... \quad g = (q-4)2^{q-3} + 1 \quad [\textbf{MSW88}].$$

It is remarkable that only for $g = 3$ the duals are not known, although considerable attempts were made. So we have

PROBLEM 4.2. Find other combinatorially regular polyhedra; in particular Klein's $\{7,3\}_8$ and Dyck's $\{8,3\}_6$.

5. Combinatorially regular tori

For completeness we also mention Schwörbel's general result on combinatorially regular tori:

THEOREM 5.1. [**S88**] *Each regular toric map (i.e., of genus 1, oriented) which is a topological cell complex, can be realized as a polyhedral embedding in E^3 with convex or nonconvex faces.*

In other words: All abstract oriented regular polyhedra of genus 1 are embeddable as c–polyhedra or nc–polyhedra. In more detail:

We give the list of the 6 types of regular toric maps from [**CM80**]:

$\{4,4\}_{a,0}$	and $\{4,4\}_{a,a}$	for $a \geq 3$
$\{3,6\}_{a,0}$	and $\{6,3\}_{a,0}$	for $a \geq 3$
$\{3,6\}_{a,a}$	and $\{6,3\}_{a,a}$	for $a \geq 2$

which are realizable.

In Figure 2 we show $\{4,4\}_{3,3}$.

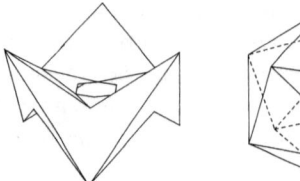

FIGURE 2. A combinatorially regular torus of type $\{4,4\}$

Some of the "small" tori, i.e., the first in the series, were already discovered before, namely $\{3,6\}_{2,2}$ by Brehm (1978, unpubl.) and independently by Grünbaum and Shephard [**GS84**]. Further, $\{3,6\}_{3,0}$ and $\{6,3\}_{3,0}$ are investigated in [**SW88**]. Finally, $\{4,4\}_{3,0}$ is the so-called triangular picture frame and is folklore. Schwörbel's theorem occurred in his Diplom thesis and was never published. Czaszar's and Szilassi's tori are equivelar, but not combinatorially regular.

Different from equivelarity, combinatorial regularity is not easy to see. The algebraic and combinatorial symmetries usually do not coincide with geometric symmetries.

Typically the geometric symmetry group is a rather small subgroup of the automorphism group of the underlying regular map.

Most symmetries can only be observed combinatorially, e.g. via Petrie polygons (cf. e.g. [**CM80**]). So they are often called hidden symmetries.

6. Minimality for small genus

The minimal number of vertices or faces that a polyhedron with $g = 0$ can have, is 4, and is attained by the tetrahedron.

The corresponding number for $g = 1$, the torus, is 7 and is attained by Czaszar's (vertices) and Szilassi's (faces) tori. These polyhedra are equivelar.

Brehm [**B81**] constructed polyhedra of genus 2 and 3 with 10 vertices, and Bokowski and Brehm [**BB89**] of genus 4 with 11 vertices, which is also combinatorially the best possible. These polyhedra are not equivelar.

So it is a natural question to ask for equivelar polyhedra of genus $g \geq 2$ with a minimal number of vertices v_g or faces f_g.

From (1.1)–(2.1), using elementary combinatorial arguments, we obtain $v_g \geq 12$ and $f_g \geq 12$ for $g = 2,3,4,5,6$ and $v_g \geq 14$ and $f_g \geq 14$ for $g \geq 7$.

Only for a few cases it is known if these bounds are attained; namely, $v_2 = 12$ in [**MSW82**], $v_3 = 12$ in [**B87, B89**] and [**GS84**].

Here we give a short proof for $f_2 = 12$ by constructing a $\{7,3;2\}$, i.e., a heptagon–dodecahedron with 3–valent vertices and of genus 2.

PROOF FOR $f_2 = 12$. The Szilassi torus has two hexagons which lie in the convex hull of this torus. We reflect this torus at the affine hull of one of these faces and obtain a polyhedron of genus 2 with 12 hexagons and 22 vertices. Six of these 22 vertices are 4–valent; they are the vertices of the hexagon of reflection. All other vertices are 3–valent.

A slight moving of the six reflected hexagons turns each 4–valent vertex into two 3–valent vertices and leaves the combinatorial structure at all other vertices unchanged. But from the restriction to planar faces it is not obvious, that this can be done such that all hexagons become heptagons and so we give a precise construction.

We denote the 4–valent vertices in a cyclic order by $a_1, \ldots, a_6$ and the hexagons incident to a_i and a_{i+1} by A_i (here $a_7 = a_1$).

The edges incident with A_{i-1} and A_i and hence with a_i are denoted by e_i. Now let $C_i(\varepsilon)$, $i = 1, \ldots, 6$ be the semidisk with center a_i and radius $\varepsilon > 0$ in aff A_i and bounded by aff e_i. We choose $\varepsilon > 0$ so small that the 3-valent vertices remain 3–valent by the following construction. Let $x_i(\varepsilon) \in C_i(\varepsilon) \backslash$ aff e_i, $i = 1, \ldots, 6$. Let A_i' be the reflection in A_i. We replace A_i' by A_i'', such that A_i'' contains the following three non-collinear points: $x_i(\varepsilon), x_{i+1}(\varepsilon)$ and z_i, where $z_i \neq a_i, a_{i+1}$ is a vertex of A_i'. This defines the new faces A_i'', $i = 1, \ldots, 6$. Any A_i'' meets A_{i-1} and A_i in a 3–valent vertex $y_i \in$ aff e_i. In general $y \neq a_i$, but both are in aff e_i. By construction the x_i are 3–valent and $x_i \notin$ aff e_i. Further the twelve vertices x_i and y_i are joined by edges in an alternating cycle. So we obtain a polyhedron with twelve heptagons and 28 3–valent vertices and hence with 42 edges.   □

Clearly this proof is not helpful for an explicit construction.

7. Minimality for large genus

The construction of the series of equivelar polyhedra in Theorem 3.2 (and Theorem 5.1) is described in detail in [**MSW83**], and we do not repeat these details.

Here it is relevant that at each step of the inductive proof the number of vertices (or faces) doubles, so that it grows with a power of 2. But the genus grows even faster. This has the consequence, as described in [**MSW83**], that asymptotically

$$(7.1) \qquad v_g = O(g/\log g) \quad \text{and} \quad f_g = O(g/\log g).$$

In particular there is an equivelar polyhedron of genus 577 with 576 vertices and another one of genus 4077 with 4076 convex faces, i.e.,

$$(7.2) \qquad V_{577}^2 \leq 576 \quad \text{and} \quad f_{4077}^e \leq 4076.$$

The asymmetry comes from the restriction to convex faces in Theorem 3.2, so that the series $\{p, 3\}$ do not occur.

The asymptotic bounds in (7.1) are much weaker than the corresponding bound $O(\sqrt{g})$ for abstract polytopes from Ringel and Youngs [**RY68**].

On the other hand (7.1) and (7.2) are until now the best known bounds for polyhedra with large genus g, even without the restriction of equivelarity.

So we have the last problem:

PROBLEM 7.1. Improve (7.1) and (or) (7.2) for equivelar or general polyhedral embeddings.

References

[B82]   Barnette, D. W. *Nonconvex vertices of polyhedral 2–manifolds.* Discrete Math. **41** (1982), 123–130

[BG84]   Betke, U.; Gritzmann, P. *Polyedrische 2–Mannigfaltigkeiten mit wenigen nichtkonvexen Ecken.* Monatsh. Math. **97** (1984), 1–21

[B89]   Bokowski, J. *A geometric realization without selfintersections does exist for Dyck's regular map.* Discrete Comp. Geom. **6** (1989), 583–589

[BB89]   Bokowski, J.; Brehm, U. *A polyhedron of genus 4 with minimal number of vertices and maximal symmetry.* Geom. Dedicata **29** (1989), 53-64

[BW84]   Bokowski, J.; Wills, J. M. *Regular polyhedra with hidden symmetries.* Math. Intellig. **10** (1984),27–32

[B81]   Brehm, U. *Polyeder mit 10 Ecken vom Geschlecht 4.* Geom. Dedic. **11** (1981), 119–124

[B87]   Brehm, U. *Maximally symmetric polyhedral realization of Dyck's regular map.* Mathematika **34** (1987), 229–236

[BDN02]   Brehm, U.; Datta, B.; Nilakantan, N. *The edge-minimal polyhedral maps of Euler characteristic -8.* Contrib. Algebra and Geometry **43** (2002), 583–596

[BW93]   Brehm, U.; Wills, J. M. *Polyhedral manifolds,* pp. 535–554 in: Handbook of Convex Geometry, North Holland 1993

[C49]   Coxeter, H. S. M. *A polyhedron without diagonals.* Acta Sci. Math. **13** (1949), 140–142

[C73]   Coxeter, H. S. M. *Regular Polytopes.* Dover, New York 1973

[CM80]   Coxeter, H. S. M.; Moser, W. O. J. *Generators and Relations for Discrete Groups.* Springer, Berlin, 4th edit. 1980

[D04]   Datta, B. *A class of equivelar polyhedral 2–manifolds.* to appear

[DN01]   Datta, B.; Nilakantan, N. *Equivelar polyhedra with few vertices.* Discrete Comp. Geom. **26** (2001), 429–461

[EEK82]   Edmonds, A. L.; Ewings, J. H.; Kulkarni, R. S. *Regular tessellations of surfaces and $(p, q, 2)$ triangle groups.* Ann. of Math. **116** (1982), 113-132

[G67]   Gott, J. R. *Pseudopolyhedron.* Amer. Math. Monthly **74** (1967), 497–504

[G83]   Gritzmann, P. *The toroidal analogue of Eberhard's theorem.* Mathematika **30** (1983), 274–290

[GS84]   Grünbaum, B.; Shephard, G. C. *Polyhedra with transitivity properties.* C. R. Math. Dep. Acad. Sci. Canad. **6** (1984), 961–966

[GS88]   Grünbaum, B.; Shephard, G. C. *Duality of polyhedra.* in: Shaping Space, eds. G. Fleck and M. Senechal, Birkhäuser, Boston 1988

[MS02]   McMullen, P.; Schulte, E. *Abstract Regular Polytopes.* Cambridge Univ. Press, Cambridge 2002

[MSW82]   McMullen, P.; Schulz, Ch.; Wills, J. M. *Equivelar polyhedral manifolds in E^3.* Israel J. Math. **41** (1982), 331–346

[MSW83]   McMullen, P.; Schulz, Ch.; Wills, J. M. *Polyhedral 2–manifolds in E^3 with unusually large genus.* Israel J. Math. **46** (1983), 127–144

[MSW88]   McMullen, P.; Schulte, E.; Wills, J. M. *Infinite series of combinatorially regular polyhedra in three-space.* Geom. Dedic. **26** (1988), 299-307

[RY68]   Ringel, G.; Youngs, I. T. W. *Solution of the Heawood map-coloring problem.* Proc. Nat. Acad. Sci. USA **60** (1968), 438-445

[SSW02]   Scholl, P.; Schürmann, A.; Wills, J. M. *Polyhedral models of Felix Klein's group.* The Math. Intellig. **24** (2002), 37–42

[SW85]   Schulte, E.; Wills, J. M. *A polyhedral realization of Felix Klein's map $\{3,7\}_8$ on a Riemann surface of genus 3.* J. London Math. Soc. **32** (1985), 539–547

[SW86]   Schulte, E.; Wills, J. M. *On Coxeter's regular skew polyhedra.* Discrete Math. **60** (1986), 253–262

[SW88]   Schwörbel, J.; Wills, J. M. *The two Pappus tori.* Geom. Dedic. **28** (1988), 359–362

[S88]   Schwörbel, J. *Die kombinatorisch regulären Tori.* Diplom thesis, Universität Siegen, Germany, 1988

[S86]   Szilassi, L. *Regular toroids.* Structural Topology **13** (1986), 69–80

[W83]   Wills, J. M. *Semi-Platonic manifolds,* pp. 45–60 in: Convexity and its Applications, Birkhäuser, Basel 1983

[W76]   Wilson, S. E. *New techniques for the construction of regular maps.* Ph. D. Thesis, University of Washington, Seattle, USA, 1976

DEPARTMENT OF MATHEMATICS, UNIVERSITY OF SIEGEN, D–57068 SIEGEN, GERMANY
E-mail address: wills@mathematik.uni-siegen.de

Combinatorics of Sections of Polytopes and Coxeter Groups in Lobachevsky Spaces

Askold Khovanskii

Coxeter classified all discrete isometry groups generated by reflections that act on a Euclidean space or on a sphere of an arbitrary dimension (see [5]). His fundamental work became classical long ago. Lobachevsky spaces (classical hyperbolic spaces) are as symmetric as Euclidean spaces and spheres. However, discrete isometry groups generated by reflections, with fundamental polytopes of finite volume (see [19]), are not classified for Lobachevsky spaces. In 1985, M.N. Prokhorov and myself proved the following theorem.

THEOREM 0.1. [6, 14] *In a Lobachevsky space of dimension > 995 there are no discrete isometry groups generated by reflections, with fundamental polytope of finite volume.*

For groups with compact fundamental polytopes, an analogous result had been previously obtained by E.B. Vinberg ([20, 21]). According to his theorem, such groups do not exist in Lobachevsky spaces of dimensions > 29. (Most likely, it is possible to reduce the number 29 considerably, all the more so the number 995. But nobody knows how to do that). The result of E.B. Vinberg came after the work of V.V. Nikulin ([12, 13]) who worked out the case of arithmetic groups. Nikulin estimated the average number of l-dimensional faces on k-dimensional faces of simple n-dimensional polytopes (the definition of a simple polytope is given below in this section) and applied his estimate to groups generated by reflections. In fact, a compact fundamental polytope of such a group is always simple, and Nikulin's estimate is applicable to it. Prokhorov and myself followed Nikulin's plan. We performed non-overlapping parts of the necessary work for the realization of this plan. Prokhorov proved the following theorem.

THEOREM 0.2. [14] *In a Lobachevsky space of dimension > 995, there are no discrete groups generated by reflections, with fundamental polytope of finite volume, satisfying Nikulin's estimate.*

It is known that if a polytope of finite volume is the fundamental polytope of a group generated by reflections in a Lobachevsky space, then this polytope is always almost simple, and, therefore, is simple at the edges (the definitions of

2000 *Mathematics Subject Classification.* Primary 14M25; Secondary 52B05, 14F43.
Partially supported by OGP grant 0156833 (Canada).

almost simple polytopes and polytopes simple at the edges are given below in this section). I proved the following theorem.

THEOREM 0.3. [6] *Nikulin's estimate holds for polytopes simple at the edges.*

COROLLARY 0.4. [6] *Nikulin's estimate holds for almost simple polytopes.*

Theorem 0.1 follows from Theorem 0.2 and Corollary 0.4. Nikulin proved his estimate using a very hard theorem — the theorem on the h-vector — a necessary and sufficient condition on a collection of integers to be the h-vector of a simple polytope (a variant of Nikulin's proof is given in Section 2). R. Stanley's proof of the necessity part of the theorem on the h-vector uses nontrivial results from algebraic geometry (see Section 3 for his proof). Several years after the publication of [6], some elementary proofs of the theorem on the h-vector were found (see [10], [17]). But they are also far from being simple.

In [6], I found a simple elementary proof of the statement in Corollary 0.4 (see the remark in Section 9 after Theorem 9.1). This statement is necessary for the proof of Theorem 0.1. It contains Nikulin's original estimate as a partial case. But I failed to find a simple proof of Theorem 0.3 — my proof of it is based on the theorem on the h-vector.

The *Klein model* of a Lobachevsky space is the interior U of the unit ball in a Euclidean space. Polytopes in this model are intersections of Euclidean polytopes with the region U. If a polytope is bounded in the Lobachevsky space, then in the Klein model it lies entirely in the region U. If a polytope in the Lobachevsky space has a finite volume, then in the Klein model it can intersect the horizon ∂U by vertices only. Recall the definition of a *simple polytope* and of a *polytope simple at the edges*. These definitions play a central role in this article.

DEFINITION 1. A convex n-dimensional polytope is said to be *simple*, if every vertex is incident with exactly n facets.

A neighborhood of any vertex of a simple n-dimensional polytope can be transformed into a neighborhood of the origin in the positive n-dimensional octant $(R^+)^n$ by an affine transformation. Hence *exactly n edges meet at each vertex of a simple n-dimensional polytope.*

DEFINITION 2. A convex n-dimensional polytope is said to be *simple at the edges*, if every edge is incident with exactly $(n-1)$ facets.

Let us also give the definition of an almost simple polytope.

DEFINITION 3. A convex n-dimensional polytope is said to be *almost simple*, if it looks like the cone over a product of simplices at each of its vertices.

It is clear that each simple polytope is almost simple, and each almost simple polytope is simple at the edges. At each of its vertices, a simple n-dimensional polytope looks like the cone over an $(n-1)$-dimensional simplex. At each of its vertices, an n-dimensional polytope simple at the edges looks like the cone over an $(n-1)$-dimensional simple polytope.

While preparing my talk for the "Coxeter Legacy" conference, I found out that my article [6], which had been written 19 years ago, is hard to read. The reason is that the journal "Functional Analysis and its Applications", where I published my article, had a restricted space. That is why I had to abridge the article considerably.

Luckily, I found an unabridged variant of the article, which helped me a lot in preparation of my talk and in writing this article.

My student V.A. Timorin found another proof of Theorem 0.3 (see [18]). His arguments are parallel to mine for the most part, but they do not at all use the combinatorics of sections of polytopes, which my proof relies upon. On one hand, this shortens the proof essentially. On the other hand, the facts from article [6] related to combinatorics of sections of polytopes, are interesting by themselves. They explain the geometric meaning of Nikulin's estimate.

This article is devoted to combinatorics of sections of polytopes and to a generalization of Nikulin's estimate. It is a considerably expanded and revised version of article [6].

I am grateful to the organizers of the conference for the invitation to give a talk, to G. Kalai and V.A. Timorin for useful discussions, to T.V. Belokrinitskaia for the help with the preparation of the Russian version of this article, and to V.A. Timorin for the help with the English translation.

1. Statements of Nikulin's theorem and of the theorem on the h-vector

Nikulin's estimate deals with the *average number of l-dimensional faces on k-dimensional faces of an n-dimensional polytope*. This average number is defined as follows. First, for every k-dimensional face (k-face for short) of the polytope, we compute the number of all l-faces on it. Then we take the arithmetic mean of these numbers over all k-faces of the polytope. By the *total number of l-faces on k-faces* we mean the number of pairs consisting of a k-face and an l-face of it. Thus the average number of l-faces on k-faces equals the total number of l-faces on k-faces divided by the number of k-faces.

Let us start with the 3-dimensional case. The following classical theorem is well known.

THEOREM 1.1. *The average number of edges on faces of a convex 3-polytope is strictly less than 6.*

Before proving this theorem, let us discuss the simplest example.

EXAMPLE 1. Consider a prism whose base is a convex n-gon. The upper and the lower bases of the prism contain n edges each. Its side faces are n quadrangles. Hence the total number of edges on faces of the prism is $n+n+4n = 6n$. The prism has $(n+2)$ faces. Hence the average number of edges on its faces is $6n/(n+2) < 6$.

The example given above shows that the estimate from the theorem cannot be improved, i.e., that the number 6 in its statement cannot be replaced by a smaller number.

PROPOSITION 1.2. *The estimate from Theorem 1.1 holds for simple 3-polytopes.*

PROOF. Denote by f_0, f_1 and f_2 the number of vertices, edges and faces of the polytope, respectively. We have

$$f_0 - f_1 + f_2 = 2,$$
$$3f_0 = 2f_1.$$

The first of these identities is the Euler formula. The second identity follows from the fact that exactly 3 edges meet at each vertex of the polytope. From these equalities we obtain that $2f_1/f_2 = 6 - 12/f_2$. The claim is thus proved, since the total number of edges on faces of the polytope equals $2f_1$, and the number f_2 of its faces is positive.   □

Theorem 1.1 can be proved in the same way as Proposition 1.2. We only need to replace the equality $3f_0 = 2f_1$ with the inequality $3f_0 \leq 2f_1$. The latter means that at least 3 edges meet at each vertex of the polytope. Using the Euler formula, we obtain the inequality $2f_1/f_2 \leq 6 - 12/f_2$ that implies the theorem. Nikulin generalized Proposition 1.2 to the multidimensional case. Namely, he proved the following theorem.

Nikulin's theorem *The average number of l-dimensional faces on k-dimensional faces of a simple n-dimensional polytope is strictly less than*

$$\binom{n-l}{n-k} \frac{\binom{[n/2]}{l} + \binom{[(n+1)/2]}{l}}{\binom{[n/2]}{k} + \binom{[(n+1)/2]}{k}}$$

for $0 \leq l < k \leq (n+1)/2$, $1 < k$.

According to Theorem 0.3, Nikulin's estimate holds for polytopes simple at the edges. Theorem 0.3 includes Theorem 1.1, since any convex 3-polytope is simple at the edges. If in the proof of Theorem 0.3 we confine ourselves to the 3-dimensional case, then we obtain a proof of Theorem 1.1 not using the Euler formula (see Section 10). The following statement is a supplement to Nikulin's theorem:

PROPOSITION 1.3. *1) For each triple of integers l, k, n such that $0 \leq l < k$, $(n+1)/2 < k \leq n$, $1 < n$, the average number of l-dimensional faces on k-dimensional faces of a simple n-polytope can be arbitrarily large.*

2) For each triple of integers l, k, n satisfying the conditions of Nikulin's theorem, his estimate is the best possible.

Here are the simplest examples.

EXAMPLE 2. For the triple $l = 0, k = n = 2$, the first part of the claim is obvious, since there exist convex polygons with any number of vertices. For the triple $l = 1, k = 2, n = 3$, Nikulin's estimate gives the number 6. As the example discussed above shows, this estimate is the best possible.

Nikulin's estimate is rather cumbersome. Below, its geometric meaning is discussed (see Section 6). For now, let me just give the following remark.

REMARK 1. For fixed positive integers l and k, as $n \to \infty$, Nikulin's estimate tends to the number $2^{k-l}\binom{k}{l}$, which is equal to the number of l-dimensional faces of the k-dimensional cube. Since the n-dimensional cube is a simple polytope for every n, and its k-dimensional faces are cubes, Nikulin's estimate is asymptotically exact.

The proof of Nikulin's theorem is based on the theory of simple polytopes, which is closely related to the theory of toric varieties. Recall the classical description of the f-vectors of simple polytopes. For a convex n-dimensional polytope Δ, the f-*vector* is the integer vector $(f_0, \ldots, f_n)$, whose component f_i equals the number of i-faces of the polytope Δ for each $0 \leq i \leq n$ (in particular, $f_n = 1$). The polynomial $f(t) = f_0 + f_1 t + \cdots + f_n t^n$ is called the f-*polynomial* of the polytope Δ. The polynomial $h(t) = f(t-1)$ is called the h-*polynomial* of the polytope Δ. The vector $(h_0, \ldots, h_n)$, whose components are the coefficients of the polynomial h (i.e. $h(t) = h_0 + h_1 t + \cdots + h_n t^n$), is called the h-*vector* of the polytope Δ. The identity

$f(t) = h(t + 1)$ shows that *the h-vector determines the f-vector*; for every m with $0 \leq m \leq n$, we have $f_m = \sum\limits_{0 \leq i \leq n} \binom{i}{m} h_i$.

What integer vectors are the h-vectors of simple n-dimensional polytopes? A complete answer to this question is given by the following remarkable theorem.

Theorem on the h-vector (McMullen, Stanley, Billera, Lee) *For every simple n-dimensional polytope, the components $h_0, \ldots, h_n$ of its h-vector satisfy the following conditions:*

(1) *(Dehn–Sommerville duality) For each $0 \leq i \leq n$, we have*

$$h_i = h_{n-i};$$

(2) *All components of the h-vector are nonnegative, and the numbers h_0 and h_n are 1;*

(3) *The h-vector is unimodal, i.e., $1 = h_0 \leq \cdots \leq h_{[n/2]}$;*

(4) *The sequence of numbers $h_1 - h_0, h_2 - h_1, \ldots, h_{[n/2]} - h_{[n/2]-1}$ has a bounded rate of growth: for $i = 0, \ldots, [n/2] - 1$, we have the inequalities $h_{i+1} - h_i < Q_i(h_i - h_{i-1})$, where Q_i are some explicit functions of an integer argument. The functions Q_i are not simple, but we will not need their explicit form.*

For each integer vector $h = (h_0, \ldots, h_n)$ satisfying conditions 1)–4), there exists a simple n-dimensional polytope, whose h-vector is equal to h.

The Dehn–Sommerville duality was discovered in the beginning of the last century (see [15]). In its entire form, the theorem on the h-vector was first conjectured by McMullen (see [9]). Stanley proved the necessity of McMullen's conditions on the h-vector (see [16]). Stanley's proof is based on a nontrivial technique from algebraic geometry (see [16] and Section 3). For every integer vector h satisfying conditions 1)–4), Billera and Lee gave an example of a simple polytope, whose h-vector equals h. Thus they concluded the proof of the theorem on the h-vector (see [2]).

Nikulin's estimate is a direct corollary from the theorem on the h-vector (see Section 2). In fact, to deduce this estimate, we only need parts 1) and 2) of the theorem on the h-vector, together with a couple of elementary lemmas given in Section 2.

Parts 1) and 2) of the theorem on the h-vector can easily be proved using a Morse theoretic type argument. Namely, one can use a generic linear function on the polytope (see Section 4). Thus one obtains a simple proof of Nikulin's estimate. Similar arguments can be also employed to prove a generalization of Nikulin's estimate necessary for the Lobachevsky geometry. But this generalization makes use of the theorem on the h-vector in corpore (to be more precise, we will need part 3) of this theorem, which is the most difficult).

A simple argument based on a generic linear function on the polytope came to my mind when I was thinking about Stanley's proof. In Section 3, we will discuss the idea of Stanley's proof.

2. Derivation of Nikulin's estimate from the theorem on the h-vector

To deduce Nikulin's estimate from the theorem on the h-vector, we will need Lemmas 2.1 and 2.4. They are quite elementary. Lemma 2.1 is very intuitive. The proof of Lemma 2.4 is simple, but a little cumbersome. It is based on claims 2.3 and 2.4

Let $A = (A_1, \ldots, A_m)$ and $B = (B_1, \ldots, B_m)$ be fixed vectors from $\mathbb{R}^m$, and suppose that all components of the vector A are strictly positive. Consider the set $\Omega \subset R^m$ defined by the relations $\alpha \in \Omega \Leftrightarrow \alpha_1 \geq 0, \ldots, \alpha_m \geq 0$ and $\alpha_1 + \cdots + \alpha_m > 0$. Denote by $\langle x, y \rangle$ the standard inner product of vectors $x, y \in \mathbb{R}^m$.

LEMMA 2.1. *The maximum C of the function $F(\alpha) = \frac{\langle \alpha, B \rangle}{\langle \alpha, A \rangle}$ on the region Ω, $\alpha \in \Omega$, is equal to*

$$\max_{1 \leq i \leq m} \frac{B_i}{A_i}.$$

Furthermore, the maximum C is attained on any vector $\alpha = (\alpha_1, \ldots, \alpha_m)$ such that its j-th components α_j vanish for all indices j such that $\frac{A_j}{B_j} < C$.

PROOF. Let Ω_1 be a subset of Ω defined by the condition $\alpha_1 + \cdots + \alpha_m = 1$. Multiplying the vector α by a positive number, we can arrange that $\alpha \in \Omega_1$. With the vector α, associate the point $(\langle \alpha, A \rangle, \langle \alpha, B \rangle)$ in the plane with coordinates a, b. The image of the set Ω_1 under this correspondence coincides with the convex polygon Δ that is the convex hull of the points $(A_i, B_i), i = 1, \ldots, m$. The polygon Δ lies in the right half-plane $a > 0$. The function $\frac{b}{a}$ on this half-plane is continuous, its level sets are rays beginning at 0. Furthermore, this function depends monotonely on the angle between this ray and the positive ray on the a-axis. The lemma is now geometrically evident. □

PROPOSITION 2.2. *1) Let A_1, A_2 be positive numbers, and B_1, B_2, μ_1, μ_2 non-negative numbers. Suppose that $\frac{B_1}{A_1} < \frac{B_2}{A_2}$, $0 < \mu_1 + \mu_2$, $\mu_1 \leq \mu_2$. Then*

$$\frac{B_1 + B_2}{A_1 + A_2} \leq \frac{\mu_1 B_1 + \mu_2 B_2}{\mu_1 A_1 + \mu_2 A_2}.$$

2) Assume additionally that $B_1 < B_2$ and that there are numbers λ_1 and λ_2 such that $\mu_1 + \mu_2 \leq \lambda_1 + \lambda_2$ and $\lambda_1 < \mu_1 < \mu_2 < \lambda_2$. Then

$$\frac{B_1 + B_2}{A_1 + A_2} < \frac{\lambda_1 B_1 + \lambda_2 B_2}{\mu_1 A_1 + \mu_2 A_2}.$$

PROOF. Part 1) of Proposition 2.2 follows from Lemma 2.1. Indeed, according to Lemma 2.1, $\frac{B_1 + B_2}{A_1 + A_2} < \frac{B_2}{A_2}$. By the same Lemma 2.1,

$$\frac{B_1 + B_2}{A_1 + A_2} \leq \frac{\mu_1(B_1 + B_2) + (\mu_2 - \mu_1)B_2}{\mu_1(A_1 + A_2) + (\mu_2 - \mu_1)A_2} = \frac{\mu_1 B_1 + \mu_2 B_2}{\mu_1 A_1 + \mu_2 A_2}.$$

Part 2) follows from part 1). Indeed, since $\lambda_2 - \mu_2 \geq \mu_1 - \lambda_1 > 0$, we have $(\lambda_2 - \mu_2)B_2 \geq (\mu_1 - \lambda_1)B_2 > (\mu_1 - \lambda_1)B_1$. Hence $\lambda_1 B_1 + \lambda_2 B_2 > \mu_1 B_1 + \mu_2 B_2$. It remains to use the inequality from part 1). □

For i and j such that $0 \leq i \leq n$ and $0 \leq j \leq (n+1)/2$, denote by $\varphi(j, i)$ the number

$$\frac{\binom{i}{j} + \binom{n-i}{j}}{\binom{[n/2]}{j} + \binom{[(n+1)/2]}{j}}.$$

PROPOSITION 2.3. *1) The numbers $\varphi(0, i)$, $\varphi(1, i)$ and the numbers $\varphi(j, [n/2])$, $\varphi(j, [(n+1)/2])$ equal 1;*

2) we have $\varphi(j, i) = \varphi(j, n - i)$;

3) for a fixed i such that $0 \leq i < [n/2]$, the numbers $\varphi(j, i)$ increase strictly as j runs from 1 to $[(n+1)/2]$.

Parts 1) and 2) are obvious. Let us prove part 3). We need to verify that for $0 \leq i < [n/2]$ and $1 \leq j \leq (n-1)/2$, we have the inequalities $\varphi(j, i) < \varphi(j+1, i)$. This is easy to do using part 2) of Proposition 2.2. It suffices to set $B_1 = \binom{i}{j}$, $B_2 = \binom{n-i}{j}$, $A_1 = \binom{[n/2]}{j}$, $A_2 = \binom{[(n+1)/2]}{j}$. The following inequality holds:

$$\frac{\binom{i}{j}}{\binom{[n/2]}{j}} < \frac{\binom{n-i}{j}}{\binom{[(n+1)/2]}{j}}.$$

Indeed, the left hand side of this inequality is less than 1, but the right hand side is greater than 1. Furthermore, we have

$(j+1)\binom{i}{j+1} = \lambda_1\binom{i}{j}$, where $\lambda_1 = \min(i - j, 0)$;

$(j+1)\binom{n-i}{j+1} = \lambda_2\binom{n-i}{j}$, where $\lambda_2 = n - i - j$;

$(j+1)\binom{[n/2]}{j+1} = \mu_1\binom{[n/2]}{j}$, where $\mu_1 = [n/2] - j$;

$(j+1)\binom{[(n+1)/2]}{j+1} = \mu_2\binom{[(n+1)/2]}{j}$, where $\mu_2 = [(n+1)/2] - j$.

The conditions of part 2) of Proposition 2.2 are satisfied, since $\lambda_1 + \lambda_2 \geq (n - 2j) = \mu_1 + \mu_2$ and $(n - i - j) > [(n+1)/2] - j \geq [n/2] - j > \min(i - j, 0)$. Using part 2) of Proposition 2.2, we obtain the desired inequality.

LEMMA 2.4. *Let $0 \leq l < k \leq (n+1)/2$. For each $0 \leq i \leq n$, set $A_i = \binom{i}{k} + \binom{n-i}{k}$, $B_i = \binom{i}{l} + \binom{n-i}{l}$. Then*

$$\max_i \frac{B_i}{A_i} = \frac{\binom{[n/2]}{l} + \binom{[(n+1)/2]}{l}}{\binom{[n/2]}{k} + \binom{[(n+1)/2]}{k}}.$$

PROOF. It suffices to verify that for each $0 \leq i \leq n/2$ and l, k subject to the conditions of the lemma, we have

$$\frac{B_i}{A_i} \leq \frac{\binom{[n/2]}{l} + \binom{[(n+1)/2]}{l}}{\binom{[n/2]}{k} + \binom{[(n+1)/2]}{k}}.$$

This inequality is equivalent to the inequality $\varphi(l, i) < \varphi(k, i)$ from part 3) of Proposition 2.3.   □

Let us turn to the proof of Nikulin's estimate.

PROOF. **of Nikulin's estimate.** The estimate of the average number of l-faces on k-faces of a simple n-polytope follows immediately from the theorem on the h-vector. Indeed, firstly, each l-face of a simple n-polytope is contained in exactly $\binom{n-l}{n-k}$ k-dimensional faces of the polytope. Secondly, the number f_m of m-faces of a simple polytope is determined by its h-vector for every m, namely, $f_m = \sum_i \binom{i}{m} h_i$. The desired average number equals

$$\binom{n-l}{n-k} \frac{\sum_i \binom{i}{l} h_i}{\sum_i \binom{i}{k} h_i}.$$

By the Dehn–Sommerville duality, this number equals

$$\binom{n-l}{n-k} \frac{\sum_{0 \leq i \leq n/2} \left(\binom{i}{l} + \binom{n-i}{l} \right) \tilde{h}_i}{\sum_{0 \leq i \leq n/2} \left(\binom{i}{k} + \binom{n-i}{k} \right) \tilde{h}_i}, \tag{1}$$

where $\tilde{h}_i = h_i$ for $0 \le i < n/2$ and $\tilde{h}_{[n/2]} = \frac{1}{2}h_{[n/2]}$ for even n. Now apply Lemma 2.1 for $m = 1 + [n/2]$, $A_i = \binom{i-1}{k} + \binom{n-i+1}{k}$, $B_i = \binom{i-1}{l} + \binom{n-i+1}{l}$ and $\alpha_i = \tilde{h}_{i-1}$ (Lemma 2.1 is applicable, since according to part 2) of the theorem on the h-vector, the numbers $h_0, \dots, h_{[n/2]}$ are nonnegative, and their sum is positive). According to Lemma 2.4, the maximum C of the ratio A_i/B_i is attained for $i - 1 = [n/2]$. According to Lemma 2.1, the value (1) is strictly less than C, since $h_0 > 0$. Nikulin's estimate is proved. $\square$

REMARK 2. The proof of Nikulin's estimate made use of only parts 1) and 2) of the theorem on the h-vector of a simple polytope.

The remark motivates the following plan: try to find a simpler proof for the symmetry of the h-vector and for the non-negativity of its components. This proof should be simpler than a complete proof of the theorem on the h-vector. This would allow to simplify the proof of Nikulin's estimate, and possibly this would allow to generalize it.

PROOF. **of Proposition 1.3** There exists a sequence Δ^N of simple n-dimensional polytopes such that the h-vector components h_i^N of these polytopes have the following property: for any $i < [n/2]$, the limit $\lim_{N \to \infty} h_i^N/h_{[n/2]}^N$ is equal to zero. It suffices to define Δ^N as the polytope dual to an n-dimensional cyclic polytope with N vertices. From formula (1) for the average number of l-faces on k-faces of simple n-polytopes, it follows that the sequence Δ^N provides an example for both parts of Proposition 1.3. $\square$

3. The theorem on the h-vector and Morse theory

The proof of the necessity of conditions on the h-vector of a simple polytope is based on the theory of Newton polytopes, which relates geometry of polytopes to algebraic geometry of toric varieties.

Let Δ be a convex integer polytope in $\mathbb{R}^n$, i.e. the vertices of the polytope belong to the lattice $\mathbb{Z}^n$. With each integer point $m \in \mathbb{Z}^n$, we can associate the monomial $\chi_m : (\mathbb{C}^*)^n \to \mathbb{C}$ defined by the formula $\chi_m(z_1, \dots, z_n) = z_1^{m_1} \cdots z_n^{m_n}$.

Denote by A the finite set $A = \Delta \bigcap \mathbb{Z}^n$ of integer points. The *Veronese map* $V_\Delta : (\mathbb{C}^*)^n \to \mathbb{C}P^{N-1}$, where N is the number of points in the set A, is defined as the map taking each point $z \in (\mathbb{C}^*)^n$ to the point with homogeneous coordinates $[\chi_{m_1}(z) : \cdots : \chi_{m_N}(z)]$, where $m_1, \dots, m_N$ are the points of the set A taken in an arbitrary order (the Veronese map is defined up to a permutation of the set A, however, the property of this map we are interested in does not depend on the choice of the ordering).

The toric variety M_Δ is the normalized projective closure of the image $V_\Delta((\mathbb{C}^*)^n)$ of the group $(\mathbb{C}^*)^n$ under the Veronese map (if the polytope Δ is "not too small", then the projective closure is automatically normal and so it does not need to be normalized). The natural action of the group $(\mathbb{C}^*)^n$ extends to M_Δ. With respect to this action, M_Δ splits into a finite number of orbits.

If the polytope Δ is simple, then the algebraic variety M_Δ is called *quasi-smooth variety* (i.e. an orbifold). Quasi-smooth varieties possess many properties of smooth algebraic varieties. In particular, the main results of Hodge theory persist for these varieties. In the sequel, in our heuristic arguments, we will assume that M_Δ is a smooth manifold.

Stanley's proof of the necessary conditions on the h-vector is based on the following fact. It turns out that *the number h_i of a simple integer polytope Δ coincides with the $2i$-th Betti number of the manifold M_Δ*. After this observation, all necessary conditions on the h-vector of a simple integer polytope Δ follow from the theory of toric varieties. Indeed, the non-negativity of the numbers h_i becomes obvious, the Dehn–Sommerville duality follows from Poincaré duality $\dim H^i = \dim H^{2n-i}$. The unimodality of the numbers h_i follows from the hard Lefschetz theorem, the inequalities $h_{i+1} - h_i < Q_i(h_i - h_{i-1})$ follow from the fact that the cohomology ring of the manifold M_Δ is generated by the elements of the vector space $H^2(M_\Delta)$. (The function Q_i appears in the Macaulay theorem from commutative algebra (see [7]) describing the Hilbert functions of the quotients of the polynomial ring in several variables.)

The necessary conditions on the h-vector for simple but non-integer polytopes can easily be reduced to the integer case. To perform this reduction, one can do a small perturbation of the facets of the polytope to make them rational. Then all vertices of the polytope become rational as well, and the combinatorial type of the polytope remains unchanged, since the original polytope was simple. After that, we can make all vertices integer by a suitable dilation of the polytope (multiplying by the common denominator of all vertices).

As we have seen, to prove Nikulin's estimate, it suffices to use only the positivity of the numbers h_i and their symmetry $h_i = h_{n-i}$. Positivity of Betti numbers of the manifold M_Δ and Poincaré duality are responsible for these properties. Thus we use neither the existence of the ring structure on the cohomology space of M_Δ, nor the hard Lefschetz theorem.

Morse theory helps frequently to compute Betti numbers. One of the simplest proofs of Poincaré duality is also based on this theory. Hence it is natural to try to use Morse theory for a proof of parts 1) and 2) of the theorem on the h-vector. To this end, we need to consider a simple enough function on M_Δ. To construct such a function, we can use the moment map (see [1]). The moment map $M : M_\Delta \to R^n$ has the following property. First, it takes the manifold M_Δ to the polytope Δ. Second, it establishes a one-to-one correspondence between the orbits of the group $(\mathbb{C}^*)^n$ in M_Δ and the faces of the polytope Δ. Namely, each orbit of (real) dimension $2i$ is mapped by the moment map to the interior of the corresponding i-dimensional face of the polytope.

A linear function on the polytope is said to be *generic*, if on no edge of the polytope does it restrict to a constant.

DEFINITION 4. The *index* of a generic linear function at a vertex of a simple n-dimensional polytope is the number of edges containing this vertex and such that the function decreases along them (in this case, the function increases along the $(n - i)$ remaining edges containing the vertex of index i).

It is easy to verify the following claim.

PROPOSITION 3.1. *Let L be a generic linear function on the polytope Δ. Then the function $L \circ M : M_\Delta \to \mathbb{R}$ on the manifold M_Δ is a Morse function, and its critical points are exactly the zero-dimensional orbits of M_Δ. The Morse index of a critical point $A \in M_\Delta$ equals twice the index of the vertex $M(A)$ of the polytope Δ with respect to the linear function L.*

The connection between the Morse index of the function $L \circ M$ at the point A (a zero-dimensional orbit of M_Δ) and the index of the linear function L at the vertex $M(A)$ of the polytope Δ, admits the following explanation. Let the index of the function L at the vertex $M(A)$ be equal to i. By definition, the vertex $M(A)$ must belong to a face Γ_1 of dimension i and to a face Γ_2 of dimension $(n-i)$ such that the maximum (respectively, minimum) of the function L on Γ_1 (respectively, on Γ_2) is attained at $M(A)$. The pre-images of the faces Γ_1 and Γ_2 under the moment map M are $2i$-dimensional and $2(n-i)$-dimensional submanifolds of M_Δ, respectively, such that the function $L \circ M$ restricted to these submanifolds attains the maximum (respectively, the minimum) at the point A.

The existence of such submanifolds shows that the Morse index of the point A equals $2i$. Thus the function $L \circ M$ on M_Δ has critical points of even indices only. Hence all odd Betti numbers of the manifold M_Δ are zero, and the number $\dim H^{2i}(M_\Delta)$ is equal to the number of vertices of the polytope Δ, where the function L has index i. However, as we have mentioned before, the number $\dim H^{2i}(M_\Delta)$ equals h_i. Hence the following theorem must be true, whose statement is absolutely elementary (it involves neither algebraic geometry nor topology).

THEOREM 3.2. *For any generic linear function L on a simple polytope Δ, the number $h_i(L)$ of vertices of the polytope Δ, where the index of the function L is i, does not depend on the function L and coincides with the number h_i of the polytope Δ.*

This theorem has a very simple elementary proof, which is given in the next section.

REMARK 3. The elementary proof of Theorem 3.2, together with Proposition 3.1, gives the simplest computation of Betti numbers for the manifold M_Δ: all odd Betti numbers are zero, and $\dim H^{2i}(M_\Delta) = h_i$, where h_i is the i-th component of the h-vector of the polytope Δ for $0 \le i \le n$.

REMARK 4. It turned out that Theorem 3.2 and its elementary proof in the next section had been known (see [4]) before the article [6], and some close arguments were used even earlier (see [3], [8]). However, neither the connection of Theorem 3.2 with the theory of toric varieties and Morse theory, nor the elementary deduction of Nikulin's estimate from Theorem 3.2 (see Corollary 4.3 and Section 5) had been known.

4. Generic linear function on a simple polytope[1]

Let us give an elementary proof of Theorem 3.2 and discuss its geometric corollaries.

PROOF. **of Theorem 3.2** Consider the set of faces of a simple n-dimensional polytope Δ (we mean the set of faces of all dimensions, including vertices, as well as the polytope Δ itself). Let us map this set into the set of vertices of the polytope.

---

[1]Linear functions on general convex polytopes, and, in particular, the problem of maximizing such functions, are studied in linear programming. Linear programming has a substantial practical value. Its creator, a distinguished mathematician and economist L.V. Kantorovich (1912–1986), was awarded the Nobel prize in economics, largely for his classical work on linear programming. In the last years of his life, Leonid Vital'evich was working in economics, but he preserved a life-long interest in mathematics. He had read the article [6] and was very enthusiastic about it.

To each face, we assign the vertex, where the restriction of L to this face attains the maximum. It is clear that the pre-image of a vertex A under this map contains exactly $\binom{i}{k}$ faces of dimension k, where i is the index of the function L at the vertex A. Each k-dimensional face belongs to the pre-image of some vertex, hence for each k, $0 \leq k \leq n$, we obtain the equality $f_k = \sum_i \binom{i}{k} h_i(L)$. The collection of all these equalities is equivalent to the identities $h_i = h_i(L)$, which prove the theorem.

Indeed, the equalities we obtained mean that the polynomial

$$\sum_{0 \leq i \leq n} h_i(L)(t+1)^i$$

coincides with the polynomial

$$\sum_{0 \leq i \leq n} f_i t^i.$$

But, by definition, the polynomial $\sum\limits_{0 \leq i \leq n} h_i(t+1)^i$ also has this property. Hence $h_i = h_i(L)$, as desired. $\qquad \square$

COROLLARY 4.1. *For every n-dimensional simple polytope, all numbers h_i are non-negative for $0 \leq i \leq n$, and the numbers h_0 and h_n are equal to 1.*

Indeed, the number of vertices of index i is nonnegative, and every generic linear function on the polytope has exactly one minimum and exactly one maximum.

COROLLARY 4.2. *For every n-dimensional simple polytope, we have the Dehn–Sommerville duality, i.e. $h_i = h_{n-i}$.*

Indeed, for any generic linear function L, according to Theorem 3.2, we have $h_i = h_i(L)$. For the computation of the numbers h_{n-i}, we can use the function $-L$. According to Theorem 3.2, we have $h_{n-i} = h_{n-i}(-L)$. From the definition of the index we see that the numbers $h_i(L)$ and $h_{n-i}(-L)$ are equal. Corollary 4.2 is thus proved.

COROLLARY 4.3. *The estimate from Nikulin's theorem holds.*

Indeed, the proof of Nikulin's theorem given in Section 2 uses parts 1) and 2) of the theorem on the h-vector together with the elementary lemma from Section 2. Corollaries 4.1 and 4.2 prove parts 1) and 2) of the theorem on the h-vector. Hence we obtain a simple elementary proof of Nikulin's estimate. In Section 5, we will rewrite this proof separately, without using the notion of the h-vector.

Let us now discuss other corollaries of Theorem 3.2. Corollary 4.1 can easily be strengthened. The following holds:

COROLLARY 4.4. *Under the assumptions of Corollary 4.1, all the numbers h_i are strictly positive.*

PROOF. For any fixed vertex of the polytope, we can choose a linear function L so that the index of the function L at this vertex is any given number from 0 to n. But according to the theorem, the numbers $h_i(L)$ do not depend on the choice of L and are equal to h_i, which proves Corollary 4.4. $\qquad \square$

COROLLARY 4.5. *Consider an arbitrary n-dimensional simple polytope and an arbitrary affine hyperplane containing no vertices of the polytope. Under these conditions, there exists a face of the polytope of dimension $[n/2]$ that is disjoint from the hyperplane.*

PROOF. Perturbing the affine hyperplane slightly, if necessary, we can arrange that it is a level hypersurface $L = c$ of a linear function L generic with respect to the polytope. According to the proof of Corollary 4.4, there exists a vertex A such that the index of the function L at it equals $[n/2]$. There exists an $[n/2]$-face Γ_1 of the polytope such that the maximum of the function L restricted to this face is attained at the vertex A, and there exists an $(n - [n/2])$-dimensional face Γ_2 of the polytope such that the minimum of the function L restricted to this face is attained at the vertex A. If $L(A) < c$, then the hyperplane $L = c$ is disjoint from the face Γ_1; if $L(A) > c$, then the hyperplane $L = c$ is disjoint from the face Γ_2.   □

In Sections 7 and 8, we will generalize Corollary 4.5 and give an estimate for the number and the ratio of the faces of different dimensions disjoint from a generic hyperplane section.

5. An elementary proof of Nikulin's estimate

Let us rewrite the proof of Nikulin's inequalities from Corollary 4.3 without using the notion of the h-vector.

Let Δ be a simple n-dimensional polytope, let l and k be integers satisfying the inequalities $0 \leq l < k \leq (n + 1)/2$, and let m be the number of vertices of the polytope Δ. Denote by $V_1, \ldots, V_m$ the vertices of this polytope taken in any order.

Fix a generic linear function L on the polytope Δ. To each face of dimension j, where j is any nonnegative integer not exceeding n, assign the vertex where the function L restricted to the face attains its maximum. We obtain the equality

$$f_j = \sum_{1 \leq i \leq m} \binom{\mathrm{ind}(V_i)}{j},$$

where $\mathrm{ind}(V_i)$ is the index of the function L at the vertex V_i. Analogously, to any face, assign the vertex where L attains the minimum. Then we obtain the equality

$$f_j = \sum_{1 \leq i \leq m} \binom{n - \mathrm{ind}(V_i)}{j}.$$

Therefore, we have

$$2f_j = \sum_{0 \leq i \leq m} \binom{\mathrm{ind}(V_i)}{j} + \binom{n - \mathrm{ind}(V_i)}{j}.$$

Let us take into account that each l-dimensional face of a simple n-dimensional polytope is contained exactly in $\binom{n-l}{n-k}$ of its k-dimensional faces. We obtain that the average number of l-dimensional faces on k-dimensional faces of the polytope Δ equals

$$\binom{n-l}{n-k} \frac{\sum_{1 \leq i \leq m} \left(\binom{\mathrm{ind}(V_i)}{l} + \binom{n-\mathrm{ind}(V_i)}{l} \right)}{\sum_{1 \leq i \leq m} \left(\binom{\mathrm{ind}(V_i)}{k} + \binom{n-\mathrm{ind}(V_i)}{k} \right)}.$$

Set

$$A_i = \binom{\mathrm{ind}(V_i)}{k} + \binom{n - \mathrm{ind}(V_i)}{k}, \quad B_i = \binom{\mathrm{ind}(V_i)}{l} + \binom{n - \mathrm{ind}(V_i)}{l}.$$

Using Lemma 2.4 and Corollary 4.4, by which among the vertices of the polytope there is a vertex V_j such that $\mathrm{ind}(V_j) = [n/2]$, we obtain that

$$\max_i \frac{B_i}{A_i} = \frac{\binom{[n/2]}{l} + \binom{[(n+1)/2]}{l}}{\binom{[n/2]}{k} + \binom{[(n+1)/2]}{k}}.$$

Using Lemma 2.1 for the number m being the number of vertices of the polytope Δ, for the numbers A_i and B_i introduced above and for $\alpha_i \equiv 1$, we obtain a proof of Nikulin's theorem. We only need to notice that for $1 < k$, the average number of faces under this estimate is strictly less than $\max_i \frac{B_i}{A_i}$, since among the vertices of the polytope there exist points of the maximum and points of the minimum of the function L, at which the corresponding ratio is strictly less than the maximal one. The proof of Nikulin's inequalities is completed.

6. Sections of a simplex and a geometric meaning of Nikulin's estimates

In this Section, we will need several simple formulas concerning the combinatorics of hyperplane sections of a simplex. First we present these formulas and show that they play a certain role in Nikulin's estimate. After that, we discuss a plan of a proof of a certain generalization of Nikulin's estimate.

Consider a section of an $(n-1)$-dimensional simplex by an affine hyperplane $L = c$ not passing through its vertices. Suppose that i vertices of the simplex lie on one side of the hyperplane, and $(n - i)$ vertices lie on the other side, where i is any number such that $0 \leq i \leq n$. Then:

1) for $j > 0$, the number f_j^c of $(j - 1)$-faces of the simplex disjoint from the hyperplane $L = c$, is equal to

$$\binom{i}{j} + \binom{n-i}{j}.$$

Indeed, on one side of the hyperplane there are $\binom{i}{j}$ such faces, and $\binom{n-i}{j}$ such faces are on the other side;

2) for $0 < k \leq \max(i, (n - i))$, twice the number f_{k-1} of $(k - 1)$-faces of the simplex divided by the number f_{k-1}^c satisfies the equality

$$\frac{2f_{k-1}}{f_{k-1}^c} = \binom{n}{k} \frac{2}{\binom{i}{k} + \binom{n-i}{k}};$$

3) for l, k such that $0 < l < k$, the number $f_{l-1,k-1}^c$ of pairs of faces $\Gamma_1 \subset \Gamma_2$ of the simplex, where Γ_1 is a face of dimension $(l - 1)$ disjoint from the hyperplane $L = c$, and Γ_2 is a $(k - 1)$-face, satisfies the equality

$$f_{l-1,k-1}^c = \binom{n-l}{n-k} \left(\binom{i}{l} + \binom{n-i}{l} \right).$$

Indeed, the number of $(k - 1)$-faces of the simplex containing a fixed $(l - 1)$-face, is $\binom{n-l}{n-k}$. It now remains to use the equality from 1);

4) for l,k such that $0 < l < k \leq \max(i, (n - i))$, the number $f_{l-1,k-1}^c$ divided by the number f_{k-1}^c satisfies the equality

$$\frac{f_{l-1,k-1}^c}{f_{k-1}^c} = \binom{n-l}{n-k} \frac{\binom{i}{l} + \binom{n-i}{l}}{\binom{i}{k} + \binom{n-i}{k}}.$$

PROPOSITION 6.1. *a) For $0 < k \leq (n+1)/2$, the maximal value of the ratio $\frac{2f_{k-1}}{f_{k-1}^c}$ for a generic section $L = c$ of the simplex equals*

$$\frac{2\binom{n}{k}}{\binom{[n/2]}{k} + \binom{[(n+1)/2]}{k}}.$$

b) For $0 < l < k \leq (n+1)/2$, the maximal value of the ratio $\frac{f_{l-1,k-1}^c}{f_{k-1}^c}$ for a generic section $L = c$ of the simplex equals

$$\binom{n-l}{n-k} \frac{\binom{[n/2]}{l} + \binom{[(n+1)/2]}{l}}{\binom{[n/2]}{k} + \binom{[(n+1)/2]}{k}}.$$

Indeed, this is readily seen from formulas 2) and 4) for these ratios and from Lemma 2.4.

THEOREM 6.2 (on a geometric meaning of Nikulin's estimate). *For l and k such that $0 \leq l < k \leq (n+1)/2$ and $1 < k$, the average number of l-dimensional faces on k-dimensional faces of a simple n-dimensional polytope Δ is strictly less than the maximum over all $(n-1)$-dimensional sections of a simplex by a generic hyperplane $L = c$:*
  1) for $l = 0$: of the ratio $\frac{2f_{k-1}}{f_{k-1}^c}$;
  2) for $0 < l$: of the ratio $\frac{f_{l-1,k-1}^c}{f_{k-1}^c}$.

PROOF. This theorem follows from Nikulin's estimate and Proposition 6.1. □

Theorem 6.2 can be proved directly, by translating the proof of Nikulin's inequalities from Section 5 to the language of sections of a simplex. The reason that we can perform this translation is the following. Near each vertex, a simple n-dimensional polytope looks like a cone over an $(n-1)$-dimensional simplex. A level hypersurface of a linear function L passing through a vertex of the polytope, gives rise to a section of this $(n-1)$-dimensional simplex. If the index of the function L at the vertex is i, then i vertices of the simplex lie on one side of this section, and $(n-i)$ vertices lie on the other side. This observation allows to perform the desired translation.

Our further plan is as follows. In Section 9, we consider sections of simple $(n-1)$-dimensional polytopes by generic hyperplanes and solve the same problems for them as those we solved in Proposition 6.1 for a simplex. Then, in Section 10, we prove generalized Nikulin's estimates for n-dimensional polytopes simple at the edges. We will use the fact that a polytope simple at the edges looks like a cone over some simple $(n-1)$-dimensional polytope near each of its vertices.

We now proceed to the realization of this plan.

7. An estimate for the number of faces of a section

To estimate the average number of l-dimensional faces on k-dimensional faces of an n-dimensional polytope, we need to deal with $(n-1)$-dimensional polytopes, with their $(l-1)$-dimensional and $(k-1)$-dimensional faces and with sections of these polytopes. To avoid the persisting "-1" in dimensions of polytopes and their faces throughout the remaining part of the text, and since the problems on sections of simple polytopes are interesting on their own right, we change the notation for dimensions. We will speak of s-dimensional faces on r-dimensional faces of q-dimensional polytopes and of hyperplane sections of these q-dimensional polytopes.

Thus let $\Delta \subset \mathbb{R}^q$ be a simple q-dimensional polytope, and let a hyperplane not passing through the vertices of the polytope Δ be fixed. Perturbing the hyperplane slightly, we can arrange that it will be a level hypersurface $L = c$ of a generic linear function L on the polytope Δ. Let O and Π be the sets of vertices of the polytope Δ where the function L is less than c and greater than c, respectively. The set of all vertices of the polytope is the union of the subsets O and Π, since the hyperplane does not pass through the vertices of the polytope.

THEOREM 7.1. *The number f_j^c of j-dimensional faces of the polytope Δ disjoint from the hyperplane $L = c$, is given by the formula*

$$f_j^c = \sum_{b \in O} \binom{\mathrm{ind}(b)}{j} + \sum_{b \in \Pi} \binom{q - \mathrm{ind}(b)}{j}. \tag{2}$$

PROOF. The set of j-dimensional faces disjoint from the hyperplane $L = c$, splits into two subsets: the subset of faces where the function L is strictly greater than c, and the subset of faces where the function L is strictly less than c.

The number of faces in the first set equals the first summand in formula (2) for f_j^c. To prove this, we associate each face from this set with the vertex, where the restriction of the function L to the face attains the maximum (an analogous argument was used in the proof of Theorem 3.2). Associating the faces from the second set with the minimum points of the function L, we see that the number of faces in the second set is equal to the second summand in formula (2). The Theorem is thus proved. □

We will need formula (2) in the next sections for the proof of the generalized Nikulin estimate.

Let us discuss here one interesting corollary of this formula, which is not relevant for this generalization. The corollary allows to give upper bounds for all components of the h-vector of the section of Δ by a generic affine plane of codimension l in terms of the h-vector of the polytope Δ. This, in turn, allows to estimate the numbers of faces of all dimensions of any (not necessarily generic) affine section of the polytope Δ.

To describe this estimate, we will need the following operation S, taking each reciprocal polynomial with nonnegative coefficients to a reciprocal polynomial with nonnegative coefficients, whose degree is one less. By definition, the polynomial $S \circ p_m(t)$ can be constructed from a polynomial $p_m(t)$ of degree m in the following way: the polynomial $S \circ p_m(t)$ is the unique reciprocal polynomial of degree $(m-1)$ such that for $k \geq (m-1)/2$, its coefficient with the monomial t^k coincides with the coefficient with the monomial t^k in the Laurent series for the rational function $p_m(t)(t-1)^{-1}$ at ∞.

THEOREM 7.2. *All coefficients of the h-polynomial of any generic hyperplane section of a simple polytope Δ do not exceed the corresponding coefficients of the polynomial $S \circ h(t)$, where $h(t)$ is the h-polynomial of the polytope Δ.*

PROOF. Denote by $f_j^{<c}$ and by $f_j^{>c}$ the number of j-dimensional faces of the polytope Δ, lying beneath and, respectively, above the level hypersurface $L = c$. Denote by $h_j^{<c}$ and by $h_j^{>c}$ the number of vertices of index j with respect to the function L lying beneath the level hypersurface $L = c$, and, respectively, the number of vertices of index $(q - j)$ lying above this hypersurface. Denote by $\tilde{f}_j$ and $\tilde{h}_j$ the j-th components of the f-vector and the h-vector, respectively, of a section of the polytope Δ.

For each j, we have the obvious relation $f_j = f_j^{<c} + \tilde{f}_{j-1} + f_j^{>c}$. Using these relations, we can rewrite formula (2) in the form

$$h(t) = h^{<c}(t) + \tilde{h}(t)(t - 1) + h^{>c}(t),$$

where $h(t), \tilde{h}(t)$ are the h-polynomials of the polytope Δ and of its section, respectively, and $h^{<c}(t), h^{>c}(t)$ are generating polynomials for the sequences $h_j^{<c}$ and $h_j^{>c}$.

The identity (2) means that

$$\tilde{h}(t) = (h(t) - h^{<c}(t) - h^{>c}(t))(t - 1)^{-1}. \tag{3}$$

All coefficients $h_j^{<c}, h_j^{>c}$ of the polynomials $h^{<c}(t), h^{>c}(t)$ are nonnegative. Near the point ∞, we have $(t - 1)^{-1} = t^{-1} + t^{-2} + \cdots$. Hence the identity (3) implies Theorem 7.2.   □

DEFINITION 5. A section of a simple q-dimensional polytope by a generic hyperplane is said to be *successful*, if it intersects all faces of dimension $> q/2$ (or, in other words, if it intersects all faces of codimension $\leq (q - 1)/2$).

From formula (2) it is readily seen that a section $L = c$ is successful if and only if at all vertices of index $< q/2$, the values of the function L are less than c, and at all vertices of index $> q/2$ the values are greater than c.

PROPOSITION 7.3. *The upper bounds for the h-vector components of a generic hyperplane section of a simple q-dimensional polytope from Theorem 7.2 are simultaneously attained if and only if the section is successful.*

Indeed, according to formula (3), for all estimates to be simultaneously attained it is necessary that the polynomials $h^{<c}(t)$ and $h^{>c}(t)$ have degree $\leq q/2$. This happens if and only if the section $L = c$ is successful.

DEFINITION 6. A section of a simple q-dimensional polytope by a generic affine plane of dimension l is called *successful*, if it intersects all codimension $\leq l/2$ faces of the polytope.

THEOREM 7.4. *For any generic l-dimensional affine section of a simple polytope Δ, none of the coefficients of the h-polynomial of this section exceeds the corresponding coefficients of the polynomial $S^{(q-l)} \circ h(t)$, where $h(t)$ is the h-polynomial of the polytope Δ and $S^{(q-l)}$ is the $(q-l)$-th iteration of the operation S. The upper bounds for the h-vector components of the section are simultaneously attained if and only if the section is successful.*

PROOF. The inequalities from Theorem 7.4 follow from Theorem 7.2, since any section of dimension l can be obtained by taking a hyperplane section, then a hyperplane section of this section, and so on.

The case when all inequalities turn simultaneously into equalities can be worked out in the same way as in Proposition 7.3.   □

From Theorem 7.4 we see how to estimate the f-vector of any generic l-dimensional section of a simple q-dimensional polytope Δ in terms of the f-vector of the polytope.

Let us construct the f-vector of some abstract simple l-dimensional polytope, such that the number $\tilde{f}_{(l-j)}$ of its faces of dimension $(l-j)$ is equal to $f_{(q-j)}$ for $j \leq l/2$. The numbers of faces of smaller dimensions of such a polytope can be recovered using the Dehn–Sommerville duality (from the theorem on the h-vector it is readily seen that there exist simple polytopes with such f-vector.)

COROLLARY 7.5. *For any k, the number of k-faces of a section of the polytope Δ by a generic affine plane of dimension l does not exceed the component $\tilde{f}_k$ of the f-vector thus constructed.*

COROLLARY 7.6. *The estimate from Corollary 7.5 holds for a section of the polytope Δ by an arbitrary affine plane of dimension l, which can even be non-generic for the polytope Δ.*

PROOF. Let a simple polytope Δ be given by the linear inequalities $0 \leq L_i$, and suppose that we study a section of the polytope Δ by an affine plane P of dimension l. Consider the one-parameter family of polytopes $\Delta(u)$, given by the inequalities $0 \leq L_{i,u} = L_i + \epsilon_i(u)$, where $\epsilon_i(u)$ are generic linear functions of u that are positive for $u > 0$.

For small positive values of u, all polytopes $\Delta(u)$ are combinatorially equivalent and have the same h-vector. If the functions $\epsilon_i(u)$ are generic, then the polytopes lying in the plane P of codimension l and given there by the inequalities $0 \leq L_{i,u}$ with small positive u are simple. We can apply Corollary 7.5 to those polytopes. Polytopes $P \cap \Delta(u)$ corresponding to different small $u > 0$ have parallel facets, and they give rise to the same partition $\Delta^*(u)$ of the dual space P^*. The polytopes $\Delta(u)$ degenerate to the section $\Delta \cap P$ for $u = 0$.

It is clear that for such degeneration, the number of faces in each dimension does not increase (the partition $\Delta^*(u)$ of P^* dual to the polytope $\Delta(u)$, is a subdivision of the partition Δ_0^*). The corollary is thus proved. □

PROBLEM 7.7. Let the h-vector of a simple q-dimensional polytope be given. What can be the h-vector of a section of the polytope by a generic affine plane of dimension l?

Theorem 7.4 gives an upper estimate for the components of the h-vector of the section. I think that this estimate is sharp but I cannot prove this. For a proof, we need to construct a simple q-dimensional polytope with a given h-vector, such that there exists a generic affine plane of dimension l, intersecting all faces of the polytope of codimension $\leq l/2$.

REMARK 5. According to the famous Upper Bound Conjecture, the l-dimensional polytope dual to a cyclic polytope with N vertices has the maximal number of faces in any dimension among all simple l-dimensional polytopes having N facets. This conjecture is proved (see [3,4],[8],[11]). The estimate from the Upper Bound Theorem follows from the partial case of Corollary 7.6 when the polytope Δ is an $(N-1)$-dimensional simplex. The arguments from Section 7 are very close to the arguments that were used to prove the Upper Bound Conjecture.

8. The ratio of faces disjoint from a section

Let us return to the realization of our plan (see the end of Section 6). Let the h-vector of some simple q-dimensional polytope be fixed. Consider an arbitrary simple polytope Δ with the given h-vector and fix an arbitrary generic hyperplane section $L = c$ of this polytope.

Denote by f_j, f_j^c the number of j-dimensional faces of the polytope Δ and, respectively, the number of j-dimensional faces of the polytope Δ disjoint from the hyperplane $L = c$. We are interested in the following

PROBLEM 8.1. 1) Give an upper estimate for the ratio f_r / f_r^c with any r, $1 \le r \le q/2$, in terms of the h-vector.

2) Give an upper estimate for the ratio f_s^c / f_r^c with any s, r, $0 \le s < r \le q/2$ in terms of the h-vector.

To state the results on Problem 8.1, let us introduce the following notation. For any vector $h = (h_0, \ldots, h_q)$ with positive (not necessarily integer) components h_i and with the symmetry property $h_i = h_{q-i}$, set:

1) $F_j(h) = \sum_{0 \le i \le q} h_i \binom{i}{j}$;

2) $\Phi_j(h) = \sum_{0 \le i < q/2} 2h_i \binom{i}{j} + Q_j$, where $Q_j = h_{q/2} \binom{q/2}{j}$ for even q and $Q_j = 0$ for odd q.

THEOREM 8.2. *For every generic hyperplane section $L = c$ of a simple q-dimensional polytope Δ with the h-vector h, the following inequalities hold:*

1) For any r such that $1 \le r \le q/2$,

$$\frac{f_r}{f_r^c} \le \frac{F_r(h)}{\Phi_r(h)};$$

2) for any s, r such that $0 \le s < r \le q/2$,

$$\frac{f_s^c}{f_r^c} \le \frac{\Phi_s(h)}{\Phi_r(h)}.$$

For a successful section $L = c$ of the polytope Δ, all these inequalities are equalities.

Conversely, if for at least one r satisfying the conditions of part 1), or for at least one pair s, r satisfying the conditions of part 2), the inequality turns to equality, then the section $L = c$ of the polytope Δ is successful.

REMARK 6. For any generic hyperplane section of the polytope Δ the ratio $\frac{f_0}{f_0^c}$ is equal to one, since a generic hyperplane intersects no vertices of the polytope Δ.

The proof of Theorem 8.2 is based on the solution of Problem 8.3, which is posed below. With a vector $h = (h_0, \ldots, h_q)$ having positive integer components h_i and having the symmetry property $h_i = h_{q-i}$, associate the collection of sets $V_0, \ldots, V_q$ containing, respectively, $h_0, \ldots, h_q$ elements. The number of elements in a finite set A will be denoted by $\aleph(A)$.

PROBLEM 8.3. Find partitions $V_i = O_i \cup \Pi_i$ of the sets V_i such that
1) for given r, $1 \le r \le q/2$, the ratio

$$\frac{\sum_{0 \le i \le q} h_i \binom{i}{r}}{\sum_{0 \le i \le q} \aleph(O_i) \binom{i}{r} + \aleph(\Pi_i) \binom{q-i}{r}}$$

is maximal;

2) for given s, r, $0 \leq s < r \leq q/2$, the ratio

$$\frac{\sum\limits_{0 \leq i \leq q} \aleph(O_i)\binom{i}{s} + \aleph(\Pi_i)\binom{q-i}{s}}{\sum\limits_{0 \leq i \leq q} \aleph(O_i)\binom{i}{r} + \aleph(\Pi_i)\binom{q-i}{r}} \tag{4}$$

is maximal.

REMARK 7. The question from part 1) of Problem 8.3 can be posed even for $r = 0$. But in this case the ratio does not depend on the choice of a partition and is identically equal to 1.

A collection of partitions $V_i = O_i \cup \Pi_i$ of the sets V_i is called *successful*, if:
1) for $i < q/2$, the sets O_i and V_i coincide, and the set Π_i is empty,
2) for $i > q/2$, the set O_i is empty, and the sets Π_i and V_i coincide,
3) for $i = q/2$, the partition of the set V_i into subsets O_i and Π_i is arbitrary.
The following theorem provides a complete solution of Problem 8.3.

THEOREM 8.4. *1) A successful collection of partitions of the sets $V_0, \ldots, V_m$ maximizes the ratio from part 1) for any r. The desired maximum is $\frac{F_r(h)}{\Phi_r(h)}$.*

2) A successful collection of partitions of the sets $V_0, \ldots, V_m$ maximizes the ratio from part 2) for any s and r. The desired maximum is $\frac{\Phi_s(h)}{\Phi_r(h)}$.

3) If the partitions $V_i = O_i \cup \Pi_i$ of the sets V_i maximize the ratio from part 1) for some r or maximize the ratio from part 2) for some s and r, then this collection of partitions is successful.

Let us deduce Theorem 8.2 from Theorem 8.4. A generic linear function L defines a partition of the set V of vertices of the polytope Δ into subsets $V_0, \ldots, V_q$: the subset V_i contains h_i elements and is defined as the set of vertices having index i with respect to the function L. By fixing a level $L = c$, we partition each set V_i into subsets O_i and Π_i, consisting of vertices where $L < c$ and $L > c$, respectively.

To deduce Theorem 8.2 from Theorem 8.4, it now suffices to compare formula (2) from Theorem 7.1 with Problem 8.3.

The first part of Problem 8.3 is very simple. The following lemma is obvious:

LEMMA 8.5. *The following value*

$$\sum_{0 \leq i \leq q} \aleph(O_i)\binom{i}{r} + \aleph(\Pi_i)\binom{q-i}{r}$$

attains the minimum on successful partitions and only on them.

According to Lemma 8.5, successful partitions and only they give a solution for part 1) of Problem 8.3.

For the proof of the remaining parts of Theorem 8.4 it is convenient to use the *fractional linear programming*. Fractional linear programming is the maximizing of the ratio L_1/L_2 of two linear functions on a convex polytope (it is assumed that the function L_2 vanishes nowhere on the polytope). Fractional linear programming is not very different from linear programming. Indeed, consider an arbitrary projective transformation of the space $\mathbb{R}P^q \supset \mathbb{R}^q$ mapping the hyperplane $L_2 = 0$ to the hyperplane at infinity. The convex polytope gets transformed into another convex polytope, and the fractional linear function L_1/L_2 becomes linear. Thus the original problem transforms to a problem of linear programming.

Hence the set of points where the maximum of a fractional linear function is attained, is a face of the polytope. (In particular, the maximum is attained at a vertex of the polytope. Lemma 2.1 from Section 2 is based on this fact.) In the case of general position, this face is necessarily a vertex of the polytope.

Let us formulate a continuous variant of part 2) of Problem 8.3. Let $h = (h_0, \ldots, h_q)$ be a vector with positive (not necessarily integer) components h_i, and with the symmetry property $h_i = h_{q-i}$. Consider the parallelepiped Δ in the space $\mathbb{R}^{[(q+1)/2]}$, defined by the inequalities $0 \leq x_i \leq 2h_i$ for $0 \leq i < q/2$ (the number $[(q+1)/2]$ is the number of indices i satisfying the inequalities $0 \leq i < q/2$). For each integer j, define a linear function on $\mathbb{R}^{[(q+1)/2]}$ by the formula

$$L_j = \sum_{0 \leq i < q/2} \left(x_i \binom{i}{j} + (2h_i - x_i)\binom{q-i}{j} \right) + Q_j,$$

where $Q_j = h_{q/2}\binom{q/2}{j}$ for even q and $Q_j = 0$ for odd q.

Consider the following problem of fractional linear programming.

PROBLEM 8.6. Maximize the function L_s/L_r on the parallelepiped Δ, where s, r are fixed numbers satisfying the inequalities $0 \leq s < r \leq q/2$.

THEOREM 8.7. For any s, r satisfying the conditions of Problem 8.6, the strict maximum of the function L_s/L_r is attained at the vertex Γ of the parallelepiped such that its i-th coordinate x_i is $2h_i$ for $0 \leq i < q/2$. This maximum $L_s(\Gamma)/L_r(\Gamma)$ is equal to $\Phi_s(h)/\Phi_r(h)$.

Theorem 8.7 allows to conclude the proof of Theorem 8.4 started in Lemma 8.3. For a fixed collection of partitions $V_0 = O_0 \cup \Pi_0, \ldots, V_q = O_q \cup \Pi_q$ and for each $0 \leq i < q/2$, set $x_i = \chi(O_i) + \chi(\Pi_{q-i})$. Then the value $L_s(x)/L_r(x)$ at the point x, where $x \in \mathbb{R}^{[(q+1)/2]}$ is a vector with coordinates x_i, equals the value of ratio (4) for the given collection of partitions of the sets V_i.

Furthermore, $x_i = 2h_i$ if and only if $\chi(O_i) = h_i$ and $\chi(\Pi_{q-i}) = h_i$. Hence, after we prove Theorem 8.7, Theorem 8.4 will be proved completely.

The proof of Theorem 8.7 uses the following property of binomial coefficients.

LEMMA 8.8. Suppose that $s < r$. Then the ratio $\psi(m) = \binom{m}{s}/\binom{m}{r}$ strictly decreases as m runs from r to ∞, and the ratio $\varphi(m) = \binom{r}{m}/\binom{s}{m}$ strictly increases as m runs from 0 to s.

Indeed, the denominator of the ratio

$$\psi(m) = (r!/s!)(1/(m-s)(m-s-1)\cdots(m-r+1))$$

increases as m increases, and the numerator of the ratio

$$\varphi(m) = (r!/s)!(s-m)\cdots(r-m+1)$$

increases as m increases.

PROOF. of Theorem 8.7.

Step 1. The value $L_s(\Gamma)/L_r(\Gamma)$ at the vertex Γ is bigger than $\binom{[q/2]}{s}/\binom{[q/2]}{r}$. Indeed, $L_s(\Gamma)/L_r(\Gamma) > L_{s,r}(\Gamma)/L_r(\Gamma)$, where

$$L_{s,r}(\Gamma) = \sum_{r \leq i < q/2} 2h_i \binom{i}{s} + Q_s.$$

According to Lemma 8.8 applied to the function $\psi(m)$, the numbers $\binom{i}{s}/\binom{i}{r}$ increase as i increases from $i = r$. To complete step 1, it remains to use Lemma 2.1 (it is applicable, since the numbers h_i are strictly positive for $r \leq i \leq [n/2]$).

Step 2. The value $L_s(V)/L_r(V)$ at any vertex V adjacent to the vertex Γ (i.e. at a vertex V that is connected with Γ by an edge), is strictly less than the corresponding value at the vertex Γ. Indeed, all coordinates of the vertex V, except one, coincide with coordinates of the vertex Γ. Let the index of this special coordinate be i. If the vertex Γ gets replaced by the vertex V, then the value of the function L_s increases by $2h_iB$, where $B = \binom{q-i}{s} - \binom{i}{s}$, and the value of the function L_r increases by the number $2h_iA$, where $A = \binom{q-i}{r} - \binom{i}{r}$. By Step 1, the number $\binom{[q/2]}{s}/\binom{[q/2]}{r}$ is at most $L_s(\Gamma)/L_r(\Gamma)$.

Let us show that $B/A < \binom{[q/2]}{s}/\binom{[q/2]}{r}$. The inequality $B/\binom{[q/2]}{s} < A/\binom{[q/2]}{r}$ follows from Lemma 8.8 applied to the function $\varphi(m)$. Indeed, $\binom{q-i}{s}/\binom{[q/2]}{s} < \binom{q-i}{r}/\binom{[q/2]}{r}$, since $(q - i) > [q/2]$ and $s < r$. Furthermore,

$$-\frac{\binom{i}{s}}{\binom{[q/2]}{s}} \leq -\frac{\binom{i}{r}}{\binom{[q/2]}{r}}.$$

(For $i < r$ and for $i = [q/2]$, this inequality turns to equality. For $r \leq i < [q/2]$ and $s < r$, this inequality is strict and is equivalent to the relation $\binom{[q/2]}{s}/\binom{i}{s} < \binom{[q/2]}{r}/\binom{i}{r}$, which also follows from Lemma 8.8 applied to the function $\varphi(m)$.)

Summing up the two obtained inequalities, we arrive at the desired result.

Step 3. Fractional linear programming and the result of Step 2 prove that the function L_s/L_r attains its maximum at the vertex Γ. Theorem 8.7, together with Theorem 8.4 and 8.2, is proved. $\qquad\square$

9. Extremal property of sections of a simplex

In the statement of Theorem 9.1, which is the central result of this section, we use the notation introduced in Section 8.

THEOREM 9.1. *For every generic hyperplane section $L = c$ of a simple q-dimensional polytope Δ, the following inequalities hold, which turn to equalities for a successful section of a q-dimensional simplex:*

1) For $1 \leq r \leq q/2$,

$$\frac{f_r}{f_r^c} \leq \frac{\binom{q+1}{r+1}}{\binom{[(q+1)/2]}{r+1} + \binom{[(q+2)/2]}{r+1}};$$

2) for $0 \leq s < r \leq q/2$,

$$\frac{f_s^c}{f_r^c} \leq \frac{\binom{[(q+1)/2]}{s+1} + \binom{[(q+2)/2]}{s+1}}{\binom{[(q+1)/2]}{r+1} + \binom{[(q+2)/2]}{r+1}}.$$

If for at least one r from the inequality of part 1) we have the equality, then the polytope Δ is a simplex, and its section $L = c$ is successful. If for some pair s, r the inequality from part 2) is the equality, then the components h_i of the h-vector of the polytope Δ are equal to each other for $s \leq i \leq (q - s)$, and the section $L = c$ of the polytope Δ is successful.

In the proof of Theorem 9.1, we will use the fact that the h-vector of a simple polytope is unimodal (see Section 3). This is a part of the theorem on the h-vector.

REMARK 8. For some simple polytopes, the unimodality of the h-vector is obvious, and does not require the use of the theorem on the h-vector. For example, the h-vector of a direct product of simplices possesses this property, since the h-polynomial of the direct product of polytopes is the product of their h-polynomials. This fact, together with the theorem on reduction from Section 10, proves Nikulin's estimate for almost simple polytopes (i.e. it proves Corollary 0.4) without using the theorem on the h-vector.

We will need Problem 9.2 posed below. A solution to this problem is given by Theorem 9.3, which implies Theorem 9.1 immediately. Consider the set of vectors $h = (h_0, \ldots, h_q)$ such that the components h_i of these vectors are:
1) positive,
2) symmetric, i.e. $h_i = h_{q-i}$,
3) unimodal, i.e. $0 \leq h_0 \leq \cdots \leq h_{[q/2]}$.
By the symmetry condition, a vector h is determined by its components $h_0, \ldots, h_{[q/2]}$, the number of which is $[q/2]+1$. Consider a simplex Δ in $\mathbb{R}^{[q/2]+1}$ defined by the inequalities $0 \leq h_0 \leq \cdots \leq h_{[q/2]}$ and the equation $h_0 + \cdots + h_{[q/2]} = v$, where v is an arbitrary positive number.

PROBLEM 9.2. Maximize on the simplex Δ:
1) for $1 \leq r \leq q/2$, the function $\frac{F_r(h)}{\Phi_r(h)}$,
2) for $0 \leq s < r \leq q/2$, the function $\frac{\Phi_s(h)}{\Phi_r(h)}$.

First note that the maximum in Problem 9.2 does not depend on the choice of the constant v, since the functions we maximize are homogeneous of degree 0. A complete answer to this problem is given by the following

THEOREM 9.3. 1) For $1 \leq r \leq q/2$, the maximum of the function $\frac{F_r(h)}{\Phi_r(h)}$ is attained for $h_0 > 0, h_0 = \cdots = h_{[q/2]}$, and is equal to

$$\frac{\binom{q+1}{r+1}}{\binom{[(q+1)/2]}{r+1} + \binom{[(q+2)/2]}{r+1}}.$$

2) For $0 \leq s < r \leq q/2$, the maximum of the function $\frac{\Phi_s(h)}{\Phi_r(h)}$ is attained for $h_0 + \cdots + h_s > 0, h_s = \cdots = h_{[q/2]}$, and is equal to

$$\frac{\binom{[(q+1)/2]}{s+1} + \binom{[(q+2)/2]}{s+1}}{\binom{[(q+1)/2]}{r+1} + \binom{[(q+2)/2]}{r+1}}.$$

Let us deduce Theorem 9.1 from Theorem 9.3. According to Theorem 8.2, for a fixed h-vector h of the polytope Δ, the ratios $\frac{f_r}{f_r^c}$ and $\frac{f_s^c}{f_r^c}$ do not exceed the ratios $\frac{F_r(h)}{\Phi_r(h)}$ and $\frac{\Phi_s(h)}{\Phi_r(h)}$, and the equality is attained for successful sections only. To conclude the deduction of Theorem 9.1 from Theorem 9.3, it remains to note that a q-dimensional simplex is the only simple polytope such that all components of its h-vector are equal. Indeed, since $h_0 = 1$, all components of the h-vector must be equal to 1. It follows that $f_q = q + 1$, i.e. that the polytope is a simplex.

For the proof of Theorem 9.3, we will need the simple Lemmas 9.4 and 9.5 on sums of binomial coefficients. We will also use classical Abel's lemma, which is a discrete variant of the integration by parts, together with its application to the functions we maximize (Lemma 9.6). Finally, we will need a simple general fact from fractional linear programming (Lemma 9.7) and two simple lemmas dealing with the functions we maximize (Lemmas 9.8 and 9.9).

LEMMA 9.4 (on a sum of binomial coefficients). *The following formula holds:*

$$\sum_{i \leq j \leq k} \binom{j}{m} = \binom{k+1}{m+1} - \binom{k+1}{j+1}.$$

PROOF. Computing the sum of a geometric series, we obtain the identity

$$\sum_{i \leq j \leq k} (1+t)^j = ((1+t)^{k+1} - (1+t)^{i+1})/t.$$

Equating the coefficients with t^m in this identity, we obtain the required equality.
□

LEMMA 9.5. *The following formula holds:*

$$\sum_{i \leq k \leq (q-i)} \binom{\min(k, q-k)}{r} = \binom{[(q+1)/2]}{r+1} + \binom{[(q+2)/2]}{r+1} - 2\binom{i+1}{r+1}.$$

PROOF. The desired sum can be rewritten in the form

$$\sum_{i \leq k \leq [(q-1)/2]} \binom{k}{r} + \sum_{i \leq k \leq [q/2]} \binom{k}{r}.$$

Applying the previous lemma to each sum, we obtain the required equality.

To make the use of the unimodality condition for the h-vector simpler, let us transform the functions we maximize. Recall the following discrete variant of the integration by parts. For every sequence $a_0, \ldots, a_n$, define sequences $(\Delta a)_i$ and $(Sa)_i$, where $(\Delta a)_i = a_i - a_{i-1}$ for $0 < i \leq n$, and $(\Delta a)_0 = a_0$, and $(Sa)_i^n = \sum_{i \leq j \leq n} a_j$ for $0 \leq i \leq n$.
□

Abel's lemma *For any pair of sequences $a_0, \ldots, a_n$ and $b_0, \ldots, b_n$, the following equality holds:*

$$\sum_{0 \leq i \leq n} a_i b_i = \sum_{0 \leq i \leq n} (\Delta a)_i (Sb)_i^n.$$

LEMMA 9.6. *We have the following equalities:*
1) $F_j(h) = \sum_{0 \leq i \leq [q/2]} (\Delta h)_i \sum_{i \leq k \leq q-i} \binom{k}{j}$;

2) $\Phi_j(h) = \sum_{0 \leq i \leq [q/2]} (\Delta h)_i \sum_{i \leq k \leq (q-i)} \binom{\min(k, q-k)}{j}$.

This follows from the Abel lemma for $n = [q/2]$ and the sequence $a_i = h_i$. The sequence b_i for the function $F_j(h)$ is defined by the following relations: $b_i = \binom{i}{j} + \binom{q-i}{j}$ for $i < q/2$, and $b_{q/2} = \binom{q/2}{j}$ for even q, and for the function $\Phi(h)$ by the relations: $b_i = 2\binom{i}{j}$ for $i < q/2$, and $b_{q/2} = \binom{q/2}{j}$ for even q.

LEMMA 9.7. *Let all components of the vectors $A = (A_1, \ldots, A_n)$ and $B = (B_1, \ldots, B_n)$ be strictly positive and $\frac{B_1}{A_1} > \cdots > \frac{B_n}{A_n}$. Then the numbers*

$$D_i = \frac{\sum_{i \leq j \leq n} B_j}{\sum_{i \leq j \leq n} A_j}$$

satisfy the inequalities $\frac{B_1}{A_1} > D_1 > \cdots > D_n = \frac{B_n}{A_n}$.

PROOF. Suppose that for some $i > 1$ we proved that $\frac{B_i}{A_i} \geq D_i$. Then, by Lemma 2.1,

$$D_{i-1} = \frac{B_{i-1} + \left(\sum\limits_{i \leq j \leq n} B_i\right)}{A_{i-1} + \left(\sum\limits_{i \leq j \leq n} A_i\right)}$$

is strictly bigger than D_i and strictly smaller than $\frac{B_{i-1}}{A_{i-1}}$. The lemma is thus proved. □

LEMMA 9.8. *1) For $1 \leq r \leq q/2$, the numbers*

$$D_i = \frac{\sum\limits_{i \leq k \leq q-i} \binom{k}{r}}{\sum\limits_{i \leq k \leq (q-i)} \binom{\min(k,q-k)}{r}}$$

strictly decrease as i runs from 0 to $[q/2]$.
2) We have

$$\max_{0 \leq i \leq [q/2]} D_i = D_0 = \frac{\binom{q+1}{r+1}}{\binom{[(q+1)/2]}{r+1} + \binom{[(q+2)/2]}{r+1}}. \tag{5}$$

PROOF. 1) The numbers $\frac{\binom{i}{r} + \binom{n-i}{r}}{2\binom{i}{r}}$ strictly decrease as i increases on the semi-segment $r \leq i < q/2$. Moreover, for even q, all these numbers are bigger than $\frac{\binom{q/2}{r}}{\binom{q/2}{r}} = 1$. It now remains to use Lemma 9.7.

2) The numbers D_i strictly decrease as i increases on the segment $0 \leq i \leq r$. Indeed, as i increases, the numerator

$$\sum_{i \leq k \leq q-i} \binom{k}{r} = \sum_{r \leq k \leq q-i} \binom{k}{r}$$

strictly decreases, whereas the denominator

$$\sum_{i \leq k \leq (q-i)} \binom{\min(k, q-k)}{r} = \sum_{r \leq k \leq (q-r)} \binom{\min(k, q-k)}{r}$$

remains unchanged. Formula (5) for the number D_0 follows from Lemmas 9.4 and 9.5. The lemma is thus proved. □

LEMMA 9.9. *For $0 \leq s < r \leq q/2$,*
1) the numbers

$$D_i = \frac{\sum\limits_{i \leq k \leq (q-i)} \binom{\min(k,q-k)}{s}}{\sum\limits_{i \leq k \leq (q-i)} \binom{\min(k,q-k)}{r}}$$

strictly decrease as i runs from s to $[q/2]$,
2) we have

$$\max_{0 \leq i \leq [q/2]} D_i = D_0 = \cdots = D_s = \frac{\binom{[(q+1)/2]}{s+1} + \binom{[(q+2)/2]}{s+1}}{\binom{[(q+1)/2]}{r+1} + \binom{[(q+2)/2]}{r+1}}. \tag{6}$$

PROOF. 1) For $r \leq j < q/2$, the number $\binom{j}{s}$ enters the numerator exactly twice: for $k = j$ and for $k = q - j$. Analogously, the number $\binom{j}{r}$ enters the denominator exactly twice. If q is even, then $\binom{q/2}{s}$ and $\binom{q/2}{r}$, respectively, appear once in the numerator and in the denominator, respectively.

The numbers $\frac{2\binom{j}{s}}{2\binom{j}{r}}$ strictly decrease as j increases on the semi-interval $r \leq j <$ $q/2$, moreover, for even q, all these numbers are bigger than $\frac{\binom{q/2}{s}}{\binom{q/2}{r}}$ (see Lemma 8.7 for the function $\psi(m)$). It now remains to use Lemma 9.7. The numbers D_i decrease strictly as i increases on the segment $s \leq i \leq r$. Indeed, as i increases, the numerator

$$\sum_{i \leq k \leq (q-i)} \binom{\min(k, q-k)}{s}$$

decreases strictly, whereas the denominator

$$\sum_{i \leq k \leq (q-i)} \binom{\min(k, q-k)}{r} = \sum_{r \leq k \leq (q-r)} \binom{\min(k, q-k)}{r}$$

remains unchanged.

2) It is readily seen that

$$D_0 = \cdots = D_s = \frac{\sum\limits_{s \leq k \leq (q-s)} \binom{\min(k,q-k)}{s}}{\sum\limits_{r \leq k \leq (q-r)} \binom{\min(k,q-k)}{r}}.$$

Formula (6) for numbers $D_0 = \cdots = D_s$ follows from Lemma 9.4. Part 1) shows that this formula gives the maximal value of the numbers D_i. The lemma is thus proved.   □

We are now ready to prove Theorem 9.3.

PROOF. **of Theorem 9.3** 1) According to Lemma 9.6, the function we maximize can be written in the form

$$\frac{F_r(h)}{\Phi_r(h)} = \frac{\sum\limits_{0 \leq i \leq [q/2]} (\Delta h)_i \sum\limits_{i \leq k \leq q-i} \binom{k}{r}}{\sum\limits_{0 \leq i \leq [q/2]} (\Delta h)_i \sum\limits_{i \leq k \leq (q-i)} \binom{\min(k,q-k)}{r}}.$$

On the simplex under consideration, all numbers $(\Delta h)_i$ are nonnegative. Using Lemma 9.8 and Lemma 9.7, we see that the maximum of the function is attained for $(\Delta h)_0 > 0, (\Delta h)_1 = \cdots = (\Delta h)_{[q/2]} = 0$. This means that the maximum is equal to D_0 and is attained for $h_0 > 0, h_0 = \cdots = h_q$.

2) According to Lemma 9.6, the function we maximize can be written in the form

$$\frac{\Phi_s(h)}{\Phi_r(h)} = \frac{\sum\limits_{0 \leq i \leq [q/2]} (\Delta h)_i \sum\limits_{i \leq k \leq (q-i)} \binom{\min(k,q-k)}{s}}{\sum\limits_{0 \leq i \leq [q/2]} (\Delta h)_i \sum\limits_{i \leq k \leq (q-i)} \binom{\min(k,q-k)}{r}}.$$

On the simplex under consideration, all numbers $(\Delta h)_i$ are nonnegative. Using Lemma 9.9 and Lemma 9.7, we obtain that the maximum of the function is attained for $(\Delta h)_0 + \cdots + (\Delta h)_s > 0, (\Delta h)_{s+1} = \cdots = (\Delta h)_{[q/2]} = 0$. This means that the maximum is equal to $D_0 = \cdots = D_s$ and is attained for $0 \le h_s, h_s = \cdots = h_{q-s}$. Theorem 9.3 and Theorem 9.1 are thus proved.   □

10. A generalization of Nikulin's Theorem

The problem of estimating the average number of l-dimensional faces on k-dimensional faces of a convex n-dimensional polytope, not necessarily simple, can be reduced to a series of problems on a possible mutual disposition of an $(n-1)$-dimensional convex polytope and a hyperplane. For such reduction, we need to consider a generic linear function on the polytope and, in the spirit of Morse theory, to study the level hypersurfaces of this linear function passing through vertices of the polytope.

Let Δ be a convex n-dimensional polytope, not necessarily simple. Denote by $f_{l,k}$ the number of all pairs consisting of an l-dimensional face of the polytope Δ and a k-dimensional face containing it. The average number of l-dimensional faces on k-dimensional faces of the polytope Δ is the ratio $f_{l,k}/f_k$, where f_k is the number of k-dimensional faces of the polytope. Let us show how to reduce the problem of estimating this number to an $(n-1)$-dimensional problem.

Fix a generic linear function L. With each vertex A of the polytope Δ, associate the pair consisting of a polytope $\Delta(A)$ and its hyperplane section L_A. This pair is defined up to a projective transformation. Here is the definition of this pair. Near each vertex A, the polytope Δ looks like a convex n-dimensional cone. This cone is sectioned by the hyperplane defined by the equation $L(x) = L(A)$. The pair $\Delta(A), L_A$ is defined as the projectivization of the pair consisting of the cone and the hyperplane described above.

In the theorem on reduction stated below, we will also assume that the polytope Δ is simple at the edges. The theorem is valid even without this assumption. However, it helps to shorten the proof, and in the sequel we will not need polytopes that are not simple at the edges.

Denote by $f_k(\Delta(A))$ the number of all k-dimensional faces of the polytope $\Delta(A)$, by $f_k(\Delta(A), L_A)$ the number of all k-dimensional faces of the polytope $\Delta(A)$ disjoint from the hyperplane L_A, and by $f_{l,k}(\Delta(A), L_A)$ the number of all pairs consisting of an l-dimensional face of the polytope $\Delta(A)$ disjoint from the hyperplane L_A and any k-dimensional face containing it. The set of all vertices of the polytope Δ will be denoted by V.

Theorem on reduction *Let Δ be a polytope simple at the edges, and L a generic linear function on it. Then we have the following inequalities:*

(1) *for $1 < k \le (n+1)/2$,*

$$\frac{f_{0,k}}{f_k} \le \max_{A \in V} \frac{2f_{k-1}(\Delta(A))}{f_{k-1}(\Delta(A), L_A)},$$

(2) *for $0 < l < k \le (n+1)/2$,*

$$\frac{f_{l,k}}{f_k} \le \max_{A \in V} \frac{f_{l-1,k-1}(\Delta(A), L_A)}{f_{k-1}(\Delta(A), L_A)}.$$

PROOF. 1) The numerator $f_{0,k}$, as well as the denominator f_k, of the ratio $f_{0,k}/f_k$ are representable as sums of nonnegative numbers over the vertices of the polytope Δ. To this end, to each pair $\Gamma_0 \in \Gamma_k$ consisting of a vertex Γ_0 and a k-dimensional face Γ_k containing it, assign the vertex Γ_0. To each k-dimensional face

Γ_k assign the two vertices, where the function L restricted to the face Γ_k attains its maximum and minimum.

Summing up the associated objects over the vertices, we obtain :

$$f_{0,k} = \sum_{A \in V} f_{k-1}(\Delta(A)),$$

$$f_k = \sum_{A \in V} \frac{1}{2} f_{k-1}(\Delta(A), L_A).$$

The number $f_{k-1}(\Delta(A), L_A)$ is strictly positive for each vertex A, since for $k \leq (n+1)/2$, a generic hyperplane does not intersect at least one $(k-1)$-dimensional face of the n-dimensional polytope $\Delta(A)$ (see Corollary 4.5). To conclude the proof of part 1), it suffices to use Lemma 2.1.

2) Analogously to part 1), the numerator $f_{l,k}$, as well as the denominator f_k, of the ratio $f_{l,k}/f_k$ are representable as sums of nonnegative numbers over the vertices of the polytope Δ. To this end, with each pair $\Gamma_l \subset \Gamma_k$ consisting of a face Γ_l and a k-dimensional face Γ_k containing it, associate two vertices, where the function L restricted to the face Γ_l attains the maximum and the minimum. With each k-dimensional face Γ_k, associate two vertices, where the function L restricted to the face Γ_k attains the maximum and the minimum. Summing up the associated objects over all vertices, we obtain:

$$f_{l,k} = \sum_{A \in V} \frac{1}{2} f_{l-1,k-1}(\Delta(A), L_A),$$

$$f_k = \sum_{A \in V} \frac{1}{2} f_{k-1}(\Delta(A), L_A).$$

To conclude the proof of part 2), it suffices to use Lemma 2.1. □

We are now ready to prove the central result of the article, which generalizes Nikulin's estimate to the case of polytopes simple at the edges.

Theorem *The average number of l-dimensional faces on k-dimensional faces of an n-dimensional polytope simple at the edges is strictly less than*

$$\binom{n-l}{n-k} \frac{\binom{[n/2]}{l} + \binom{[(n+1)/2]}{l}}{\binom{[n/2]}{k} + \binom{[(n+1)/2]}{k}}$$

for $0 \leq l < k \leq (n+1)/2$, $1 < k$.

PROOF. A slightly weaker result, where the strict inequalities are replaced with non-strict inequalities, is a direct corollary of the theorem on reduction and Theorem 9.4, which allows to estimate the numbers $\frac{f_{k-1}(\Delta(A))}{f_{k-1}(\Delta(A), L_A)}$ and $\frac{f_{l-1,k-1}(\Delta(A), L_A)}{f_{k-1}(\Delta(A), L_A)}$.

It remains to explain why the inequalities are strict. The point is that among the vertices of the polytope Δ there are two vertices, where the function L attains the maximal and the minimal values. The sections L_A of polytopes $\Delta(A)$ are certainly not successful for these vertices, since the hyperplane L_A does not intersect the polytope $\Delta(A)$. Hence the ratios from Theorem 9.4 are certainly not extremal for these vertices. Adding this remark to the proof of the theorem on reduction, we obtain that all inequalities are in fact strict. □

To conclude, let us revisit the classical Theorem 1.1, according to which the average number of edges on faces of a 3-polytope is strictly less than 6. The theorem on reduction reduces the problem of estimating this number to the following simple two-dimensional problem: estimate from above the ratio of the number of pairs consisting of a vertex of a polygon and an edge containing this vertex, to the number of edges of the polygon disjoint from a generic line.

If the polygon has m edges, then the number of pairs is $2m$, and the number of edges disjoint from a generic line is m, if the line does not intersect the polygon, and is $(m-2)$ in the opposite case. The desired ratio is either $2m/m = 2$ or $2m/(m-2)$. Since the number of edges of a polygon is at least three, the ratio does not exceed 6, which proves Theorem 1.1 (note that in accordance with Theorem 9.1, the maximal value of the desired ratio is attained in the case of a triangle intersected by the line). The argument just given does not even use the Euler formula and is most likely the simplest proof of Theorem 1.1.

References

[1] Atiyah, M.F. *Convexity and commuting Hamiltonians.* Bull. London Math. Soc. **14** (1982), 1–15

[2] Billera, L. J.; Lee, C. W. *A proof of the sufficiency of McMullen's conditions for f-vectors of simplicial convex polytopes.* J. Comb. Theory, Ser. A **31** (1981), 237–255

[3] Bondesen A., Brondsted A. *A dual proof of the upper bound conjecture for convex polytopes.* Math. Scand. **46** (1980), 95–102

[4] Brondsted A. *An introduction to convex polytopes.* Springer-Verlag 1983

[5] Coxeter, H. S. M., *Discrete groups generated by reflections.* Ann. Math. **35** (1934), No. 3, 588–621

[6] Khovanskii, A.G. *Hyperplane sections of polyhedra, toric varieties and discrete groups in Lobachevsky space.* Funct. Anal. and its Appl. **20** (1986), No. 1, 41–50

[7] Macaulay, F.S. *Some properties of enumeration in the theory of modular systems.* Proc. London Math. Soc. **26** (1927), 531–555

[8] McMullen P. *The maximum numbers of faces of a convex polytope.* Mathematika **17** (1970), 179–184

[9] McMullen, P. *The numbers of faces of simplicial polytopes.* Israel J. Math. **9** (1971), 559–570

[10] McMullen, P. *On simple polytopes.* Invent. math. **113** (1993), 419–444

[11] McMullen P., Shephard G.C. *Convex polytopes and the upper bound conjecture.* London Mathematical Society Lecture Note Series, Cambridge University Press, London–New York **3** (1971)

[12] Nikulin, V.V. *On the arithmetic groups generated by reflections in Lobachevsky spaces. (Russian)* Izv. AN SSSR, ser. mat. **44** (1980), 637–699

[13] Nikulin, V.V. *On the classification of arithmetic groups generated by reflections in Lobachevsky spaces. (Russian)* Izv. AN SSSR, ser. mat. **45** (1981), 113–142

[14] Prokhorov, M.N. *Absence of discrete groups of reflections with a noncompact fundamental polyhedron of finite volume in a Lobachevsky space of high dimension. (Russian)* Izv. AN SSSR, ser. mat. **50** (1986), No. 2, 320–332

[15] Sommerville, D.M.Y. *The relations connecting the angle-sums and volume of polytope in space of n dimensions.* Proc. Roy. Soc. London, ser. A115 (1927), 103–119

[16] Stanley, R.P. *The number of faces of a simplicial convex polytope.* Adv. Math. **35** (1980), 236–238

[17] Timorin, V.A. *An analogue of the Hodge–Riemann relations for simple polytopes.* Russian Mathematical Surveys **54** (1999), No. 2, 381–426

[18] Timorin V.A. *On polytopes that are simple at the edges.* Funct. Anal. and its Appl. **35** (2001), No. 3, 189–198

[19] Vinberg, E.B., *Discrete groups generated by reflections in Lobachevsky spaces. (Russian)* Matem. Sb. **72** (1967), No. 3, 471–488

[20] Vinberg, E.B. *The nonexistence of crystallographic reflection groups in Lobachevsky spaces of large dimension. (Russian)* Funktsional. Anal. i Prilozhen. **15** (1981), No. 2, 67–68

[21] Vinberg, E.B. *Absence of crystallographic reflection groups in Lobachevsky spaces of large dimension. (Russian)* Trudy Moskov. Mat. Obshch. **47** (1984), 68–102

DEPARTMENT OF MATHEMATICS, UNIVERSITY OF TORONTO, TORONTO ON M5S 3G3 CANADA
E-mail address: askold@math.utoronto.ca

Donald and the Golden Rhombohedra

Marjorie Senechal

Introduction

I first met Donald Coxeter at a conference in 1981. I was in awe: Coxeter the geometer was legendary, 'Coxeter' the adjective ubiquitous. I feared he might grill me on 11-dimensional snub polytopes. But no, the only question he asked me was very down-to-earth: did I know that apples don't have cores? Of course they did, but why argue? It was October, the height of the apple season; an apple was easily procured. Professor Coxeter deftly sliced a juicy red specimen into thin horizontal sections. Lo and behold, there was no stem-to-stern core, just an elongated seed pod in the center. But as he kept slicing, I realized the core was a ruse: the real surprise lay hidden in the pod. Apple seeds, *in situ*, are arranged in a five-pointed star! The apple reveals its secret symmetry in its seeds and in its blossoms, with no hint in its outer form.

FIGURE 1. Apple seeds in an apple.

Donald Coxeter delighted in the subtle geometries of apparently simple things. I hope he would be amused, even pleased, that "Donald and the Golden Rhombohedra" sounds like the title of a children's story. He found inspiration in toys and models all his life, worlds in grains of sand. He had a special affection for the

2000 *Mathematics Subject Classification*. Primary 52C22; Secondary 52C23.

FIGURE 2. A rectangle is golden if the ratio of its edge lengths is
the "golden number" $\tau = (1 + \sqrt{5})/2$. A *golden rhombus* is the
convex hull of the midpoints of the edges of a golden rectangle; its
diagonals are in golden ratio.

stretched and squashed cubes he called the golden rhombohedra: they make an
appearance in *Regular Polytopes* [3], and he returned to them again and again. For
the golden rhombohedra are apple seeds of the polyhedron kingdom, perhaps the
mineral kingdom too.

The Golden Isozonohedra. "Thank you for your letter of March 24 and
your beautifully illustrated essay 'On the golden polytopes'," Coxeter wrote to
Koji Miyazaki in April 1977.[1] "Perhaps the name should be more specific. 'On the
golden isozonohedra' (see my 'Twelve Geometric Essays, p. 66.)'."

Coxeter's more specific name specifies the little polyhedral family completely.
A zonohedron is a centrosymmetric convex polyhedron with centrosymmetric faces;
a zonohedron is "iso" if its faces are congruent, and an isozonohedron is "golden"
if its faces are golden rhombs (Figure 2).[2]

There are five golden isozonohedra (Figure 3): the rhombic triacontahedron,
K_{30}, with icosahedral symmetry; the rhombic icosahedron, F_{20}, with the symmetry
of a pentagonal antiprism; a rhombic dodecahedron, B_{12}, with the symmetry of a
rectangular parallelepiped; and two rhombohedra, one acute A_6, and one obtuse,
O_6; both have the symmetry of a trigonal antiprism. In Coxeter's elegant symbols,
K, F, and B stand for their discoverers (Kepler, Federov, Bilinski); the subscript
indicates the number of faces.

The family members are interrelated in various ways. The triacontahedron is
the acknowledged *pater familias*; the others are, literally, descended from it by the
process of zone contraction ([8]). Zone contraction classifies all zonohedra, and
zonotopes in any dimension, into families and constructs their family trees, but
let us consider, as Coxeter did in *Regular Polytopes*, the special case of rhombic
zonohedra in dimension three. The edges of a zonohedron belong to distinct sets

---

[1]H. S. M. Coxeter to Koji Miyazaki, 22 April, 1977. Coxeter sent me a copy of this letter
many years ago.

[2]This terminology is unique to Coxeter, and does not imply that the zonohedron is isohedral.
An n-polytope is *isohedral* if its $(n-1)$-faces form a single orbit under the action of its symmetry
group. Of the five golden isozonohedra, K_{30}, A_6, and O_6 are isohedral.

FIGURE 3. K_{30}, F_{20}, B_{12}, A_6 and O_6.

of parallel edges; the band of faces bounded by such a set is called a *zone*. "By removing one of the p zones of a rhombic zonohedron and bringing together the two remaining pieces of the surface, one obtains a simpler zonohedron, with $p-1$ replacing p" ([**4**]). The process can be repeated k times, until $p-k=3$; after that, the zonohedron collapses to a zonogon. The triacontahedron has six zones; removing any one of them yields the rhombic icosahedron, which has five; removing any one of these five yields a rhombic dodecahedron. The rhombic dodecahedron's four zones belong to two classes (two orbits of its symmetry group.) Removing a zone of one class, we get the acute rhombohedron; of the other, the obtuse.

Beginning with A_6 and O_6, we can build the other golden isozonohedral by nesting them in one another. Two A_6s and two O_6s build one B_{12}; one B_{12}, three A_6s and three O_6s build one F_{20}, and one F_{20}, five A_6s and five O_6s build a K_{30}. In Coxeter shorthand,

$$B_{12} = 2A_6 + 2O_6$$
$$F_{20} = B_{12} + 3A_6 + 3O_6,$$
$$K_{30} = F_{20} + 5A_6 + 5O_6.$$

Nesting is not the inverse of zone removal, as it does not build zone upon zone.

"I am especially pleased with page 15," Coxeter continued his letter to Miyazaki, "where you decompose a B_{12} of edge-length 2 into 4 A_6 and 4 O_6 and 6 B_{12} (of edge-length 1), and decompose an F_{20} of edge-length 2 into 10 A_6 and 10O_6 and 10B_{12} and 2 F_{20} (or, equivalently, into 30 A_6 and 30 O_6 and 2 F_{20}). In fact, this decomposition can be extended inductively to an F_{20} of edge 2^n ($n = 1, 2, ...$), showing that the whole Euclidean 3-space can be filled with rhombohedra and B_{12}s and F_{20}s (or simply with rhombohedra and F_{20}s) in a manner that has the symmetry of F_{20} itself (or of a pentagonal antiprism, so that the symmetry group is $D_5 \times \{I\}$ in the notation of my 'Introduction to Geometry', p. 277). As this space-filling has only one axis of pentagonal rotation into itself, and no translation, it is a non-periodic honeycomb."

A honeycomb (tiling) with axes of 5-fold rotation must be nonperiodic: periodic tilings of R^2 and R^3 may have 2-fold, 3-fold, 4-fold, or 6-fold axes, but not

FIGURE 4. Penrose kites and darts. The tiles must be placed edge
to edge, with marked vertices matched to marked vertices, and
unmarked to unmarked.

5-fold or k-fold where $k > 6$ ([**18**]). Because molecules in crystals were long assumed to arrange themselves like point lattices, this classical result is known as the
crystallographic restriction.

"What I would be interested to know," Coxeter concluded his letter, "is whether
rhombohedra and F_{20}s fill space in a manner that is essentially non-periodic (in the
sense that no honeycomb composed of these particular bricks can have a translation
into itself). If the answer is yes, we would have here a very nice 3-dimensional
analogue for the non-periodic tilings of Roger Penrose . . ."

The golden rhombohedra can, of course, tile R^3 periodically; any rhombohedron can. (Venkov and McMullen proved, independently, that a zonotope in any
dimension n tiles R^n if, and only if, all of its zones have length 4 or 6.[**23**],[**17**].)
If a shape tiles both periodically and nonperiodically, then the nonperiodicity is a
property of the tiling, not the tiles. Implicit in Donald's question to Miyazaki (and
explicit in later correspondence, see below), was the difficult question of whether the
rhombohedra could be decorated in some way to permit only nonperiodic arrangements. Golden rhombohedra, so decorated, would then be *aperiodic tiles*. Roger
Penrose had recently found just such markings for a curious pair of quadrilaterals
in the plane.

Penrose rhombs. Martin Gardner's article on the Penrose kites and darts
had appeared in *Scientific American* just three months before ([**9**]). The stunning
cover tiling of almost but never quite repeating kites and darts, and Gardner's
account of the tilings' remarkable properties, made "Penrose" almost a household
word.[3]

By "Penrose tiles" we mean either of two pairs of decorated tiles, the fanciful
"kites and darts" discussed by Gardner, and the more prosaic "thick and thin
rhombs" with acute angles 72° and 36°, respectively. These rhombs are not golden,
but the ratio of their areas is. Every tiling by kites and darts is, at the same time,
a tiling by thick and thin rhombs and vice versa, since the tiles of either set can be
bisected into triangles and spliced *in situ* into tiles of the other. In Figure 4, the
kites and darts are decorated with colored vertices, which must be matched in the

---

[3]Hao Wang had shown, in 1960, that the existence of a set of polygons that tile the plane
"essentially nonperiodically" implied the nonexistence of an algorithm for determining whether
any given set could tile the plane in any manner. In 1964, Berger found the first such set, settling
Wang?s conjecture; his comprised 20,426 tiles. Gradually the number was reduced: 104, 92, 40,
6, 4, and finally 2 [**10**]. (The existence of an "einstein" is still an open question.)

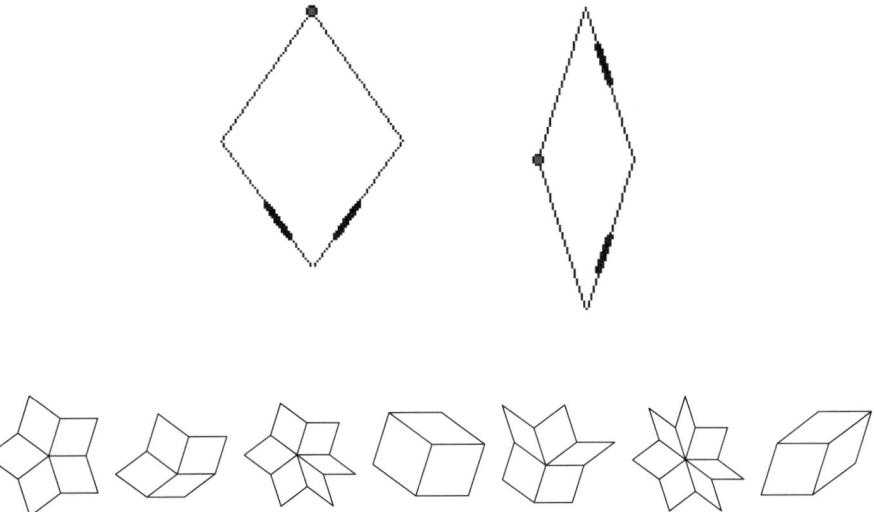

FIGURE 5. Top: matching rules for the Penrose rhombs; Bottom: exactly seven vertex configurations are allowed.

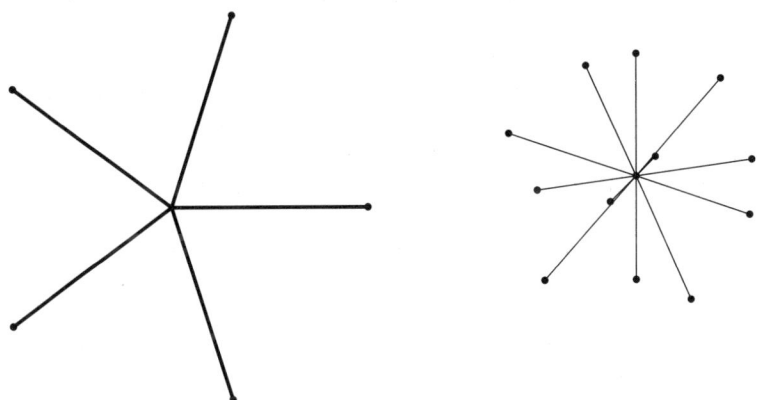

FIGURE 6. The edges of the Penrose rhombs are parallel to the vectors of a regular pentagonal star; the edges of the golden rhombohedra are parallel to the vectors of a regular icosahedral star.

tiling. In Figure 5, the rhombs are decorated with marked vertices and edges: both types of markings must be matched.

The kites and darts have attracted more popular attention, but the rhombs are more theoretically tractable and can be readily generalized: their three-dimensional analogues are the golden rhombohedra (Figure 6).

If you play with Penrose tiles for even a little while, you will quickly discover that a patch of tiles can sometimes be extended in several different ways (or none: you may hit a dead end and have to retrace your steps). You can, in principle,

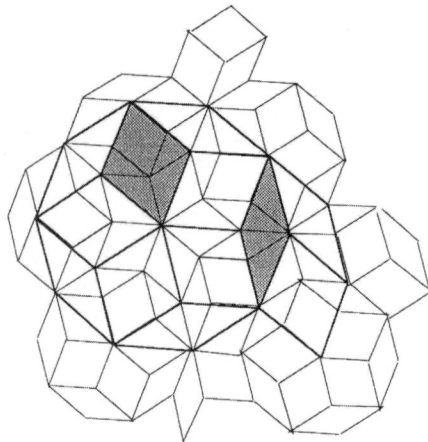

FIGURE 7. Bisected rhombs can be composed into larger rhombs *ad infinitum.*

construct uncountably many different Penrose tilings of the plane, yet all of them will have exactly the same local configurations of every size, and each of these configurations will be relatively dense in every tiling. That is, Penrose tilings are "repetitive" and all Penrose tilings belong to a single "local isomorphism class" ([19]).

The main purpose of the decorations, or "matching rules," is to forbid translational symmetry in any direction. We cannot determine, by inspection, whether the decorations accomplish that, but there are other ways to do it. If we can show that the rules force each tile in a tiling to belong to a unique, larger, tile (of either type), and that to a unique, still larger, tile, *ad infinitum*, then nonperiodicity follows from a theorem of Grünbaum and Shephard, which asserts that every tiling with unique composition is nonperiodic ([10]). Notice (Figure 7) that the rhombs are first bisected and the pieces composed into larger rhombs; it is actually the triangles that are composed.

To summarize, Coxeter asked whether matching rules could be found to enforce Miyazaki's space-filling by larger and larger rhombic icosahedra.

Quasicrystals and the Ammann rhombohedra. Coxeter did not yet know that an amateur mathematician, Robert Ammann, had proposed pairs of decorated A_6 and O_6 as aperiodic tiles in 1976 ([20]). Ammann's decorations were very simple: each rhombic face was deformed by a bump or a dent; two rhombohedra could be juxtaposed only if the bump on a face of one fitted into a dent in a face of the latter (Figure 8). The bumped and dented A_6s were mirror images, and the O_6s also.

Ammann's rules permit seven configurations of rhombohedra around an edge e (Figure 9). Viewed along e, the faces of these rhombohedra project to Penrose rhombs, their common edge e projects to a common vertex, and we obtain the vertex arrangements of (Figure 5). Continuing the Ammann tiling by fitting rhombohedra around edges parallel to e leads to an infinite undulating slab of tiles whose surfaces (upper and lower) project to a Penrose tiling. Every Penrose rhomb tiling is a blueprint for an Ammann slab, but the slab does not share the tiling's composition property.

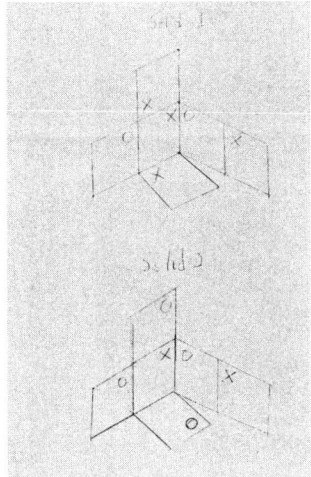
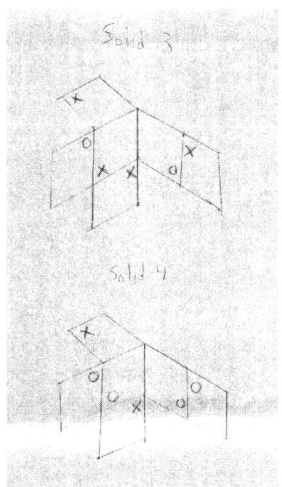

FIGURE 8. a) Ammann's nets for his four golden rhombohedral tiles (left, O_6s; right, A_6s.) Each face of each type of tile is endowed with a bump (X) and a dent (O); the marked tiles of each type are mirror-images. b) Rhombo blocks with Ammann's markings.

Ammann's 3-dimensional tiles attracted little attention at first, which may be why Coxeter hadn't heard of them when he wrote to Miyazaki. At the time, geometers were more intrigued by the several new sets of aperiodic tiles that Ammann conjured with inscrutable ease ([10],[20]). Attention shifted to nonperiodic tilings in three dimensions with the discovery of crystals with icosahedral symmetry in 1984. The icosahedron's 6 axes of 5-fold symmetry are incompatible with a point lattice: the crystallographic restriction wasn't restrictive after all. But if the molecules in a crystal aren't arranged in a lattice, how are they arranged? Since the discovery of x-ray diffraction by crystals in 1912, textbooks had taught the crystallographic restriction as hallowed fact (Figure 10).

FIGURE 9. Five of Ammann's edge configurations; compare with
Figure 5c. Ammann's rules permit the remaining two configurations, but the laws of gravity do not. Ammann's rules also admit
the golden isozonohedra B_{12}, F_{20}, and K_{30} of Figure 5.

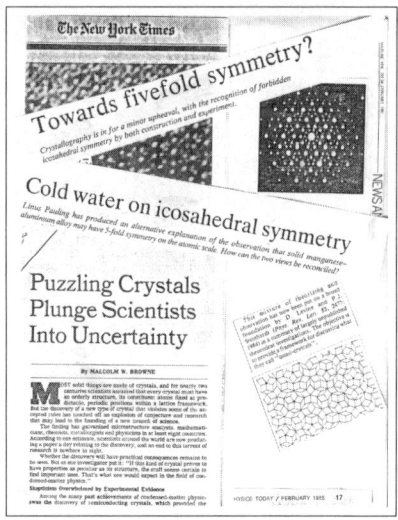

FIGURE 10. The discovery of quasicrystals in 1984 made headlines.

Diffraction photographs of random arrangements don't show sharp bright spots.
But, other than lattices, what arrangements do? A new field, the mathematics of
"aperiodic order," was born (for overviews and further references, see [**12**],[**14**]).
But the question predated these so-called quasicrystals, at least implicitly: in 1981,
the crystallographer Alan Mackay, suspecting that the repetitive order of Penrose
tilings might manifest itself through optical diffraction, made a "mask" of Penrose
tiling vertices. Sure enough, an optical transform showed sharp spots in concentric
decagons ([**16**]). Like Ammann's rhombohedra, Mackay's transform attracted little
attention at the time, but three years later its remarkable similarity to quasicrystal diffraction patterns was noted with excitement. Physicists, crystallographers,
chemists, materials scientists, and mathematicians rushed to explore the curious

Penrose tilings. Like motifs in an Escher drawing, aperiodic 3-D tiles jumped from background to foreground.

In 1986, preparing the thirteenth edition of Rouse Ball's *Mathematical Recreations and Essays*, Coxeter decided to append remarks on nonperiodic space fillings by the golden rhombohedra, A_6 and O_6 ([**2**]). By that time, he'd heard of Ammann and his tiles (though not his matching rules). As always, Coxeter reached out to anyone and everyone he thought might help him understand.

"Dear Mr. Ammann," Coxeter wrote in February, 1986, "I am interested in your 1976 discovery of a 3-dimensional analogue of Penrose's non-periodic tilings, namely a packing of acute and obtuse rhombohedra whose faces are 'golden' rhombs . . . I am anxious to know whether a triacontahedron K_{30} could possibly be surrounded by 70 A_6s and 70 O_6s in such a way as to preserve icosahedral (rotatory) symmetry and make a big K_{30} (of edge 2). Would it be possible to 'decorate' the rhombohedra with pits and pimples so as to be the infinite assembly essentially non-periodic?"[4]

The first question has recently been answered in the affirmative, as I will explain. And yes, it is "possible to 'decorate' the rhombohedra with pits and pimples" in such a way that the infinite assembly is "essentially nonperiodic." But such assemblies are not recursively generated space-fillings by K_{30}s of edges n. Moreover, the fast and furious flurry of research in the wake of the discovery of quasicrystals produced several different – not locally isomorphic – essentially non-periodic tilings of R^3 by A_6s and O_6. The very notion of aperiodic tiles is far more subtle than first was thought.

Independently of Penrose, Ammann had discovered the 2-dimensional rhomb tiles and their matching rules, also in 1976 ([**20**]). Consider five infinite families of equispaced parallel lines, orthogonal to the arms of a regular pentagonal vector star, the families translated slightly to avoid singular points where more than two lines meet. (N.G. de Bruijn called this a *pentagrid*, and so will we.) Every acute angle of a pentagrid line intersection is either 72° or 36°, and the limiting ratio of the two types of intersections is golden.

Ammann claimed that the pattern of pentagrid line intersections is a set of instructions for juxtaposing tiles, edge to edge, in infinite bands: place thick rhombs at all the 72° intersections and thin rhombs at all the 36° intersections (Figure 11).[5] The famous "Ammann bars" are a variant of the pentagrid in which the spacings between lines in each family are not equal, but "long" and "short" and obey the rules of Fibonacci sequences. (Inevitably, the $L : S$ is the ubiquitous τ.)

De Bruijn rediscovered Ammann's approach – independently and rigorously – in 1981 and showed that Penrose's matching rules could be reconstructed for those pentagrids in which translations (of the line families) satisfied a certain algebraic condition ([**5**]). In the same paper, he also showed that a regular pentagrid is a section of a periodic tiling of E^5 by unit 5-cubes, cut by a 2-plane orthogonal to $(1,1,1,1,1)$. This section meets certain k-faces of the cube tiling and thus selects a set of elements of its dual tiling. We obtain rhomb tilings by projecting these dual sets along $(1,1,1,1,1)$; the tilings are Penrose for suitably chosen sections. This

---

[4]H. S. M. Coxeter to Robert Ammann, February 19 and March 24, 1986, courtesy of Esther Ammann.

[5]Ammann, a brilliant eccentric, didn't prove things he considered obvious. It is less obvious to the rest of us that the tiles fit together without gaps or overlaps, as the tiling and the pentagrid are not dually situated, but de Bruijn later showed that Ammann's assertion is true.

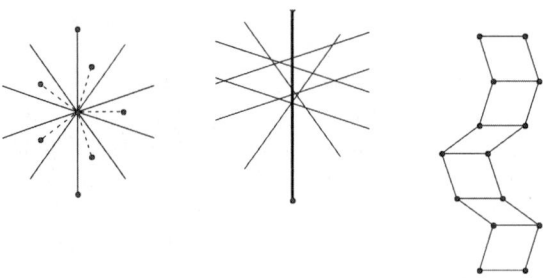

FIGURE 11. Left: A pentagrid star (dotted lines) with perpendicular line segments representing the five infinite families; Center: a portion of a line of a pentagrid (vertical), with intersections; Right: the corresponding band of Penrose tiles.

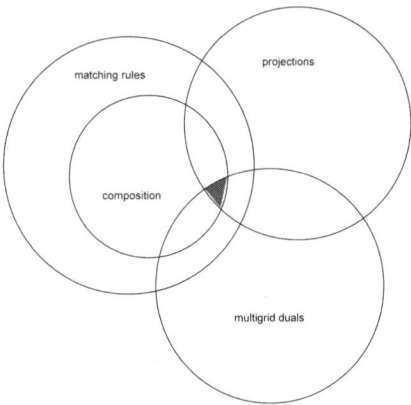

FIGURE 12. Four methods for generating nonperiodic tilings are known; Penrose tilings of the plane can be produced by all of them.

powerful technique, known as the "cut-and-project formalism," can be generalized to arbitrary dimensions and in myriad other ways.

Matching rules, composition, dualization, projection: the Penrose tilings can be generated in at least four different ways (Figure 12) ([22]). But these same methods produce, evidently, different nonperiodic tilings of R^3 by golden rhombohedra. For example,

- Many years after the fact, Ammann gave a procedure for generating his tiles ([1]). "Imagine 3-space filled by 6 sets of parallel planes, each set equally spaced and parallel to 2 of the sides of an icosahedron. Consider each of the points at which 3 such planes intersect. At each such point, construct a rhomboid, with the edge of the rhomboid perpendicular to the intersecting surfaces. . . . The points of intersection, of course, are not evenly spaced, but the resulting rhomboids can be moved around to fill

3-space." He then described his matching rules, a clever analogue of the rules for the rhombs. Ammann's rhomobohedral tilings do not, evidently, obey any law of composition. They are not, evidently, obtained through the cut-and-project formalism.

- Using the cut-and-project method, A. Katz generated tilings by A_6s and O_6s and devised matching rules for them ([**11**]). Katz's rules are "perfect": they force the tiles into a single local isomorphism class. But the two rhombohedra must be marked in twenty two different ways.

- Joshua Socolar and Paul Steinhardt, generalizing Ammann bars to three dimensional "quasi-periodic hexagrids," found they could not avoid singular points by shifting the families of parallel planes ([**22**]). The tiles dual to singular points are K_{30}, F_{20}, B_{12}, according as six, five, or four planes meet; all tiles dual to intersections of three planes turn out to be A_6s. While a tiling by K_{30}, F_{20}, B_{12} and A_6 is, by the nesting property, *ipso facto* a tiling by golden rhombohedra, the arrangement of A_6s and O_6s in a K_{30} is not uniquely determined by the hexagrid, so A_6s and O_6s cannot be marked to force such a tiling. Socolar and Steinhardt chose decorated K_{30}, F_{20}, B_{12}, and A_6 for their aperiodic tiles instead, and showed that the tilings obey a composition law.

- Ammann's tilings, remarkable in their simplicity, demand and deserve further examination. Socolar has shown that Ammann's matching rules force nonperiodicity ([**21**]), but his methods do not indicate whether these rules are "weak," "strong," or "perfect", that is, whether the tilings are repetitive as well as nonperiodic, and whether they belong to a single local isomorphism class, or several.

The picture gets murkier yet. Matching rules give rise to local arrangements of tiles, but the relationship is not necessarily reciprocal. The Penrose rules imply the atlas of vertex configurations shown in (Figure 5) but not vice versa: a periodic tiling by unmarked thick and thin rhombs exists whose vertex configurations are drawn from this set. The Penrose rules are completely specified by its atlas of configurations of slightly larger radius, but no radius is large enough to specify the matching rules for an apparently similar aperiodic tiling of the plane ([**7**]).[6] Local configurations in point sets are a key to their long-range order ([**13**]); tilings have been less studied from this point of view (see, however, [**6**]).

Dear Donald, Dear Tony. Let's return now to Coxeter, revising Rouse Ball. He shared his enthusiasm for the golden isozonohedra and polyhedral models with A.G. (Tony) Bomford, an Australian engineer.[7]

"Dear Tony," Coxeter wrote on March 17, 1986. " I am delighted to hear of your new hobby of cutting out accurate wooden polyhedra ...Turning to page 31 of *Regular Polytopes*, you may be amused to see that 'two acute and two obtuse' make Bilinksi's rhombic dodecahedron while 'five and five' make Fedorov's rhombic icosahedron. In the last case, the assembly can have pentagonal symmetry. You may like to demonstrate these results by cutting out ten acute and ten obtuse rhombohedra and sticking them together in various ways. I have a hunch that a triacontahedron could possibly be surrounded by 70 acute and 70 obtuse in such a

---

[6]This tiling, by marked squares and rhombs, was first studied by Ammann.

[7]Coxeter and Bomford corresponded for twenty years; see Doris Schattschneider's essay in this volume. I am grateful to the Bomford family for permission to quote from it here.

way as to preserve icosahedral (rotatory) symmetry and make a big triacontahedron (of edge 2). This would be interesting as establishing, by recursion, a honeycomb filling the whole space: a 'Kowalewski-Miyazaki quasilattice.'"

"Dear Professor Coxeter," Bomford replied on April 2. "I admire the restraint with which you mention your hunch that a triaconta could possibly be surrounded with 70 golden rhombos of each type to form a triaconta with double the side length. I had of course to quietly check that $20 + 70 + 70 = 2^3 \times 20$, but that done, I felt in no doubt at all that 'of course' everything would fit! You will perceive that I am altogether too much addicted to plausible reasoning; but Polya would have liked your hunch as an example for his book. Once I have the geometry of the golden rhombos correctly sorted out, it would be no trouble to cut ten obtuse and ten acute, and I shall certainly do it."

April 17: "Dear Tony, Polya is right: try a simpler related problem! So instead of surrounding the K_{30} with 70 acute rhombohedra A_6 and 70 obtuse rh. O_6, make use of the 60 rotations of the whole assembly into itself and look at the possibility of sticking on just one of each plus a small fraction ($70/60 = 1\ 1/6$). You may be able to see at once that it won't work. What gave me a glimmer of hope was the 'Scientific American,' Jan. '77, where the cover design has nearly, but not quite, decagonal symmetry."

November 24: "Dear Professor Coxeter, I tried to make three A_6s and O_6s. There was no problem with the A_6s. The first problem with the O_6s is that there is such a small area of opposing surface, so that it is difficult to hold them. The second problem is that the 36 deg knife edges are a severe test of the wood... I finished five of each, and send them, and you will be able to use them to see, loosely, how the bigger blocks are built up from the smaller; but the O_6s are far from perfect, and even the A_6s are a trifle big. ...I felt a little sad about the non-existence of the Great Triacontahedron – the C_{30}. I did not give up all hope at first. It occurred to me that if, after all, it were to exist, then a little Triacontahedron, with half-length sides, ought to reside centrally within an ordinary K_{30}...but I do not really think I am going to find a Little Triacontahedron in there, any more than you do."

December 27: "Dear Tony ...I scarcely know how to begin to thank you for your wonderful Christmas present of exquisitely crafted golden isozonohedra, including enough rhombohedra to demonstrate the dissections

$$B_{12} = 2A_6 + 2O_6,$$

$$F_{20} = B_{12} + 3A_6 + 3O_6,$$

$$K_{30} = F_{20} + 5A_6 + 5O_6.$$

As they are so smooth and shiny, one hesitates to risk spoiling them by using scotch tape (sticky on both sides) to make the pieces stick together during the assembly process."

"Dear Donald," wrote Bomford on March 31, 1987, " I am curious about Miyazaki's discoveries described in the proof sheet you sent on December 27. I wonder how he 'observed' that two F_{20} placed pentagonally apex to apex could be surrounded with 30 A_6 and 30 O_6 to make a Great F_{20} with side 2... As Polya might have remarked, it is one thing to make a conjecture; another to believe something and assert it; and another to demonstrate or prove it. Don't think that I doubt that he is right: it just is that I don't know how I would set about demonstrating such an assertion in print, and I would be curious to learn how he did it."

FIGURE 13. Wooden models of zonohedra, made for Coxeter by A.G. Bomford. Coxeter Collection, York University. Photo by Leda Weiss.

April 7: "Dear Tony," Donald confessed, "Miyzakai sent me photographs of his models... I accepted his statement about $2F_{20} + 30A_6 + 30O_6 = F_{20}2$ without verifying it; perhaps that was rash! I don't have the time or patience to try it myself, so please don't embarrass me by sending the necessary pieces! It took me over an hour to assemble the $5A_6 + 5O_6$ to make F_{20}; the difficulty was my not knowing whether the next piece (at each stage) should be A_6 or O_6."

Rash perhaps, but Miyazaki's claim was correct. And today, with the help of Michael Longuet-Higgins's Rhombo blocks – the plastic A_6s and O_6s, with magnetic strips beneath the long diagonals of their faces, that I've used in several figures in this paper – it is easy to build an $F_{20}2$ from $2F_{20} + 30A_6 + 30O_6$. This construction juxtaposes some tiles by translation Figure 14, so is not an Ammann tiling.

Happily, Coxeter lived to see Longuet-Higgins's construction of "a triaconta" "surrounded with 70 golden rhombos of each type to form a triaconta with double the side length ([**15**])."[8] Longuet-Higgins' increasingly larger K_{30}s is a recursively generated space-filling by A_6s and O_6s but, like Miyazaki's construction, is not an Ammann tiling. "I find your new paper fascinating: an ideal rebuttal to the anti-Platonists who claim that mathematics is a human invention like poetry," Coxeter told him.[9]

The Coxeter Project. Coxeter persisted in his questions and his quest. "Do your hierarchic rules

$$55A_6 + 34O_6 = A_6\phi^3$$
$$34A_6 + 21O_6 = O_6\phi^3$$

imply that somewhere your 'flower dodecahedron' $20A_6$ will automatically appear (in fact, infinitely often) so that there will be local icosahedral symmetry round infinitely many centres (but no such global symmetry)?" he asked Tohru Ogawa.[10]

---

[8]The correspondence quoted above shows that Coxeter and Bomford expected the smaller and larger triacontahedra would be concentric; they are not.

[9]By e-mail! May 6, 2001, Coxeterhsm@aol.com to mlonguet@uscd.edu.

[10]H. S. M. Coxeter to Tohru Ogawa, May 1, 1986.

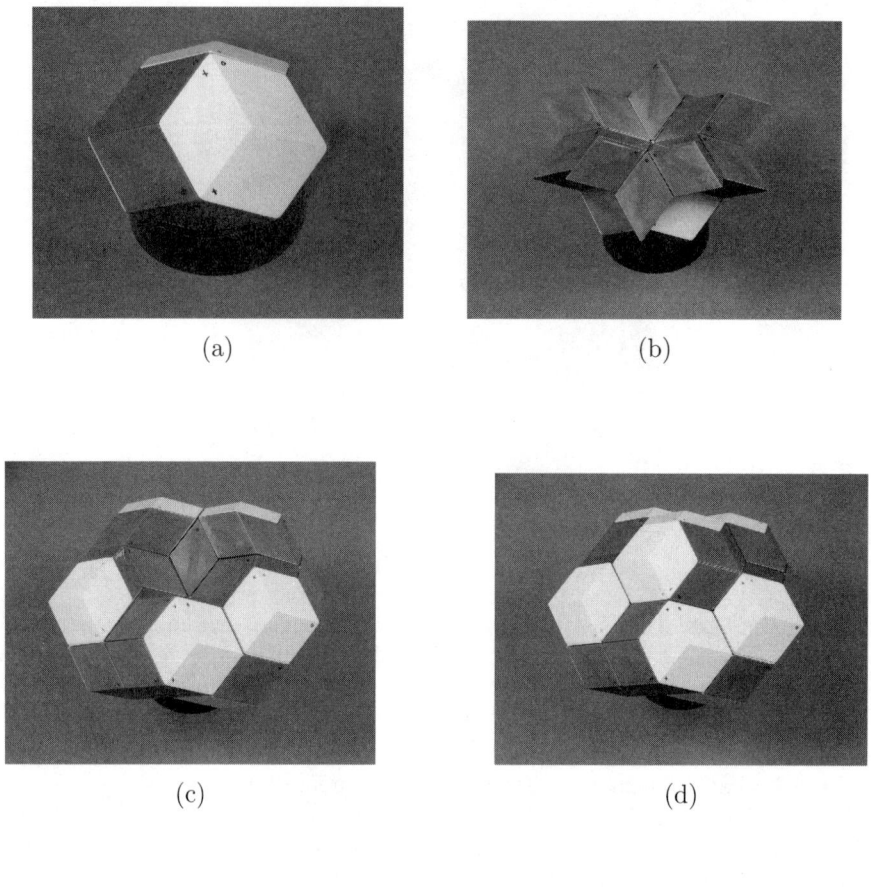

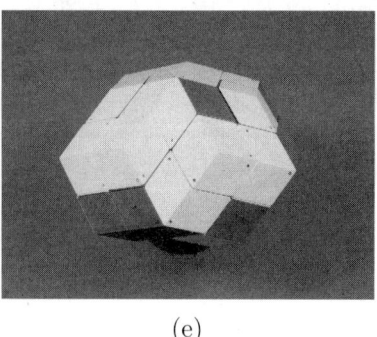

FIGURE 14. Miyazaki's recursive construction of F_{20}. a) An F_{20}; b) (a), with a "collar" of 10 A_6s; c) (b), with 10 B_{12}s fitted like gemstones around the collar; d) (c), a second F_{20} inserted into the collar's neck meets the first at its apex; e) (d) together with 10 additional O_6s completes the $F_{20}2$.

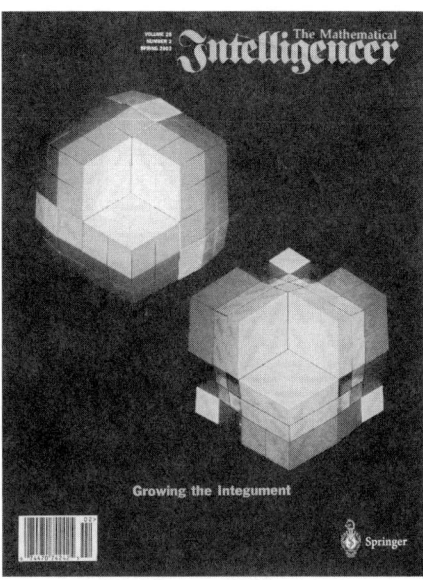

FIGURE 15. Cover page of *The Mathematical Intelligencer*, showing Longuet-Higgins's construction of $K_{30}2$. With kind permission of Springer Science and Business Media.

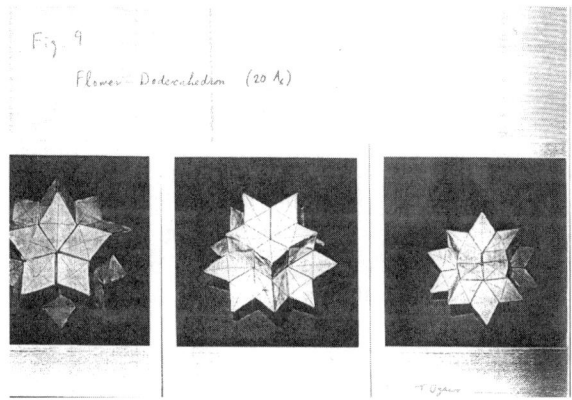

FIGURE 16. Ogawa's "flower dodecahedron." Courtesy of H. S. M. Coxeter.

"If so, is it true (as Conway observed in the 2-dimensional case) that, for any such centre, there is always another at a distance as small as n times the diameter of the flower, for some particular n?"[11]

---

[11]On his website (http://members.tripod.com/vismath2/ogawa/3d.htm), Dr. Ogawa supplies an answer: "there are infinite *local five-fold centres* in the two-dimensional Penrose tiling...[in my tiling] on the other hand, the range of icosahedral symmetry in the most strict sense is restricted to the only one case; to be referred to as a flower dodecahedron Fl_{60}. Of course, [the] three dimensional configuration can be icosahedrally symmetrical in looser senses, for example in a macroscopic sense that all of the six vectors evenly appear."

Neither wooden models, nor Rhombo blocks, nor Zometools can help us explore such questions, but powerful software might. In 1996, ten mathematicians and computer scientists, led by Egon Schulte, drafted a proposal for the Geometry Center in Minneapolis.[12]

"Three-dimensional discrete geometry lies at the heart of many fundamental problems in mathematics and other sciences," we wrote. "For example, the rapidly growing field of polytope theory is important in many branches of discrete mathematics. For polytopes, there are intriguing problems that connect geometric with combinatorial or algebraic properties. . . . The discovery of quasicrystals in physics in 1984 spurred vigorous research activity in the mathematics of long-range aperiodic order. Many aperiodic 3D structures are so complex that exploration of their local and global properties is infeasible with the visualization tools currently available.

"The complexity of geometric structures that arise in scientific and industrial applications requires better understanding. The need to build, visualize, manipulate, and deform these 3D structures in an interactive computer environment is widely recognized. The major areas that would directly benefit from such 3D software include convexity, tilings, quasicrystals and aperiodicity, lattices and periodicity, packing and covering, groups and symmetry, polyhedra and incidence structures, oriented matroids, reflection groups and hyperplane arrangements, illumination problems, molecular structures, frameworks and rigidity (ball and stick structures), and growth systems (foams, cell growth).

"Existing software either enables the researcher to visualize known structures and to give analytic (coordinate) input to build some new structures (Geomview and Mathematica) or provides graphical tools that allow one to construct objects free-hand (CAD/CAM). No existing integrated package provides the interactivity and mathematical content needed to study complex geometric problems in 3D. We propose to develop a 3D package analogous to two-dimensional programs such as The Geometer's Sketchpad, which allow the researcher to build 2D geometric figures in a 'hands-on' manner, and to explore the implications of geometric constraints.

"This involves being able to . . . . group and ungroup these objects and moving them around independently or as a group," we continued. "We need to be able to intersect and dissect the basic building blocks and attach them to each other at angles to generate larger figures in space such as 3D geometric complexes, tilings, or clusters, with the ability to also reverse the construction process and dissect figures. In building up figures from smaller units, we need to be able to specify constraints (such as perpendicularity, *parallelism*, dihedral angles, or flexible attachment) as we build."

Our e-mail correspondence tracked the proposal's dismal fate.

22 Apr 96. It would be good to settle on the project name. Naming things makes them real somehow and a good name might go a long way to persuading people to bring it into existence. . . . Several people emailed that they like the name Coxeter Project (or Project Coxeter) but not everyone has done so. Marjorie

---

[12]Heidi Burgiel, The Geometry Center; Daniel Huson, Universität Bielefeld, Germany; Nicholas Jackiw, Key Curriculum Press, Inc., Berkeley; Stuart Levy, The Geometry Center; Jesus A. de Loera, Geometry Center; Robert Moody, University of Alberta; Jiri Patera, Université de Montreal; Michelle Raymond, The Geometry Center; Doris Schattschneider, Moravian College; Egon Schulte, Northeastern University; Marjorie Senechal, Smith College.

22 Apr 1996. The Coxeter Project sounds good. It would be important to get his blessing for the name. There is a possible acronym from his initials: Hands-on Synthetic Manipulation (of 3d graphics). I am pretty sure he would not approve of that! Bob

22 Apr 1996. Coxeter might not approve of his initials being interpreted as "Hands-on Synthetic Manipulation" but I'll bet his wife would! I've heard her tell him several times that he should get with it and learn to use computers. She could talk him into giving it his blessing even if the HSM part is just a joke. Marjorie

24 Apr 1996. This is great news. Perhaps an earlier (but not well-known) precedent is Escher's use of Coxeter's name as a verb. When Escher was working on drawings for his Circle Limit hyperbolic prints he wrote to his son George that he was "Coxetering." Doris

26 Apr 1996. I just had a long talk with Dick McGehee. He is very interested in Project Coxeter and encourages us to submit a proposal to the Geometry Center. From his discussions with industrial and other scientists he knows that the kind of 3D interaction we hope to develop is urgently needed in a wide variety of problems and applications – not only in polytope theory! At the same time, the fact that Project Coxeter will address needs of the discrete geometry research community is crucial, since the Center was established to address the visualization needs of researchers. Marjorie

6 Jun 1996. Next season we plan to focus on 3-D. In a recent discussion with Dick McGehee, I learned that the Center will likely go ahead with Project Coxeter, but that (as you forwarned) they can't address any educational component or research in the pitch they make to the NSF. Dick was very interested in the possibility of a Sketchpad / Coxeter relationship to accelerate and exploit Coxeter's educational opportunities. Nick

19 Jun 1996. We are up to a bumpy start! I assume that everyone received Marjorie's message about the newest development at the Geometry Center and its consequences for our project. After the good work at our meeting and the successful proposal writing afterwards, this was a somewhat unexpected and disappointing news... From the message I got the impression that the Center's main nervousness is the next site visit by an NSF team. Until then, or at least until it is clear that the Center can "produce results" for the next visit, we cannot expect too much in the way of support. Egon.

The proposal was not approved; the Geometry Center soon closed its doors. Eight years later, I asked Nick how the 3-D Sketchpad was coming along.

27 Apr 2004. Alas, the 3D-GSP project went nowhere fast... what we really need is new interaction technology, at the hardware level. Another avenue of pursuit was to determine if any of the (relatively inexpensive) 3-D manipulation hardware available from home gaming market manufacturers were accessible to educational software development. The quick answer from Nintendo and Sega: to do anything educational would instantly destroy our credibility with our 14-year-old male customer base. Nick

Postscript. If Coxeter shared our disappointment in the demise of the Coxeter Project, he never told us. And his enthusiasm for the golden isozonohedra never flagged. At the "Symmetry 2000" conference in Stockholm in September of that year, he presented a lively overview of "The rhombic triacontahedron" and its relatives, including stellations, vividly illustrated with Rhombo blocks.

Dear Donald,

Many years ago you wrote, "As for the analogous figures in four or more dimensions, we can never fully comprehend them by direct observation. In attempting to do so, however, we seem to peep through a chink in the wall of our physical limitations, into a new world of dazzling beauty." What I would be interested to know is, which n-polytopes, on the other side of the wall, are golden?[13] Whom do the higher-dimensional relatives of golden isozonohedra resemble? Do families build nests? And non-periodic honeycombs? Must they?

Acknowledgements. I am grateful to the Bomford family, Norman Johnson, Michael Longuet-Higgins, Peter McMullen, and Asia Ivić Weiss for valuable information and permission to use it, and to the Coxeter Collection at York University for permission to use Figure 13. Special thanks to Doris Schattschneider for reading an early draft of this paper and sending me copies of the Coxeter-Bomford correspondence, and to Stan Sherer, for expertly photographing my Rhombo block constructions and for his patience as $F_{20}2$ collapsed again and again.

References

[1] Ammann, R. *Aperiodic Tiling and Crystal Structure*, unpublished and undated manuscript, 8 pages including illustrations, probably 1991.

[2] Rouse Ball, W. W. and Coxeter, H. S. M. *Mathematical Recreations and Essays*, 13th edition, Dover Publications, N.Y., 1987.

[3] Coxeter, H. S. M. *Regular Polytopes*, 2nd edition, MacMillan, New York, 1963.

[4] Coxeter, H. S. M. *The rhombic triacontahedron*, Symmetry 2000, Hargittai. I. and Laurent, T. C., editors, Part 1, London, Portland Press Ltd., 2002.

[5] De Bruijn, N. G. (1981), *Algebraic theory of Penrose's non-periodic tilings of the plane*, Proceedings of the Koninklijke Nederlandse Akademie van Wetenschappen Series A, **84** (Indagationes Mathematicae, **43**), 38–66.

[6] Dolbilin, N. and Schattschneider, D. *The Local Theorem For Tilings*, Quasicrystals and Discrete Geometry (J. Patera, editor). Fields Institute Monographs, Vol. 10, AMS, Providence, RI, 1998, 193-199.

[7] Burkov, S. E. *Absence of weak local rules for the planar quasicrystalline tiling with 8-fold symmetry*, Comm. Math. Phys. **119** (1988), 667–675.

[8] Engel, P., Michel, L., and Senechal, M. *Lattice Geometry*, preprint, http://www.ihes.fr, 2004.

[9] Gardner, M. *Mathematical Games*, Scientific American, January, 1977, 110–121.

[10] Grünbaum, B. and Shephard, G. S. *Tilings and Patterns*, W. Freeman, New York, 1987, Chapter 11.

[11] Katz, A. *Theory of matching rules for the 3-dimensional Penrose tiling*, Communications in Mathematical Physics, **118** (1988), 263–288.

[12] Lagarias, J. C., in *Mathematical quasicrystals and the problem of diffraction*, Directions in Mathematical Quasicrystals, Baake, M. and Moody, R. V., editors, CRM Monograph Series, **13**, Amer. Math. Soc.: Providence, RI, 2000, 61–93.

[13] Lagarias, J. C. and Pleasants, P. A. B. *Repetitive Delone sets and quasicrystals*, Ergod. Th. Dyn. Sys. **23** (2003), 831–867 .

[14] Lee, J.-Y., Moody, R. V., and Solomyak, B. *Consequences of Pure Point Diffraction Spectra for Multiset Substitution Systems*, Discrete and Computational Geometry, **29** (2003), 525–560.

---

[13]Norman Johnson, Coxeter's former student, has pointed out, in a private communication, that analogues of A_6 and O_6 exist in every dimension. "For $n > 2$ these come in two varieties, acute and obtuse. In either case all the facets of a golden n-rhombotope are golden $(n-1)$-rhombotopes. There may well be more interesting 4-polytopes (polychora) with golden facets."

[15] Longuet-Higgins, M. *Nested Triacontahedral Shells, or How to Grow a Quasicrystal*, The Mathematical Intelligencer, **25** (2003), 25–43.

[16] Mackay, A. *Crystallography and the Penrose pattern*, Physica A, **114** (1982), 609–613.

[17] McMullen, P. *Convex bodies which tile space by translation*, Mathematica **27** (1980), 113–121.

[18] Senechal, M. *Quasicrystals and Geometry*, Cambridge University Press, Cambridge, corrected paperback edition, 1996.

[19] Senechal, M. *Crystals and Quasicrystals*, Handbook of Discrete and Computational Geometry, J. E. Goodman and J. O'Rourke, editors, 2nd edition, Chapman and Hall/CRC, 2004, Chapter 62, 1377–1393.

[20] Senechal, M. *The Mysterious Mr. Ammann*, The Mathematical Intelligencer **26** (2004), 10–21.

[21] Socolar, J. E. S. *Weak matching rules for quasicrystals*, Communications in Mathematical Physics **129** (1990), 599–619.

[22] Socolar, J. E. S. and Steinhardt, P. *Quasicrystals. II. Unit-cell configurations*, Physical Review B **34** (1986), 617–647.

[23] Venkov, B. A. *On a class of euclidean polytopes* (in Russian), Vestnik Leningr. Univ. Ser. Mat.-Fiz. Him., **9** (1954), 11–31.

Louise Wolff Kahn Professor in Mathematics and History of Science and Technology, Smith College, Northampton, MA 01060 USA

E-mail address: senechal@smith.edu

Configurations of Points and Lines

Branko Grünbaum

1. Introduction

Configurations — as the word is interpreted throughout this paper — are simple enough that in their geometric aspect could be explained to any third-grader, but easily lead to problems that are beyond the reach of all presently available tools. While one could reasonably place configurations within the purview of elementary geometry, they could also be interpreted as belonging to algebraic geometry, or combinatorics, or topology. Despite — or possibly because — the simplicity of the concept, over the years it suffered from many instances of confusion and downright errors. Even if this fact may be understood as arising from misinterpretation of the nature of configurations, it is surprising that the problematic statements were in many cases not recognized and corrected for nearly a century.

Configurations of points and lines were first defined by Reye [**R1**] some 125 years ago (see also [**R2**]). Many specific configurations have been described and studied before that — but without a framework into which they could fit. Configurations enjoyed considerable popularity during the last two decades of the nineteenth century, only to be relegated to a mathematical limbo afterwards. Neither the publication of the only serious text on configurations by Levi [**L2**] in 1929, nor the attempt by Hilbert and Cohn-Vossen [**H1**] in the 1930's, generated much activity. In the second half of the twentieth century there was a slight renewal of interest, in particular through Coxeter's frequently quoted paper [**C3**]. This, and other papers of Coxeter's on configurations make the topic appear appropriate for the present volume, as it was for the conference that led to it.

During the last two decades, research on configurations became energized by several factors.

- Re-examination of the nineteenth century papers revealed that many of the basic results that have been accepted for a century or more are, in fact, not valid as stated. Details of this aspect will be presented in Section 3.
- Novel ideas made it possible to study various kinds of configurations that were beyond the reach of earlier investigations.

2000 *Mathematics Subject Classification*. Primary 52C30; Secondary 05B30.

- At the same time, computers and computer graphics contributed to the appeal of the study, by enabling enumerations and depictions of a wide variety of configurations to an extent not possible earlier.

These aspects will be discussed in the remaining parts of this survey. A large number of intriguing open problems will be presented, along with precise definitions and statements of results, and with many references to the scattered literature.

Following general definitions in Section 2, Sections 3 and 4 deal with the history of the early papers on configurations. These sections describe in some detail the errors of these basic papers. It seems appropriate to dwell on the shortcomings as a warning: Very capable people were capable of making mistakes, and the lack of clarity and precision causing them has been known to occur more recently as well. In later sections we give brief explanations of recent developments in the theory of configurations. Through this it becomes apparent that as soon as the right questions are asked, there is no more room for dismissal of configurations as trivial geometry.

2. Definitions and notation

DEFINITION 2.1. A *configuration* C is a family of *points* (or *vertices*) and *lines* such that for some positive integers p, q, n, k each of the p points is incident with q of the n lines, and each of these lines is incident with k of the points. Such a configuration C is denoted (p_q, n_k) or, if $p = n$ and hence $q = k$, more simply by (n_k). If the parameters p and n are not relevant, a configuration (p_q, n_k) is said to be of *type* $[q, k]$.

Several explanations and clarifications should help understand the scope of this definition. To begin with, the points and lines are understood to be part of some space in which these concepts, and incidence, are meaningful. Our main settings are the real *projective* plane and the real *Euclidean* plane E^2; this is the meaning unless some other interpretation is specifically mentioned. We shall invariably interpret the projective plane as the *extended Euclidean plane* E^{2+}, that is, E^2 augmented by the "ideal points", also called "points at infinity". For geometric configurations "point" and "line" mean a point of the plane and a straight line, respectively. "Incidence" is defined as the point belonging to the line; however, we allow that some points of the configuration lie on a line with which they are not considered to be incident. To avoid trivialities we shall always assume that not all points of a geometric configuration are collinear.

In a different interpretation of Definition 2.1 we obtain combinatorial configurations: here "points" are understood as any symbols, "lines" as sets of "points", and "incidence" as "point is an element of a line". Thus, combinatorial configurations are special incidence structures. Combinatorial configurations are often specified by a configuration table, in which columns represent points incident with one line. In our considerations we insist that in a combinatorial configuration two distinct points are incident with at most one line; this implies that two distinct lines are incident with at most one point.

The earliest mention of the fact that for every choice of q and k there exist geometric configurations of type $[q, k]$ seems to be in the paper of Kantor [**K2**] published in 1879. His construction is quite complicated. As is easily seen, the existence of such configurations is more easily established by considering a block

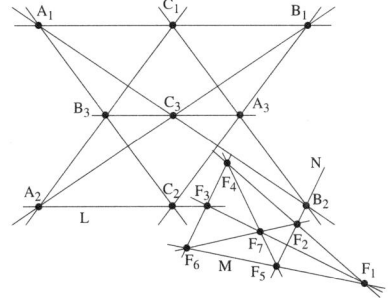

FIGURE 1. The combinatorial configuration (16_3) indicated by this diagram has a representation as shown, but it admits no realization since by the theorem of Pappus the point B_2 necessarily lies on the line L.

of size k^q of points in the integer lattice of the q-dimensional Euclidean space, and the lines through them parallel to the coordinate axes.

DEFINITION 2.2. Configurations C_1 and C_2 are said to be *isomorphic* if there is a bijection between their points, and a bijection between their lines, which preserve incidences. If there are incidence-preserving bijections between points of one configuration and lines of the other, the configurations are said to be *dual* to each other. A configuration is *selfdual* if it is isomorphic to its dual.

Clearly, every geometric configuration is isomorphic to a combinatorial one; however, as we shall see, the converse is not valid.

Discussions of the relation between geometric configurations and combinatorial ones requires a differentiation which seems largely absent in the literature. A geometric configuration C is a *realization* of the underlying combinatorial configuration if a point and a line of C are incident *if and only if* their combinatorial counterparts are incident. In contrast, C is a *representation* of the underlying combinatorial configuration provided a point and a line of C are incident *whenever* their combinatorial counterparts are incident, but points may belong to other lines as well. The essential distinction between representation and realization is illustrated by Figure 1. Since every combinatorial configuration could be represented by a collinear set of points, by convention representations by such configurations and their duals are excluded from all following considerations and statements.

Levi graphs of configurations (and more general incidence structures) are one tool that simplifies discussions and leads to parsimonious ways of defining concepts. They were introduced by Levi [**L3**] in 1942, but were brought to the attention of a wider audience only by the path-breaking paper of Coxeter [**C3**]. The Levi graph $L(C)$ of a configuration C is a bipartite graph in which the vertices of one part correspond to the points of C, and the vertices of the other part to the lines of C; two vertices of $L(C)$ are connected by an edge if and only if the corresponding line and point are incident. As an illustration, in Figure 2 is shown the Levi graph of the configuration of Figure 1. The earlier "Menger graphs" of configurations discussed by Coxeter [**C3, C2**] are still used sometimes, but they are much less useful.

The Levi graphs of combinatorial configurations are easily characterized. For configurations (n_3) these are 3-valent bipartite graphs of girth (length of shortest

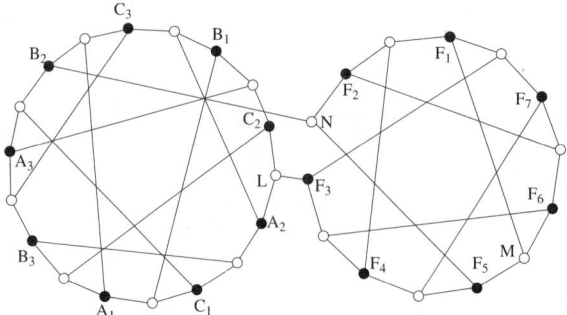

FIGURE 2. The Levi graph of the (16_3) configuration from Figure
1. (Adapted from the Levi graphs of Pappus and Fano configurations given by Coxeter [**C1**].)

circuit) at least 6, and similarly for other classes of combinatorial configurations.
The graph-theoretic connection makes it redundant to dwell at length on definitions
of concepts such as isomorphism or duality, connected or k-connected configuration,
Hamiltonian circuit in a configuration, or the girth of a configuration.

Another tool used in the study of configurations are *incidence matrices*, and
various ways of representing them; see, for example, [**L2, B3**].

Neither of these tools leads to a solution of a central problem: *Which combinatorial configurations admit representations or realizations in the (Euclidean or
projective) plane.* In Section 3 we shall mention coordinatization, one approach
that can be used to such an effect, but which has to be applied to individual configurations and is highly laborious unless the configurations are very small.

Symmetry is a central concept for the newer development of configurations. By
this we understand an isometric map of the plane onto itself that maps a configuration onto itself. All symmetries of a configuration form its *symmetry group*. A
geometric configuration is said to be of symmetry type $[h_1, h_2]$ provided its points
form h_1 orbits, and its lines h_2 orbits under its symmetry group. We shall also
say that such a configuration is $[h_1, h_2]$-*astral*. Clearly, if a configuration of type
$[q, k]$ is $[h_1, h_2]$-astral, then $h_1 \geq (k+1)/2$ and $h_2 \geq (q+1)/2$. If h_1 and h_2 have
these minimal values we shall simplify the language and say that the configuration
is *astral*. In cases where $h_1 = h_2 = h$, we shall say that the configuration is h-*astral*.
A variant of this term was introduced in [**G10**] along with a variety of examples.
Here we shall see many examples in Sections 5, 6, and 7.

3. Early results and errors, and their subsequent development

Most of the early works on configurations deal with configurations (n_3), and
in particular, with the enumeration of isomorphism classes for certain values of
n. One of the startling facts about these investigations is how hard it is to decide
nowadays whether they discuss combinatorial configurations or geometric ones, and
in the latter case, whether they consider configurations in planes or spaces over the
reals, or over complex numbers. All these possibilities were often mixed up — for
example, in the writings of Kantor, Schönflies, Schroeter, Steinitz and others listed
in the references. As we shall see soon, contributing to the confusion is the fact
that results valid for some special cases were deemed to have general validity.

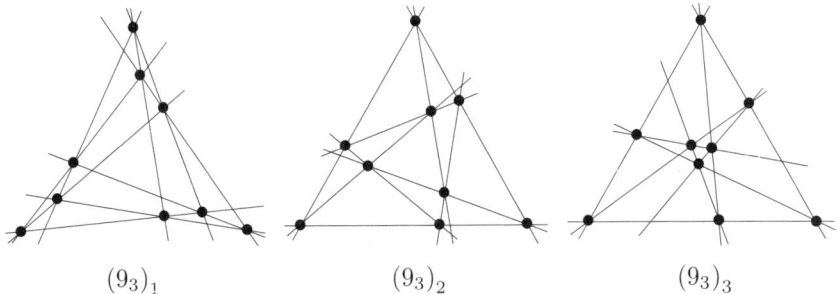

$(9_3)_1$ $(9_3)_2$ $(9_3)_3$

FIGURE 3. The three configurations (9_3). The notation is the one used by [**H1**]. It is not clear who first devised the realization with 3-fold rotational symmetry of these configurations; the ascription by Steinitz ([**S8**, p. 489], [**S10**, p. 158]) to a 1910 paper by H. A. Schwarz is mistaken.

Other investigations concerned the possibility of geometric realizations of combinatorial configurations, and of the possibilities of constructing all combinatorial configurations (n_3). It is unsettling to realize how many errors, small and large, were committed in these investigations. It is even more surprising that some of the errors remained undetected for a century.

3.1.   The first enumeration of the isomorphism classes of configurations (8_3) and (9_3) was carried out by Kantor [**K3**]. Kantor explains that *the unique configuration (8_3) can be constructed by starting with a simple quadrangle, to which another quadrangle is inscribed in such a way that it is also circumscribed to the starting one.* The mystery of how this can be done is removed later in the paper when he mentions, in passing, that if the vertices of the first quadrangle are in the real plane, then the other four vertices are imaginary. This was in a different context shown much earlier, by Möbius [**M3**]. In the review of [**K3**] by Schubert [**S5**] the description italicized above of the configuration is repeated, but there is no mention of the reality or otherwise of that configuration. A proof of the non-realizability of the configuration (8_3) was given by Möbius [**M3**] (see Coxeter [**C3**, pp. 122, 131]), Levi [**L2**, p. 99], and by Bokowski and Sturmfels [**B9**, p. 35]. More information concerning the configuration (8_3), its realizability in various planes and its symmetries, together with historical remarks and references can be found in Coxeter's papers [**C3, C5**].

Kantor [**K3**] also established that there are precisely three non-isomorphic combinatorial configurations (9_3), and that each is isomorphic to a geometric configuration; he provides diagrams of all three. We show realizations of these configurations in Figure 3. One of them, the Pappus configuration $(9_3)_1$, is an expression of one of the basic theorems of projective geometry. Hilbert and Cohn-Vossen [**H1**] and Coxeter [**C3, C5**] provide illustrations and details about the realizations and symmetries of the Pappus configuration $(9_3)_1$.

Strangely enough, Kantor never mentions the unique combinatorial configuration (7_3); it is generally known as the Fano configuration, and is important in many combinatorial contexts. The omission may possibly be explained by the fact that

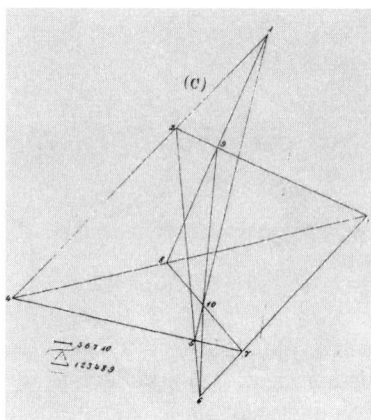

FIGURE 4. Facsimile of one of the diagrams of [**K3**], purporting
to show a (10_3) configuration. This configuration has been shown
by Schroeter [**S4**] to be not realizable in the Euclidean plane. In
the diagram, the line 9-10-6 is clearly not straight.

this combinatorial configuration is not isomorphic to any geometric configuration
in either the real or the complex plane.

A later paper of Kantor [**K4**] is devoted to the enumeration of (isomorphism
classes) of configurations (10_3). The enumeration is based on a mixture of combinatorial and geometric arguments, and is quite opaque. One of its results — that
there are ten different classes of combinatorial configurations (10_3) — is correct,
although it relies on an invalid argument. This argument, made without any justification, asserts that two configurations (n_3) are isomorphic if and only if their
families of remainder figures contain the same figures in the same numbers. (The
remainder figure of a point P in a configuration consists of those points of the configuration that are not on any configuration line through P, and of configuration
lines containing two or more of these points.) This statement was also repeated in
the review of [**K4**] by Rodenberg [**R4**], although it is false in this generality; while
true for $n \leq 10$, already for $n = 11$ there are counterexamples. Independent enumerations of the combinatorial configurations (10_3) were carried out by Martinetti
[**M1**], Schroeter [**S4**], and more recently by many others (for example, [**S11, B4**]).
The enumeration by Zacharias [**Z2, Z3**] is not correct, nor is the report on it by
Togliati [**T1**].

The other claim of [K4], that *each of the ten classes of combinatorial configurations (10₃) is realizable as a geometric configuration of points and lines in the
real plane* — is invalid. Kantor claims to derive his assertion using a statement
credited to Johann Benedict Listing (but without any reference) to the effect that
— in present-day terminology — every combinatorial (n_3) configuration can be
interpreted as having a Hamiltonian circuit. (Kantor speaks of an n-gon inscribed
and circumscribed to itself.) But the assertion attributed to Listing (which I have
been unable to locate in any work of Listing's) is invalid, not only in the stated
generality but even under quite stringent restrictions. (We shall discuss Hamiltonian circuits in more detail in Sections 3 and 4.) Kantor supports his claim of

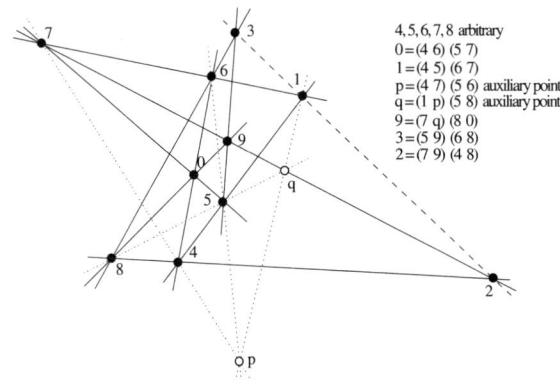

4,5,6,7,8 arbitrary
0=(4 6) (5 7)
1=(4 5) (6 7)
p=(4 7) (5 6) auxiliary point
q=(1 p) (5 8) auxiliary point
9=(7 q) (8 0)
3=(5 9) (6 8)
2=(7 9) (4 8)

FIGURE 5. Schroeter's construction in [S4] of one of the (10_3) configurations.

realizability by diagrams, purporting to show the corresponding ten geometric configurations. However, several of the diagrams do not actually show all intersection
points, and some are incorrect. A facsimile of Kantor's purported realization of one
of these configurations is reproduced in Figure 4; as clearly visible, the line 6,9,10 is
not straight. But beyond inaccuracies of the diagrams, the important fact is that,
as proved by Schroeter [S4], *one of the ten combinatorial configurations cannot be
realized by points and lines in the real plane*; this configuration is the one shown in
Figure 4. (Later investigations by Laufer [L1] and Bokowski-Sturmfels [B9] showed
that this configuration cannot be realized by points and lines in any plane over a
field, hence, in particular, not even in the complex plane.) In continuation of his
paper [S4] Schroeter shows that the other nine combinatorial configurations (10_3)
can be geometrically realized in the *real* plane. An example of his mode of construction is shown in Figure 5. In fact, the constructions Schroeter presents show
that they are realizable in the *rational* plane as well. This has been confirmed using
more modern methods by Bokowski-Sturmfels [B9].

The claim that every combinatorial configuration (n_3) can be realized by points
and straight lines (in the real or complex plane) is repeated by Martinetti [M1], in
the preliminaries to his enumeration of configurations (11_3). He makes that statement (which is repeated in the review of [M1] by Loria and Lampe [L4]) in order to
claim that he enumerates geometric configurations while his argumentation deals
only with combinatorial ones. What Martinetti finds is that there are 31 combinatorial configurations (11_3). (We shall return soon to another part of his paper
[M1].) The same 31 configurations were independently found by Daublebsky [D1]
using a different method. Going beyond this, in a supplement to [D2], Daublebsky
presents diagrams showing that all these configurations are geometric. Without giving details, Daublebsky states that the diagrams were all obtained by the method
used by Schroeter in [S4] in the construction of the nine realizable configurations
(10_3), and that they are faithful and correct. An independent proof of the geometric realizability in the Euclidean plane of the 31 configurations (11_3) was provided
by Sturmfels and White [S12, S13]; in fact, they establish the realizability of all
31 in the *rational* plane.

The configurations (12₃) have their own somewhat convoluted history. The first enumeration of the combinatorial configurations (12₃) was carried out by Daublebsky [**D2**] in 1895, using the method of remainder figures. He found that only 18 different remainder figures could possibly occur in such a configuration. Through various arguments (described only in general terms) Daublebsky arrived at the conclusion that these remainder figures could be combined to yield many hundreds of configurations (12₃). Then he "... drew a schematic diagram of each configuration on a separate piece of paper ..." and determined for each the "remainder system", that is, a list of the different remainder figures occurring in the configuration. Finally, configurations with the same remainder system were investigated to see whether they are isomorphic. This turned out to be the case in most — but not all — cases. Daublebsky presented the resulting 228 combinatorial configurations by their configurations tables (these take 23 pages!!!). He also gave some other data and provided drawings for geometric realizations of a few of the configurations. In a later paper [**D3**], Daublebsky gave results of his investigations of the groups of automorphisms of each of the 228 combinatorial configurations (12₃). The first independent enumeration of the combinatorial (12₃) configurations was carried out only in 1990, by Gropp [**G2**]. It showed that Daublebski missed one, so that there are in fact 229 such configurations. (I assume that Gropp compared his list with that of Daublebsky, and that the one additional configuration is the only discrepancy between the two lists. A statement in [**G1**] can be interpreted this way.) Gropp communicated to me the configuration table of this configuration, and it can be read off from the illustration in [**D4**] and [**G6**]. As with configurations (11₃), the 229 combinatorial configurations (12₃) have been independently enumerated (by two different algorithms) in [**B4**].

The only published proof that all 228 combinatorial configurations (12₃) found by Daublebsky are geometrically realizable was given only recently, by Sturmfels and White [**S12, S13**]. Sturmfels and White also proved that all these (12₃) configurations are realizable in the rational plane. In a private communication, B. Sturmfels showed that the "new" combinatorial configuration found by Gropp is also geometrically realizable, even in the rational plane; a diagram is shown in Dorwart and Grünbaum [**D4**].

The numbers of different combinatorial configurations (n₃) have been determined for $n \leq 14$ by Gropp [**G2**], and for $n \leq 18$ by [**B4**]. See Table 1, which includes also the number for $n = 19$ computed by the same method. There seems to be no estimate of the asymptotic growth of the number of types of combinatorial configurations (n₃) as n goes to infinity.

In contrast to the above discussion, for $n \geq 13$ there is no general information available concerning the possible realizations or representations of the configurations (n₃) as geometric configurations of points and lines in the Euclidean plane. The only known connected configuration (n₃) with $n \geq 10$ which cannot be so represented is one of Kantor's (10₃) configurations, discussed above and indicated by the incorrect drawing in Figure 4. This observation and the results mentioned earlier lead to the conjectures:

CONJECTURE 3.1. *Every connected combinatorial configuration (n₃) with $n \geq$ 11 can be represented by points and lines in the real Euclidean plane.*

TABLE 1. The number of non-isomorphic combinatorial configurations (n_3), from [**B4**] and [**B7**].

n	Number of all (n_3) configurations	Selfdual (n_3) configurations
7	1	1
8	1	1
9	3	3
10	10	10
11	31	25
12	229	95
13	2,036	365
14	21,399	1,432
15	245,342	5,799
16	3,004,881	24,092
17	38,904,499	102,413
18	530,452,205	445,363
19	7,640,941,062	1,991,320

CONJECTURE 3.2. *Every configuration (n_3) that can be represented or realized by points and lines in the real plane can also be represented or realized in the rational plane.*

It should be stressed that the distinction between representability and realizability of a configuration (n_3) is very important. In Figure 1 we indicated by a diagram a *representation* of a combinatorial configuration (16_3). This configuration (and many others) cannot be *realized* since, by the theorem of Pappus, the line L must pass through the point P with which it is not incident. However, all known examples of configurations (n_3) that are representable but not realizable are *2-connected* at most. Hence the following

CONJECTURE 3.3. *Every 3-connected configuration (n_3) that is representable in the real plane is realizable in the plane as well.*

As a sidelight to these conjectures it is appropriate to mention that the distinction between representation and realization of a configuration was very slow to be noted. For example, Schroeter [**S4**] describes very carefully the construction of the nine realizable configurations (10_3). He starts each construction by choosing a certain number of arbitrary points, and adding in some cases additional points on already constructed lines — but without noticing that some choices lead to representations that are not realizations. This is illustrated in Figures 5 and 6. Analogously, Steinitz in his fundamental theorem, which we shall discuss soon, claims to establish realizations of the configurations, or of near-configurations in which one incidence is not satisfied, — although this result is invalid. It becomes valid if representations are considered instead of realizations. However, there is a basic difference between the shortcomings of the Schroeter constructions and the Steinitz claim: The former can be made correct if assuming that the construction start with *generic* choices of points and lines, while no such correction can salvage Steinitz's assertion.

An error first committed by Schönflies [**S1**] is the claim that all combinatorial configurations (n_3) that are vertex-transitive and contain triangles are selfdual

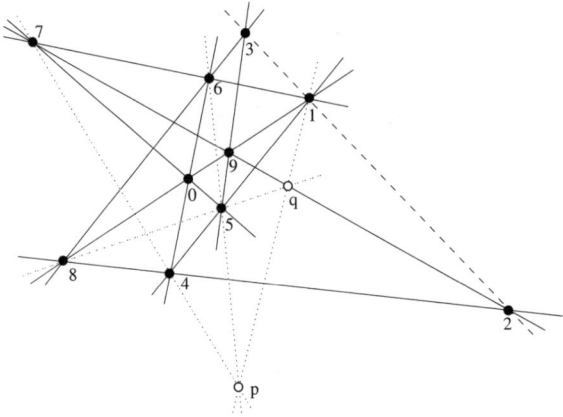

FIGURE 6. The same construction as in Figure 5 yields only a representation, since the point 1 lies on line 8-0-9.

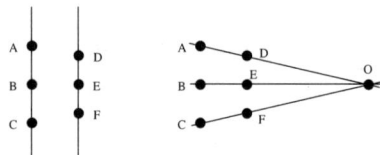

FIGURE 7. The Martinetti transformation.

("sich selbst reciprok"). In [**S2**] Schönflies states that this is a consequence of the (true) fact that if a configuration contains triplets of points that form triangles, then it must contain triplets of lines that form triangles — but this deduction is invalid; see Steinitz [**S9**, p. 307–309]. A related error is the frequently made *definition* by which configurations with the same number of points and lines are called "selfdual". As visible from Table 1, this happens to be true for configurations (n_3) with $n \leq 10$. However, already for $n = 11$ only 25 of the 31 configurations are selfdual; the fraction of self-dual configurations decreases rapidly with growing n — for $n = 16$ they form less than 1% of all configurations. This error of confusion was, unfortunately, committed also by Coxeter in [**C3**], and occurs still in some publications.

Another type of misleading terminology is still in wide use: Many authors (too many to list) call "*symmetric*" all configurations (or more general incidence structures) with the same number of points as lines. This is clearly inappropriate, since most of these objects admit no automorphisms or any other kind of incidence-preserving "symmetries". It would seem that "*balanced*" would be a far better designation — but if the use of this term is opposed because it occurs in the context of block designs, "*equinumerous*" might work as it does convey the meaning without implying properties that do not exist, and without impinging on other topics. With very few exceptions, in this paper we shall consider only equinumerous configurations.

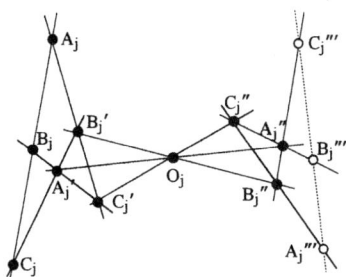

FIGURE 8. The "module" used in the Martinetti construction.
Only the ten solid dots and the ten solid lines form one module.

3.2.   A different direction in the studies of configurations (n_3) was initiated
by Martinetti [**M1**]. Speaking of combinatorial configurations but using geometric
language, he describes a simple operation which may be applied to a configuration
(n_3) to obtain a configuration $((n+1)_3)$. The operation is schematically indicated
in Figure 7. It replaces two "parallel" lines (that is, lines with no common points)
such that corresponding pairs of points (AD, BE, CF) are not contained in any line
of the configuration, by three lines that pass through these three pairs and a new
point O. If a configuration $\mathcal{C}$ can be obtained from a smaller one by this operation,
Martinetti calls $\mathcal{C}$ *reducible*; otherwise it is *irreducible*. Martinetti's main result is
the claim that for each n there are very few irreducible (n_3) configurations, and he
gives a complete description of all irreducible configurations. More precisely:

Theorem claimed by Martinetti (1887) [M1]   *A connected (n_3) combinatorial configuration is irreducible if and only if it is one of the following:*

   (i) *For $n \geq 7$, the cyclic configuration $\mathcal{C}_3(n)$ with lines $[j, j+1, j+3]$ (mod
        n), for $0 \leq j \leq n-1$;*
  (ii) *$n = 10m$ for some $m \geq 1$, and the configuration is the one described below
        and denoted $\mathcal{M}(m)$; $\mathcal{M}(1)$ is the Desargues configuration $(10_3)_1$.*
 (iii) *$n = 9$, and the configuration is the Pappus configuration $(9_3)_1$.*
 (iv) *$n = 10$, and the configuration is $(10_3)_2$ or $(10_3)_6$ in the list of (10_3)
        configurations of Schroeter [**S4**].*

Martinetti's combinatorial configuration $\mathcal{M}(m)$ can best be explained as consisting of m copies of the family of the ten points indicated by solid dots in Figure
8, and the ten solid lines shown there. The jth copy is joined to the $(j+1)$st by
identifying A_j''', B_j''', C_j'''' with A_{j+1}, B_{j+1}, C_{j+1}, respectively; all subscripts taken
(mod m).

Martinetti's proof is, not surprisingly, involved and long. The result was quoted
or mentioned many times over the next century; see, for example, Steinitz [**S8**, pp.
486–487], [**S10**, pp. 153–154], Gropp [**G1, G2, G6, G8**], Carstens et al. [**C1**].
In lecture notes for my configurations courses in 1999 and 2002 I wrote about the
proof of Martinetti's theorem the following:

> I have not checked the details, and I do not know it as a fact
> that anybody has. The statement has been accepted as true for
> these 115 years, and it may well be true. On the other hand,

> Daublebski's enumeration of the (12_3) configurations was also
> considered true for a comparable length of time ...

As it turned out, my suspicion has been vindicated by the Ph.D. thesis of
M. Boben [**B5**]; see also [**B6**]. He showed that Martinetti's list of irreducible
configurations is incomplete. The error in Martinetti's proof arises as follows. When
constructing $\mathcal{M}(m)$, he attaches m copies of the "module" in Figure 8 as indicated
above; the mth copy is attached to the first "straight", by identifying A_n''' with
A_1, and similarly for the B's and C's, thus obtaining $\mathcal{M}(m)$. However, as shown
by Boben, that attachment can also be done in "twisted" ways, two of which yield
irreducible configurations which we may denote by $\mathcal{M}^*(m)$ and $\mathcal{M}^{**}(m)$. These are
obtained by identifying A_n''' with C_1, B_n''' with B_1, and C_n'''' with A_1 for the former,
and A_n''' with C_1, B_n''' with A_1, and C_n'''' with B_1 for the latter. A separate argument
shows that the three resulting configurations are non-isomorphic for every m. With
this modification, Boben's corrected version of Martinetti's result replaces parts (ii)
and (iv) by the following:

> (ii) $n = 10m$ for some $m \geq 1$, and the configuration is one of $\mathcal{M}(m)$, $\mathcal{M}^*(m)$
> or $M^{**}(m)$ described above. For $m = 1$ these are the configurations $(10_3)_1$,
> $(10_3)_2$ and $(10_3)_6$.

3.3. A very interesting situation concerns Steinitz's Ph.D. thesis [**S6**]. In it,
Steinitz proves a remarkable result, in an inspired way. However, he does not prove
what he believes and asserts to have proved, but a considerably weaker result. It is
hard to understand how Steinitz — a profound and very painstaking researcher —
could make such a logical error; it is even harder to understand that the error was
not detected for more than a century, despite the frequent mentions of Steinitz's
theorem in the literature. The error came to light only in the presentation of T.
Pisanski at the Ein Gev conference in 2000 [**P2**].

THEOREM 3.4 (Steinitz [**S6**] (in modern and correct formulation)). *For every
connected combinatorial configuration (n_3) and every choice of one line (or of one
point), there is a selection of distinct points and lines in the plane which represent
all the incidences of the configuration except possibly the incidences of the chosen
line (or the chosen point).*

Steinitz claimed to have proved the above assertion with "realize" instead of
"represent". As stated explicitly in [**S9**], [**S8**, p. 485], or [**S10**, p. 150], Steinitz
considers as geometric configurations only those that contain no incidences besides
the ones of the combinatorial configuration. The failure of the realization claim
follows at once from considerations of point F_1 (or line M) in Figure 1; the point
B_2 has to be on line L with which it is not incident regardless of the incidences or
nonincidences of F_1 (or M).

I feel humbled to realize that although I found the configuration shown in
Figure 1 long ago (see [**D4**]), and although I lectured on Steinitz's theorem several
times, I did not detect his error.

Steinitz's proof is remarkable enough to deserve a brief description. It has a
combinatorial part, and a geometric one.

The centerpiece of the combinatorial part is the claim:

THEOREM 3.5 (Steinitz [**S6**]). *Every combinatorial configuration (n_k) admits
an orderly configuration table.*

Here a configuration table for an (n_k) configuration is said to be *orderly* if every *row* of the table contains all the points (hence each precisely once).

A statement that Theorem 3.5 holds for $k = 3$ (without any justification or hint of proof) appears in Martinetti [**M1**]. Most later authors do not mention the result — much less its proof — although many writers seem to accept it as self-evident. On the other hand, the statement in Page and Dorwart [**P1**] regarding this result is incorrect, as are the consequences deduced by them from the erroneous statement. As pointed out by Gropp [**G7**], Theorem 3.5 implies the later result known as König's theorem [**K5**], that every bipartite graph of constant valence contains a factor of degree 1.

Having established that any given (n_3) configuration has an orderly configuration table, after a few additional steps, Steinitz arrives at the result that, having chosen a point or a line, it is possible to arrange all elements (that is, points and lines) of the combinatorial configuration in such a sequence that each is incident with at most two elements preceding it in the sequence, the only exception being the chosen element, which is the last in the listing. Now, in the geometric part, since an element incident with at most two previous ones can obviously be constructed, it follows that the combinatorial configuration can be represented by points and straight lines that satisfy all the incidences, except those of the last line. Clearly, if the last three points are not incident with a (straight) line, they can always be made incident with a curve of degree 2. This is the result of Theorem 3.4.

Some additional comments seem appropriate.

First, from the proof it is obvious that the whole construction could be carried out in the rational plane, so that all the points and lines (including the last one, if it exists), are rational.

Second, Steinitz devotes more than half the dissertation [**S6**] (24 pages) to a consideration of ways in which one could guarantee that the final step in the above proof can be made using a straight line instead of a curve of degree 2. While this might be another interesting result, I have not been able to follow the exposition. (In fact, I know of nobody who claims to have understood and verified this part of [**S6**].) The opaqueness of the exposition can best be seen from the last two sentences of Steinitz's introduction to this part of the work (see [**S6**, p. 22]):

> ... Without any particular assumptions about the configurations, a method will be presented below following which one can reach a linear presentation. However, for each configuration to which we want to apply this method, an additional investigation is necessary since the method becomes illusory in certain cases.
> [My translation]

In mentioning [**S6**] in the survey [**S8**, p. 490], Steinitz is equally uninformative. Stating that his method is an extension of Schroeter's approach in [**S3, S4**], he ends the explanation by stating:

> Schroeter's method can be generalized so that it is applicable to most configurations (n_3).

It seems that the "method of Schroeter" is rooted in arguments due to Möbius in the early part of the nineteenth century, in particular in [**M3**].

Third, even if the proof is valid, and somebody were to make the exposition understandable — this would prove our Conjecture 3.1, but it would not be a proof of the analogue of Conjecture 3.1 for *realizations*, as claimed by Steinitz.

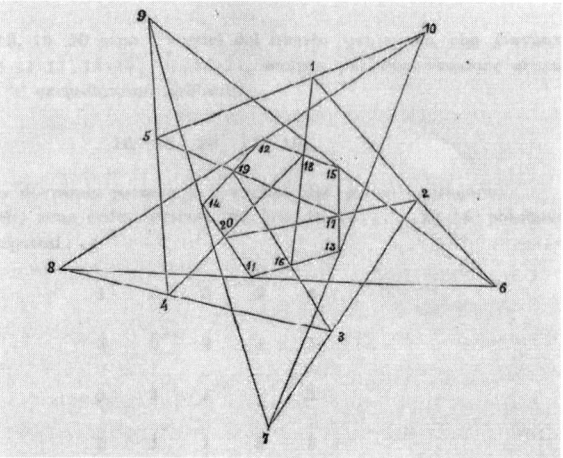

FIGURE 9. The seemingly first published graphical presentation of a polycyclic configuration, specifically a 4-cyclic (20_3) configuration. (From Visconti [**V1**].) There are four pentagons: $1, 2, 3, 4, 5$; $6, 8, 10, 7, 9$; $11, 14, 12, 15, 13$, and $16, 19, 17, 20, 18$. Each is inscribed in the preceding one, and the first one in the last.

Indeed, we know from examples such as the one in Figure 1 that some representable configurations are not realizable, hence Conjecture 3.1 cannot be generally valid for realization.

3.4. Interpretations of configurations as polygons or families of polygons go back to the very beginning of the study of configurations (n_3). The connection arises by considering segments of the lines of a configuration determined by the points of that line; such segments can be used to form one or more circuits — polygons. Utilizing the old and ever-present confusion between lines and segments, it is customary to say that one polygon (in the configuration) is *inscribed* into another polygon if the vertices of the former are configuration points (other than the endpoints) of the lines determined by the sides of the latter. The latter is also said to be circumscribed about the former. To clarify this description, consider Figure 9. As specified in the caption, the lines there form four pentagons, each inscribed into another and circumscribed about a third.

Such families of mutually inscribed/circumscribed polygons have been discussed very frequently — for example, by Kantor [**K3, K4**], Martinetti [**M1**], Schöenflies [**S1, S2**], Steinitz [**S9**], and many others. In some cases the discussions concern combinatorial configurations, in other configurations in complex or other planes. Mostly it is assumed that the inscription is "regular", by which is understood that the order of sides of a polygon and the order of the vertices of the inscribed polygon coincide. The concepts have been used to generate various families of configurations. However, several of the assertions found in these papers are not true. As they are of no particular relevance for our discussion, we shall not give details. In Sections 4, 5, and 6 we shall discuss developments that can be interpreted as streamlined families of such polygons.

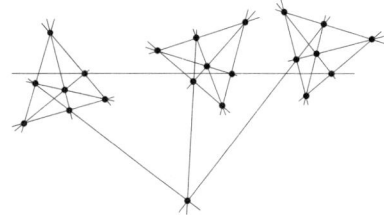

FIGURE 10. The smallest known non-Hamiltonian configuration (n_3); here n=22.

One special case deserves particular mention: In many cases an (n_3) configuration can be presented as an n-gon inscribed and circumscribed to itself. Interpreting this situation via the Levi graph it is clear that such an n-gon represents a *Hamiltonian circuit* in the configuration. The circuit passes through each point precisely once, and utilizes each line precisely once. As mentioned earlier, Kantor [K4] asserted that every configuration (n_3) has a Hamiltonian circuit. This was disproved by Steinitz [**S7**] by an example with $n = 28$. The smallest example of this kind is shown in Figure 10 (from [**D4**]). Steinitz [S7] also mentions that he verified the existence of Hamiltonian circuits for $n \leq 11$. Nothing seems to be known for $12 \leq n \leq 21$.

CONJECTURE 3.6. *All connected combinatorial (n_3) configurations with $n \leq 21$ are Hamiltonian.*

Both the $n = 28$ example of Steinitz, and $n = 22$ example in Figure 10 are only 2-connected. Until recently, I believed that 3-connected configurations (n_3) always have a Hamiltonian circuit. However, this is not the case: There is a 3-connected geometric configuration (50_3) with no Hamiltonian circuits (see [**G13**]).

CONJECTURE 3.7. *Every connected (n_4) configuration has a Hamiltonian circuit.*

4. Configurations (n_4)

The history of configurations (n_4) is much shorter than that of (n_3). It seems that among the first to publish examples of such configurations, both combinatorial and geometric, while aware of the need to distinguish between them, was Brunel [**B11**] in 1898; this work seems to have escaped attention of all later writers. In an earlier paper Brunel followed the ideas of a polygon inscribed and circumscribed to itself, which had been quite popular at the time, as a special class of (combinatorial or geometric) configurations (n_3). In [**B11**] he pursued this idea farther, by considering a polygon *doubly* "inscribed and circumscribed" to itself. In both situations we would call such polygons "Hamiltonian circuits" in the configuration. Each line of such a polygon is incident, besides the two points (vertices of the polygon) that define it as a side of the polygon, with precisely two additional vertices of the polygon. He determines that any combinatorial configuration (n_4) must satisfy $n \geq 13$.

In [**B11**] Brunel gives two constructions. In the first, he presents an orderly configuration table, and states that while the verification that this indeed determines a combinatorial configuration (35_4) is easy, the graphical representation requires some effort. From this (especially in view of his later comments) one may conclude that he had a geometric realization of this configuration. In fact, this configuration turns out to be isomorphic to the geometric configuration (35_4) mentioned in [**G14**], communicated to the authors by Ludwig Danzer. Although no reasonable diagram of this configuration seems to be available, it can be described easily enough by a construction of the kind used by Cayley and others in similar context a century and a half ago. In the case under discussion, start with seven points in general position in real 4-space; consider the 35 2-planes and 35 3-spaces they generate, and intersect this family by a 2-dimensional plane in general position to obtain the required geometric configuration (35_4).

Brunel's second construction yields combinatorial configurations (n_4) on which a cyclic group operates transitively. This includes the explicitly specified configurations for $13 \leq n \leq 16$, but the results presented are marred both by typos, and by outright errors. Without noticing their abstract isomorphism, in several cases Brunel lists isomorphic doubly selfinscribed and selfcircumscribed polygons as distinct. For example, in case $n = 13$ Brunel lists translates of $\{0, 1, 4, 6\}$ and $\{0, 1, 3, 9\}$ as the two polygons, although the permutation

$$(0)(1)(2)(3, 4)(5)(6, 9, 8, 10, 12, 7)(11)$$

maps the first polygon onto the second. But even allowing for these shortcomings, we see that Brunel anticipated the corresponding results of Merlin [**M2**], and even went a bit beyond them. A corrected list would show one cyclic configuration (or polygon) for $n = 13$ and 14, three for $n = 15$, and two for $n = 16$. This coincides with the recent list of cyclic configurations given by Betten and Betten [**B3**], to which we shall return soon. Brunel also noted that translates of $\{0, 1, 4, 6\}$ yield a configuration for all $n \geq 13$; this anticipated a result of Gropp [**G2**].

Merlin mentions in [**M2**] that configurations (n_4) have not been investigated systematically, although some isolated ones were discovered by F. Klein [**K1**], W. Burnside [**B12**], and others. He constructs a combinatorial configuration (13_4) and proves its uniqueness and minimality. He also constructs a configuration (14_4) and proves it is unique. Merlin states that there are exactly *three* distinct configurations (15_4) which, however, are not presented. In fact, he is mistaken. As shown by Betten and Betten [**B3**], there are *four* different configurations (15_4), three of which are cyclic and coincide with the three doubly selfinscribed and selfcircumscribed polygons of Brunel (who did not comment on the possibility of noncyclic configurations (15_4)). In the same context, Merlin makes two additional errors: (i) He claims that his three configurations (15_4) can be distinguished by the number of vertex-disjoint triangles present in them, which he claims to be 5, 1 and 0, respectively. In fact, all four configurations (15_4) have five such triangles, the maximal possible number. (ii) He states that his configurations (13_4), (14_4), and (15_4) have orderly configuration tables (which is correct and proved by Steinitz in [**S6**] for all configurations (n_k)), and states that it follows that there is no Hamiltonian circuit for any of them — which is wrong. Steinitz's orderliness result has no such implications, and Brunel's explicit constructions in [**B11**], of which Merlin is unaware, provide counterexamples to Merlin's claim.

By a construction analogous to the one devised by Martinetti for configurations (n_3), Merlin shows that for every $n \geq 30$ there are combinatorial configurations (n_4). In fact, it is easy to show that there are such configurations for all $n \geq 13$. As to the number $N(n)$ of distinct combinatorial configurations (n_4), the only known

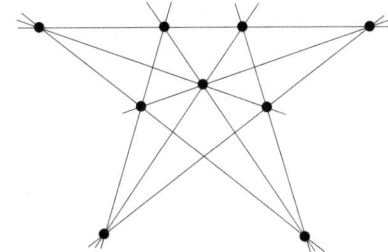

FIGURE 11. A set of nine points and nine lines that is realizable in the real Euclidean plane but is not realizable in the rational Euclidean or projective plane.

values are those given by Betten and Betten [**B3**], namely $N(13) = N(14) = 1$, $N(15) = 4$, $N(16) = 19$, $N(17) = 1972$, and $N(18) = 971171$. For $n \geq 15$ these numbers have not been independently verified.

As far as is known, none of these combinatorial configurations is geometrically realizable. Merlin [**M2**] shows that the configurations (13_4), (14_4), and (15_4) are not geometrically realizable. But he also notes that geometric configurations (n_4) do exist for infinitely many values of n. His construction uses "stacks" of configurations of type $[3, 3]$ and vertical lines through their vertices to construct configurations of type $[4, 3]$, and then stacks of duals of the projections of these into the plane to construct configurations of type $[4, 4]$. While this yields geometric configurations (n_4) for infinitely many values of n, there are infinitely many n that are not covered.

The latest result in this direction, established by a variety of mostly *ad hoc* constructions, is given in [**G12**], the last of several papers on this topic:

THEOREM 4.1. *Connected geometric configurations (n_4) exist for all $n \geq 21$, except possibly for the following ten values of n: $22, 23, 26, 29, 31, 32, 34, 37, 38, 43$.*

CONJECTURE 4.2. [**G12**] *No combinatorial configuration (n_4) with $n \leq 20$ is realizable.*

It is also highly probable that there exist no geometric configurations (n_4) for the ten values of n listed in Theorem 4.1.

CONJECTURE 4.3. *No combinatorial (n_4) configuration on which a cyclic group acts transitively has geometric realizations.*

It should be noted that in contrast to Conjecture 3.2, there exist (n_4) configurations that can be realized in the Euclidean plane but not in the rational plane. The simplest construction I know starts with the collection of nine points and nine lines shown in Figure 11. It is well known (see [**G9**, Section 5.5], [**B9**, pp. 5, 40]) that this "partial configuration" cannot be represented in the rational projective (or Euclidean) plane, but it is easily seen that it can be imbedded into an (n_4) configuration with $n \leq 44$.

5. Highly symmetric configurations: Astral configurations

A considerable part of the recent interest in geometric configurations is due to the results obtained in the study of configurations that have a large degree of geometric symmetry.

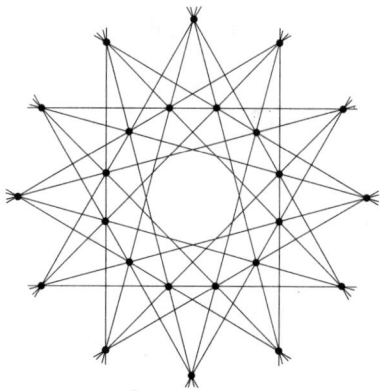

FIGURE 12. The smallest astral configuration of type [4,4]: the
configuration (24_4), which has symbol $12\#(5, 4, 1, 4)$.

Configurations (n_3) and (n_4) must have at least two orbits of points and two
orbits of lines; by the terminology introduced in Section 2, configurations with
these minimal numbers of orbits are called astral. Some instances of astral (n_3)
configurations have appeared earlier, but without any systematic considerations;
we shall return to the historical examples and references is Section 6.

In general, there are three classes of h-astral configurations (n_k):

(i) Configurations h-astral in the extended Euclidean plane E^{2+} but not contained in the Euclidean plane E^2 itself.

(ii) Configurations h-astral in E^2, with a cyclic group of symmetries; we call
them h-cyclic, or polycyclic if the value of h is not relevant.

(iii) Configurations h-astral in E^2, with a dihedral group of symmetries; we
call them h-dihedral, or polydihedral if the value of h is not relevant.

It should be noted that for even k, all h-astral configurations (n_k) are h-dihedral.
For $k = 3$ there exist astral (that is 2-astral) configurations of all three classes. We
shall discuss additional results later.

The remaining part of this section is devoted to h-astral configurations (n_4),
and in particular the astral ones among them. These configurations are the best
explored kind, hence we discuss them first. Since they are all polydihedral, we shall
simplify the terminology and call them h-astral resp. astral for short.

The first drawing of an astral (n_4) configuration (or of any (n_4) configuration!)
seems to have appeared in [**G14**]; it is the (24_4) shown in Figure 12. This was the
beginning of a development I would like to sketch now in a few words.

I had been wondering what other (n_4) configurations exist for which there are
just two transitivity classes of points (and two classes of lines) under isometric
symmetries of the configuration — that is, astral configurations. Manual drafting
of diagrams soon reached the limits of reliability. At about that time, Stan Wagon
gave a series of talks at the University of Washington, extolling the virtues and
ease of use of the *Mathematica* software as both a computational tool and a graphic
one. As it turned out, it was really easy to write a program that tested for which
spans of diagonals of an m-gon do appropriate intersection points of the diagonals,
together with the vertices of the m-gon, form a configuration (n_4), where $n = 2m$.

TABLE 2. Symbols of the 27 basic sporadic astral configurations (n_4). The full meaning of the symbols $m\#(s_1, t_1, s_2, t_2)$ will be explained below. Here it suffices to note that $n = 2m$, and that s_1 and t_2 are the spans of the diagonals of the m-gon.

30#(7,6,1,4)	30#(7,4,1,6)	30#(8,6,2,6)	30#(11,10,1,6)
30#(11,6,1,10)	30#(12,10,6,10)	30#(12,11,2,7)	30#(12,7,2,11)
30#(13,12,1,8)	30#(13,8,1,12)	30#(13,12,7,10)	30#(13,10,7,12)
30#(14,13,6,11)	30#(14,11,6,13)	30#(14,12,4,12)	42#(13,12,1,6)
42#(13,6,1,12)	42#(18,17,6,11)	42(18,11,6,17)	42#(19,18,5,12)
42#(19,12,5,18)	60#(22,21,2,9)	60#(22,9,2,21)	60#(25,24,5,12)
60#(25,12,5,24)	60#(27,26,3,14)	60#(27,14,3,26)	

(The *span* of a diagonal in a convex polygon is the number of edges spanned by the diagonal.) The first of the experimental results showed that this happens if and only if $n = 2m = 12k$ for some integer $k \geq 2$. Figure 12 shows the case $k = 2$, while Figure 13 shows six astral configurations with $k = 3$ (that is, (36_4) configurations).

The more detailed study of which pairs of spans of diagonals of a polygon with $m = 6k$ sides yield 2-astral configurations led to a puzzling situation, which is illustrated in Figure 14. (In order not to prejudice between the two spans s and t, both points (s, t) and (t, s) are indicated; the hollow circles denote pairs for which there are two different configurations for the same spans.) The valid pairs seemed to exhibit no visible regularity. But as the experimentation continued to larger values of k, illustrated in Figure 15, a clear indication emerged: Most of the pairs satisfy simple linear relations, with only a few "sporadic" pairs besides. This led to the trigonometric verification that the experimentally obtained pairs do in fact correspond to astral configurations, and to the listing of such configurations known to exist given in Theorem 5.1 (see [G11], where a slightly different notation was used; two errors in the data of [G11] have been corrected in [B1]).

THEOREM 5.1. *There exist two infinite families of astral configurations (n_4), that can be described by the symbols $(6k)\#(3k - j, 2k, j, 3k - 2j)$ where $k \geq 2$ and $1 \leq j \leq k - 1$, and $(6k)\#(2k, j, 3k - 2j, 3k - j)$ where $k \geq 2$, $1 \leq j \leq 2k - 1$ but $j \neq k$ and, if k is even, $j \neq 3k/2$. There also exist 27 basic sporadic configurations listed in Table 2, and their multiples.*

The conjecture that the astral configurations specified in Theorem 5.1 are the only ones was expressed in [G11], and established by Leah Berman in her University of Washington doctoral thesis in 2002. This can be formulated as Theorem 5.2, which was published in [B1].

THEOREM 5.2 (Berman). *The following is a complete list of astral configurations (n_4):*

(i) *Configurations having vertices on two concentric regular m-gons, where $m = n/2$, that are listed in Theorem 5.1;*

(ii) *Configurations having vertices not on two concentric regular m-gons that result by taking two concentric copies of one of the configurations in part (i), rotated with respect to each other through any angle other than a multiple of π/m.*

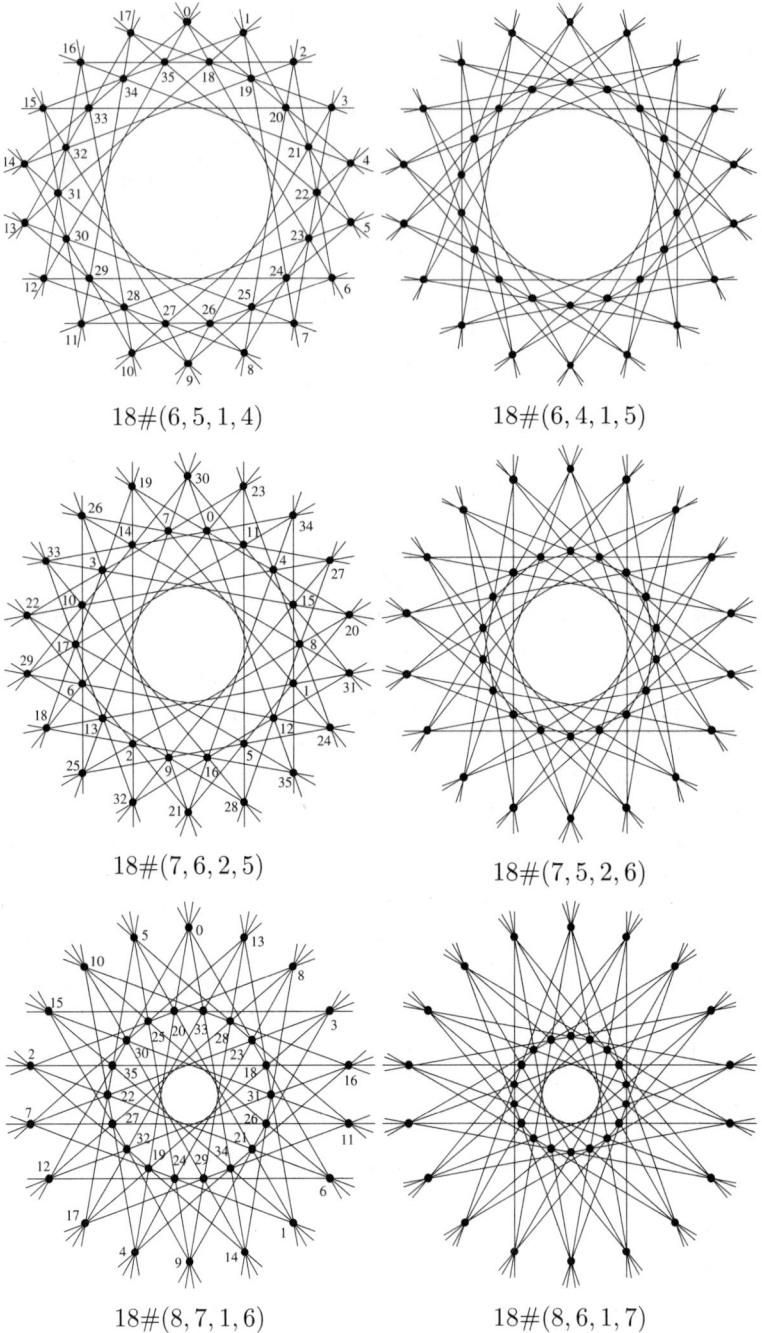

$18\#(6,5,1,4)$ $18\#(6,4,1,5)$

$18\#(7,6,2,5)$ $18\#(7,5,2,6)$

$18\#(8,7,1,6)$ $18\#(8,6,1,7)$

FIGURE 13. The six different astral configurations (36_4), with their symbols. The two configurations in each row are polars of each other. The labeling of the three configurations at left indicates an isomorphism between them.

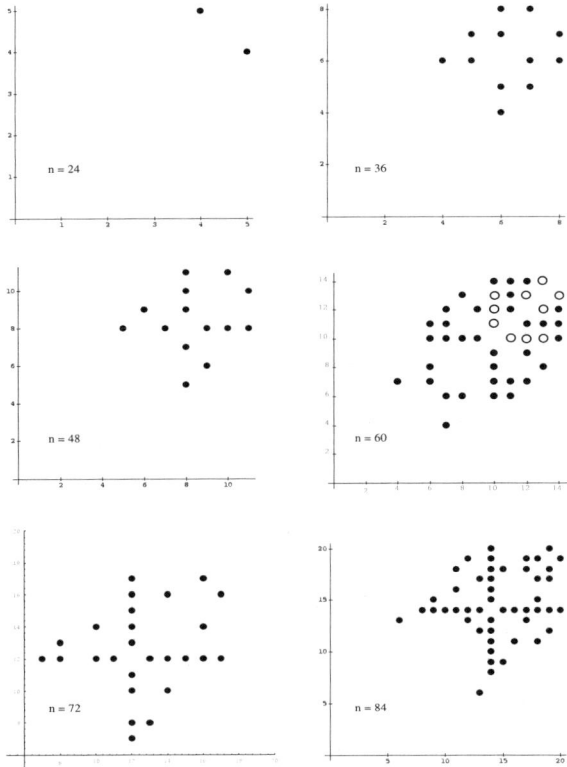

FIGURE 14. Plots of the parameters of existing astral configurations (n_4).

Berman's proof relies in a non-trivial way on the characterization of multiple intersection points of diagonals in regular polygons. Although that question appears to be a simple problem it is, in fact, surprisingly deep and delicate. It was first solved (except for minor glitches) by Bol [**B10**] but his result was apparently forgotten until the work of [**R3**], where the solution was presented in a different form. The solution was presented in a much more accessible and more general form by [**P3**] in 1998. (In the intervening years, various special cases have been rediscovered by several authors. Many references can be found in [**P3**].)

The enumeration of astral configurations (n_4) given in Theorems 5.1 and 5.2 dealt with *geometrically* distinct configurations. Very little is known about the *isomorphism* classes of these configurations. One rather unexpected result is that the six astral configurations (36_4) shown in Figure 13 are of only two non-isomorphic types. The configurations in the left column are isomorphic to each other by the labeling shown, while those on the right must therefore be isomorphic to each other, since they are polars of the ones at left. It is not hard to show that the configurations of each polar pair are not isomorphic. We venture:

CONJECTURE 5.3. *For each* $n = 12k$, *with* $k \geq 3$, *not all connected astral configurations* (n_4) *are isomorphic.*

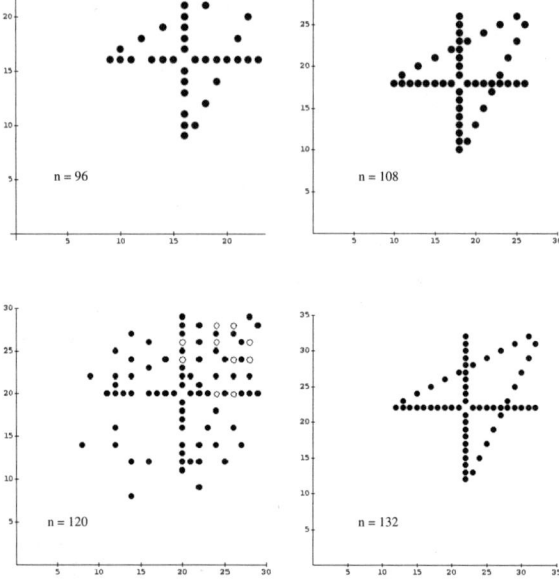

FIGURE 15. More plots of the parameters of existing astral configurations (n_4).

This is probably easy to settle by a suitable generic example. The really hard
problem in this context is finding how many non-isomorphic astral configurations
(n_4) are there for each n.

I also offer

CONJECTURE 5.4. *No astral configuration (n_4) is isomorphic to a geometric
configuration in the rational projective plane.*

The known results concerning h-astral configurations with $h \geq 3$ are far less
complete. The first example — the unique 3-astral configuration (21_4) described
in [**G14**] and shown in Figure 16 — is the smallest of all known *geometric* configurations (n_4), h-astral or not. The underlying *combinatorial* configuration has
been known for a long time (first described, it seems, by Klein [**K1**] in 1879), and
its realizations in the complex plane or in finite planes were investigated by many
people (see Burnside [**B12**], Coxeter [**C6**], and the references in [**C6**]).

The study of h-astral configurations was greatly advanced by the work of Boben
and Pisanski [**B8**], who called them *h-cyclic* or *polycyclic*. Their ideas were among
those utilized in [**G12**] in the proof of Theorem 4.1. The following account differs
in details from the approach in [**B8**]; it was developed during the presentation of
the material in courses I gave at the University of Washington, and in preparation
for a workshop in Bled (Slovenia) which unfortunately I could not attend.

The starting point for almost all h-astral configurations is a notation for *intersection points* of diagonals of span s of a regular m-gon, illustrated in Figure 17 and
explained in its caption. (One exception is described below, and shown in Figure
23.) A notation for a connected h-astral configuration $\mathcal{C}$ can be devised using a
characteristic path as follows:

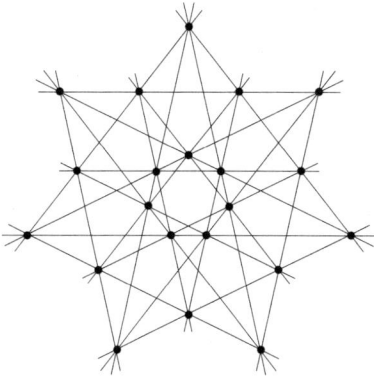

FIGURE 16. The 3-astral configuration (21_4), which is the smallest known geometric configuration (n_4). It can be described by the symbol $7\#(3,2,1,3,2,1)$, the derivation and meaning of which is described in the text.

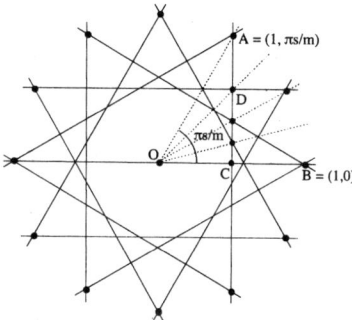

FIGURE 17. The notation for intersection points of diagonals of span s of a regular n-gon. Assuming the polygon to have unit circumcircle, the label $[[p,q]]$ for a point means that it is situatd on the diagonal of span s, and that it is the qth among the intersections of that diagonal with other diagonals of the same span, counting from the midpoint of the diagonal. For example, the point D has label $[[4,3]]$.

Choose an orientation, say counterclockwise, and orient all diagonals of the regular m-gons (that is, lines of the configuration) accordingly. Choose an arbitrary point P_0 of $\mathcal{C}$ as starting point of the path, and through it an arbitrary line L_1 of $\mathcal{C}$ for which P_0 is the earlier of the two points in the same orbit. On L_1 we take the first point (according to the order on L_1) of the other orbit of points incident with L_1, and denote it P_1. Through P_1 pass lines of an orbit other than that of L_1; we choose as L_2 that one for which P_1 is the earlier point. Continuing in this way, after a finite number h of steps we return to the first orbit of points. The chosen points determine the characteristic path. Hence the polygons are m-gons, where $m = n/h$ and the configuration is h-astral. Then the configuration $\mathcal{C}$ can be described by a symbol of the form $m\#(s_1, t_1, s_2, t_2, s_3, t_3, \ldots, s_h, t_h)$, where s_j is the span of the diagonal L_j of the jth polygon, and $[[s_j, t_j]]$ is the symbol of the point P_j on L_j. It should be noted that, in general, the point P_h will not coincide with P_0. Note that necessarily each entry is a positive integer, different from the adjacent ones (understood cyclically), and is less than $m/2$.

The characteristic path $P_0 P_1 P_2 \ldots P_h$ describes and determines the configuration. It is indicated by a heavy gray line in the examples in Figures 18 and 19. Different starting points and choices of lines lead to equivalent symbols; they result by cyclic permutations of each other that advance through an even number of places, or else by a reversal of these. On the other hand, it is easy to show that advancing an odd number of places (and reversals of these) lead to symbols of the configuration polar to the starting one. Details can be found (with slightly different notation and terminology) in [**B8**].

The entries in a symbol $m\#(s_1, t_1, s_2, t_2, s_3, t_3, \ldots, s_h, t_h)$ cannot be chosen arbitrarily. It is not hard to verify that the symbol must satisfy the conditions

$$(*) \qquad\qquad s_1 + t_1 + s_2 + t_2 + s_3 + t_3 + \cdots + s_h + t_h \qquad \text{is even}$$

and

$$(**) \qquad \frac{\cos(\pi s_1/m)}{\cos(\pi t_1/m)} \frac{\cos(\pi s_2/m)}{\cos(\pi t_2/m)} \cdots \frac{\cos(\pi s_k/m)}{\cos(\pi t_k/m)} = 1$$

If these necessary conditions are satisfied, a *combinatorial* configuration (n_4) can always be constructed, simply by following the indications of the path $P_0 P_1 P_2 \ldots P_h$. However, to assure connectedness, besides conditions $(*)$ and $(**)$ we need:

$(***)$ If $m, s_1, t_1, s_2, t_2, s_3, t_3, \ldots, s_h, t_h$ have a common factor $f > 1$, then the numbers m/f, s_1/f, t_1/f, s_2/f, t_2/f, s_3/f, t_3/f, $\ldots$, s_h/f, t_h/f fail to satisfy at least one of the conditions $(*)$ and $(**)$.

The conditions $(*)$, $(**)$, and $(***)$ are sufficient to assure the existence of a *representation* of the combinatorial configuration implied by the symbol

$$m\#(s_1, t_1, s_2, t_2, s_3, t_3, \ldots, s_h, t_h).$$

However, according to [**B8**], for *realization* (rather than representation) an additional condition is required:

$(****)$ No proper subsequence $(s_i, t_i, s_{i+1}, t_{i+1}, \ldots, s_k)$ yields a symbol

$$m\#(s_i, t_i, s_{i+1}, t_{i+1}, \ldots, s_k, t^*)$$

that satisfies conditions $(*)$, $(**)$, and $(***)$, where $1 \le t^* < m/2$. Dually, no proper subsequence $(t_i, s_{i+1}, t_{i+1}, \ldots, s_k, t_k)$ can yield a symbol

$$m\#(s^*, t_i, s_{i+1}, t_{i+1}, \ldots, s_k, t_k)$$

that satisfies conditions $(*)$, $(**)$, and $(***)$, where $1 \le s^* < m/2$.

In Figure 20 we show an example of a symbol that fails condition $(****)$ and hence leads to a representation that is not a realization.

The h-astral configurations (n_4) can be classified into three types:

- *Trivial* are configurations for which the symbol is such that the (unordered) *set* of s_j's coincides with the set of t_j's. Hence conditions $(*)$

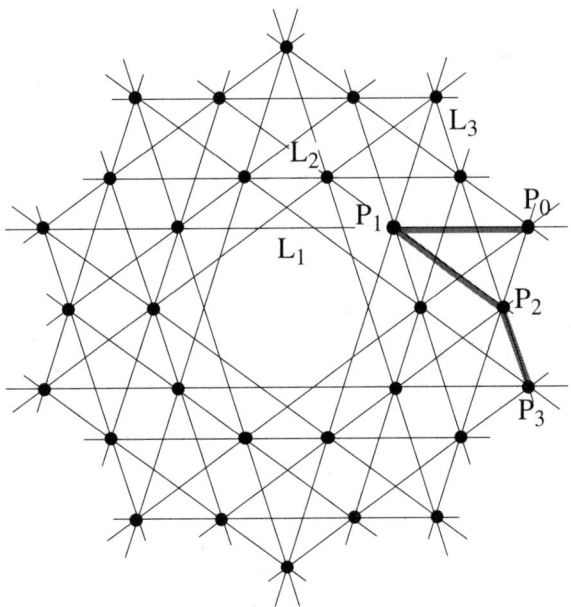

FIGURE 18. A 3-astral configuration (30_4), with symbol $10\#(4,3,1,3,1,2)$. This symbol is composed of the label $[[4,3]]$ of point P_1, label $[[1,3]]$ of point P_2, and label $[[1,2]]$ of point P_3. The spans of the lines L_j are 4,1 and 1, respectively. Note that the parameter t_j in the label $[s_j, t_j]$ of the point P_j can be interpreted as the span of L_j in the polygon formed by the points in the orbit of P_j.

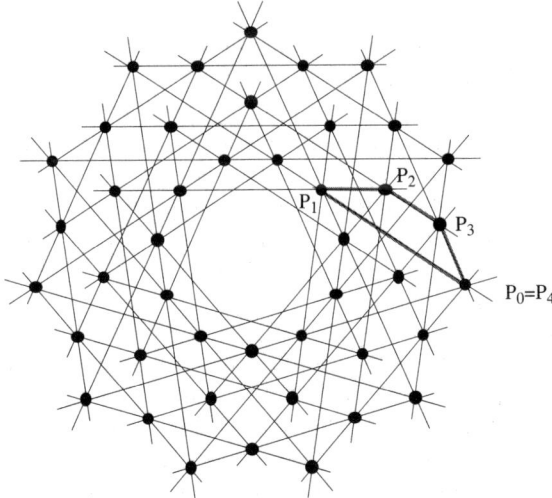

FIGURE 19. A 4-astral configuration (44_4), with symbol $11\#(4,1,3,4,2,3,1,2)$. Note that P_1 has label $[[4,1]]$, P_2 has label $[[3,4]]$, P_3 has label $[[2,3]]$, and $P_4 = P_0$ has label $[[1,2]]$.

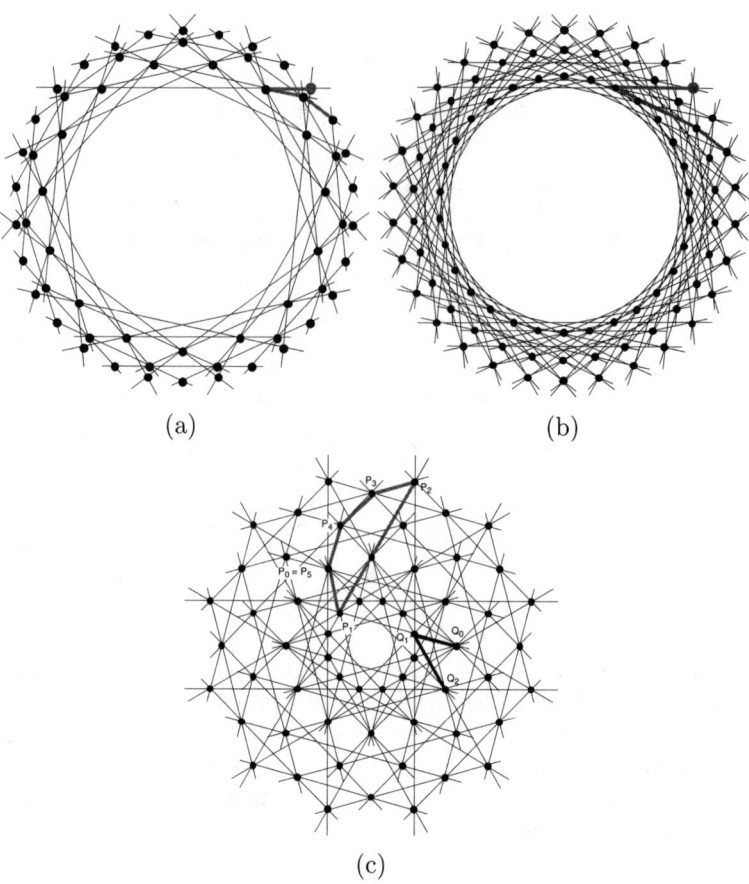

FIGURE 20. In each of the examples, a characteristic path starts at
an enlarged dot and is shown in gray. (a) The necessity of condition
(∗) is illustrated by the 3-astral symbol 15#(4, 3, 2, 3, 1, 2), which
does not lead to an (n_4) configuration. (b) The entries of the
symbol 30#(8, 4, 2, 6, 4, 6) have a common factor 2, but this symbol
leads to a connected configuration since condition (∗∗∗) is satisfied.
(c) The symbol 12#(5, 4, 1, 5, 2, 1, 3, 2, 4, 3) satisfies condition (∗),
(∗∗) and (∗ ∗ ∗) but not (∗ ∗ ∗∗). This is not a realization of
the symbol, since lines of one orbit are incident with six points
each, and points of one orbit are incident with six lines each. The
explanation is that if we take only the first four parts of the list
in parentheses, and change the terminal 5 to a 4, we obtain a
valid symbol 12#(5, 4, 1, 4) illustrated by the characteristic path
$Q_0Q_1Q_2$. This is the symbol of the configuration shown in Figure
12, which can easily be seen as formed by the two innermost orbits
of points and lines.

and $(**)$ are satisfied trivially, with no need to calculate the trigonometric functions. An example is the configuration (21_4) shown in Figure 16, with symbol $7\#(3, 2, 1, 3, 2, 1)$.

- *Systematic* are configurations having symbols which are derived by explicit formulas for the parameters, for infinitely many values of n. Thus the satisfaction of condition $(**)$ is verifiable by manipulations of the trigonometric functions, again without the necessity to actually calculate the values of these functions.

- *Sporadic* are configurations that are neither trivial nor systematic.

As visible from Theorems 5.1 and 5.2, astral configurations are either systematic or sporadic. However, for each $h \geq 3$, there are trivial h-astral configurations as well.

In unpublished work, L. Berman has identified four families of systematic 3-astral configurations $m\#(a, b, c, d, e, f)$:

(5.1) $m = 2q$, $\{a, c, e\} = \{q - p, q - 2r, p\}$, $\{b, d, f\} = \{q - 2p, q - r, r\}$

(5.2) $m = 3q$, $\{a, c, e\} = \{q + p, q - p, p\}$, $\{b, d, f\} = \{q, q, 3p\}$

(5.3) $m = 6q$, $\{a, c, e\} = \{3q - p, r, p\}$, $\{b, d, f\} = \{3q - 2p, 2q, r\}$

(5.4) $m = 10q$, $\{a, c, e\} = \{5q - p, 2p, p\}$, $\{b, d, f\} = \{|5q - 4p|, 4q, 2q\}$.

Examples of configurations in each of these families are shown in Figure 21.

In Figure 22 are shown several h-astral configuration (n_4) that exhibit various unusual phenomena. Some of these are easy to explain, but some are rather puzzling.

There are many open problems concerning h-astral configurations (n_4). A few of the more striking are:

- A complete characterization of representations of 3-astral configurations. This should take into account also configurations such as the one in Figure 23, found by L. Berman (private communication). This configuration falls outside the scope of description by symbols we used here. It is not known whether there are other configurations of this kind — it is hard to imagine that there is only a single one!

- What is the explanation for configurations such as the one in Figure 22(b), that are describable by our symbols but incompatible with condition $(****)$?

- Is it possible for an h'-astral configuration to be isomorphic to an h''-astral configuration with $h' \neq h''$?

- Are there finitely many basic sporadic h-astral configurations for each h?

- What are the possibilities of (n_4) configurations of the kind shown in Figure 24, which is $(2, 3)$-astral (in the notation introduced in Section 2)? What other unequal pairs of numbers of orbits can occur?

6. h-astral configurations (n_3)

The study of these configurations is much less advanced, and promises to be more challenging than the investigation of the (n_4) configurations. There are two sources of the variety possible for h-astral (n_3) configurations. On the one hand, in many cases there is at least one parameter that can assume a continuum of different real values. On the other hand, if $h \geq 3$, a line of the configuration can contain

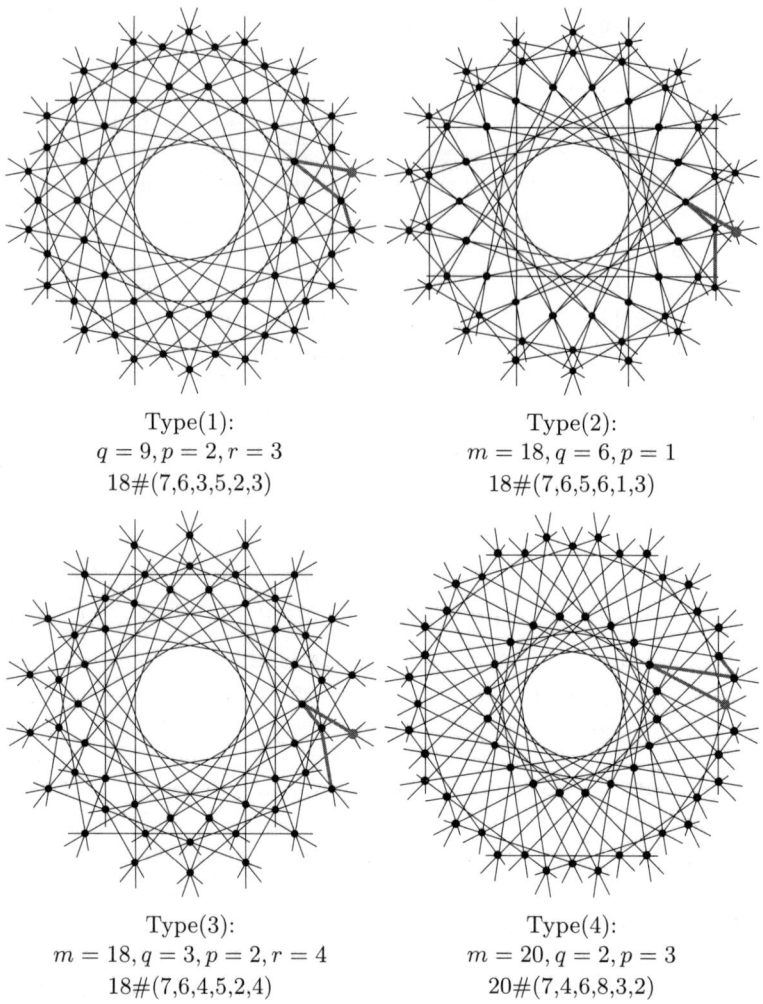

Type(1):
$q = 9, p = 2, r = 3$
18#(7,6,3,5,2,3)

Type(2):
$m = 18, q = 6, p = 1$
18#(7,6,5,6,1,3)

Type(3):
$m = 18, q = 3, p = 2, r = 4$
18#(7,6,4,5,2,4)

Type(4):
$m = 20, q = 2, p = 3$
20#(7,4,6,8,3,2)

FIGURE 21. Examples of systematic 3-astral configurations of the
four known systematic types. One characteristic path is indicated
in each, with the starting point shown by an enlarged gray dot.

points from either two or three different orbits. Even more than in the case of (n_4)
configurations, the case $h = 2$ of astral configurations is radically different from
h-astral with $h \geq 3$.

As mentioned above, the h-astral configurations (n_3) come in three varieties:

- *projectively h-astral*, that is configurations h-astral in the *extended Euclidean* (that is, *projective*) plane E^{2+}, but not in the Euclidean plane E^2
  itself.
- *h-cyclic* (chiral), that is, configurations in E^2 with a cyclic symmetry
  group.
- *h-dihedral*, that is configurations E^2 with a dihedral symmetry group.

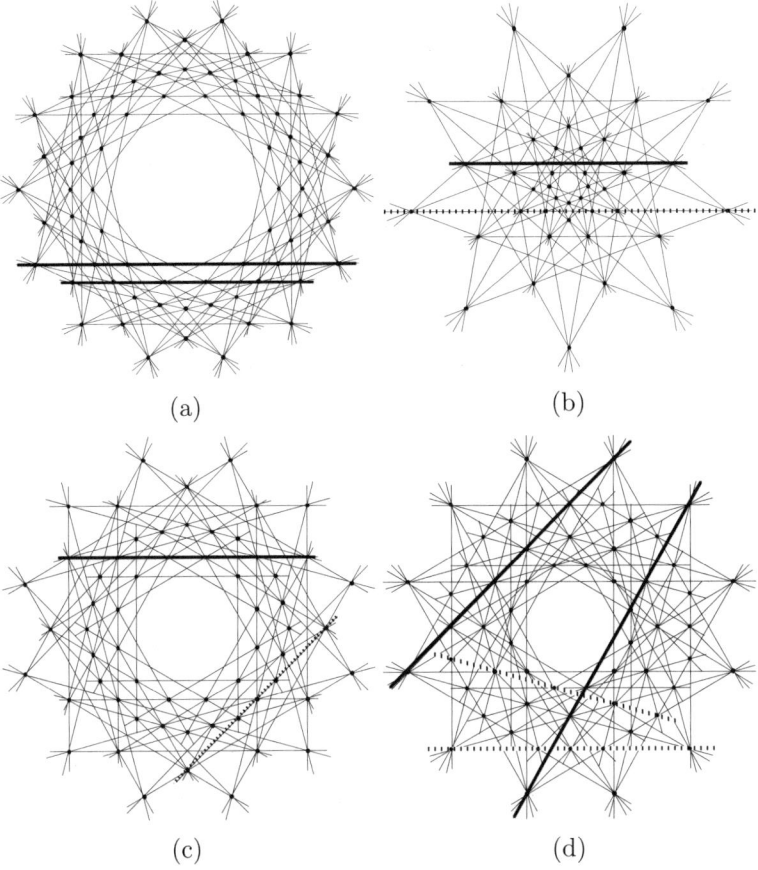

(a)   (b)

(c)   (d)

FIGURE 22. (a) A representation of the 5-astral configuration
with symbol 14#(3,1,4,3,1,3,2,4,3,2); it has two orbits of lines
incident with six points each.   One line of each such orbit
is emphasized.   One orbit arises from the trivial configuration
14#(3,1,4,3,1,4). The other orbit arises from the sporadic configuration 14#(3,2,3,1,4,5). The points in the two outermost orbits are incident with six lines each. (b) A representation of the
5-astral configuration 9#(3,1,4,3,1,3,2,4,3,2). The heavily drawn
line represents the orbit of lines through six points that arises from
the trivial 3-cyclic configuration 9#(3,1,4,3,1,4). The dashed line
arises from the 3-astral configuration with symbol 9#(3,1,3,4,3,2).
However, this symbol does not arise from the original by the process in condition (∗∗∗∗). This seems to indicate that condition
(∗∗∗∗) needs to be strengthened. (c) the representation of the 5-
astral configuration 12#(3,1,4,3,1,3,2,4,3,2) has one orbit of lines
incident with six points each, and one orbit of points incident with
six lines. It also has one orbit of lines incident with five points,
one of which is incident with five lines. (d) A representation of the
5-astral configuration 12#(4,3,1,4,3,4,3,1,4,3). It has two orbits of
lines incident with six points each, and two orbits of lines incident
with five points each; analogously for points.

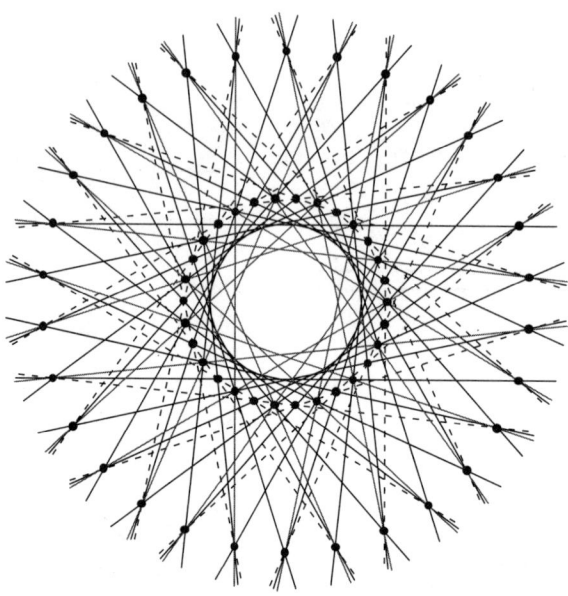

FIGURE 23. A 3-astral configuration (60_4).

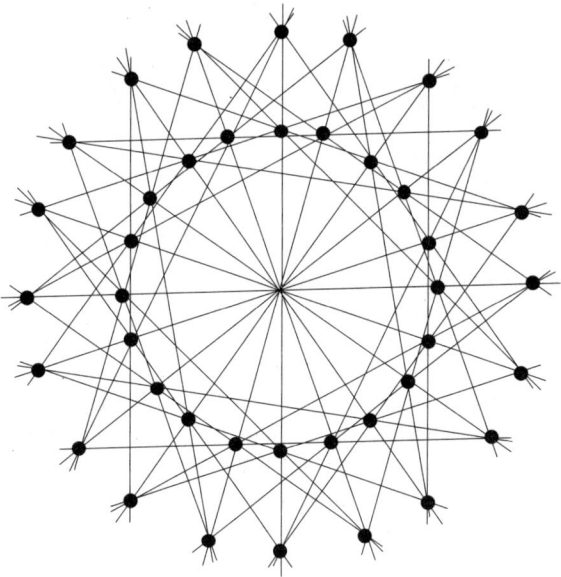

FIGURE 24. A 3-dihedral configuration (40_4), in which the points form two orbits and the lines form three orbits. It is constructed by starting with the vertices of a regular decagon, taking diagonals of span 3 and diameters. This is a configuration of type [3,4]. Taking a concentric copy, suitably rotated, one can add additional lines which yield a configuration (40_4). Here the angle of rotation is $\arccos((19\sqrt{5}-1)/44) = 19.464602895^0$.

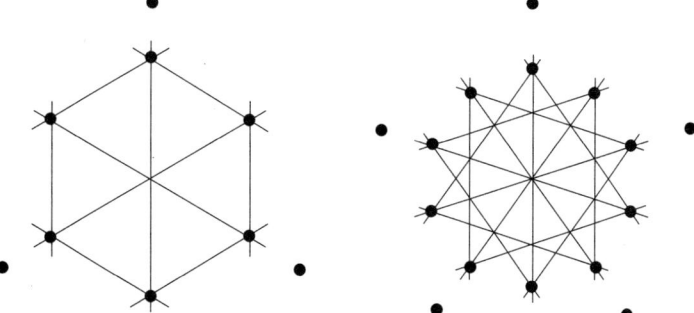

FIGURE 25. Examples of (n_3) configurations that are astral in the extended Euclidean plane E^{2+} but not contained in the Euclidean plane E^2 itself. The unattached dots indicate points at infinity.

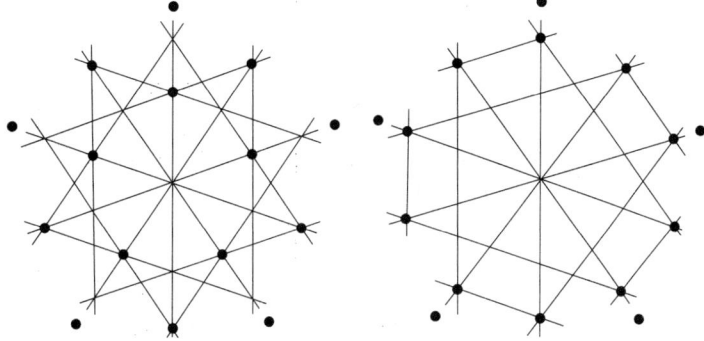

FIGURE 26. Two configuration (n_3) that are 3-astral in the projective plane E^{2+} but not in the Euclidean plane E^2. Note that the first one has only two orbits of lines, each containing points of all three orbits. The other has only two orbits of points, but three orbits of lines.

6.1. Examples of projectively astral configurations are shown in Figures 25 and 26. The first configuration in Figure 25 is a realization of the Pappus configuration; see Coxeter [**C5**]. A 3-astral realization of the Desargues configuration (10_3) in E^{2+} is given by Coxeter [**C4**, Figure 6]. It is clear that similar examples could be found for $h \geq 4$. At least for small h, the complete characterization of projectively astral configurations may be relatively simple but seems not to have been worked out.

6.2. *h-cyclic* configurations (n_3) are much more interesting. A few examples of cyclic configurations are shown in Figures 27 and 28. The first published graphical representations of h-cyclic configurations seem to be those of Visconti [**V1**]; a 4-cyclic (20_3) reproduced as Figure 9, and a 3-cyclic (30_3). It is ironic that Schönflies [**S2**] shows drawings of configurations (12_3) and (9_3) which could have been presented as 3-cyclic — but were drawn with non-regular polygons. This is the same Schönflies who a few years later determined the 230 discrete symmetry groups of the Euclidean 3-space! Zacharias [**Z1**] shows several examples of what are

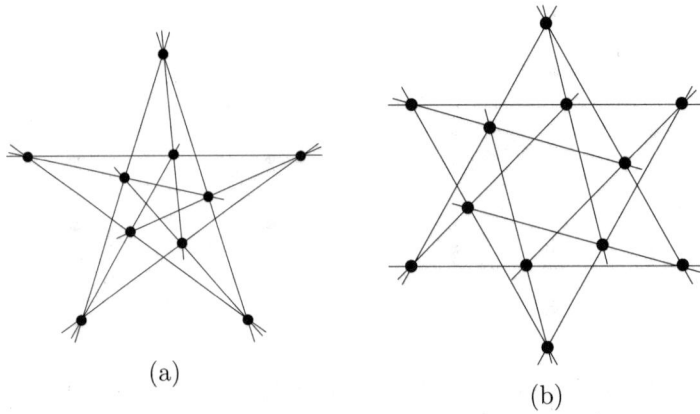

(a)

(b)

FIGURE 27. The two smallest 2-cyclic configurations (n_3): (a) (10_3) with symbol $5\#(2,2;1)$ and (b) (12_3) with symbol $6\#(2,2;1)$. The symbols are explained later in the text.

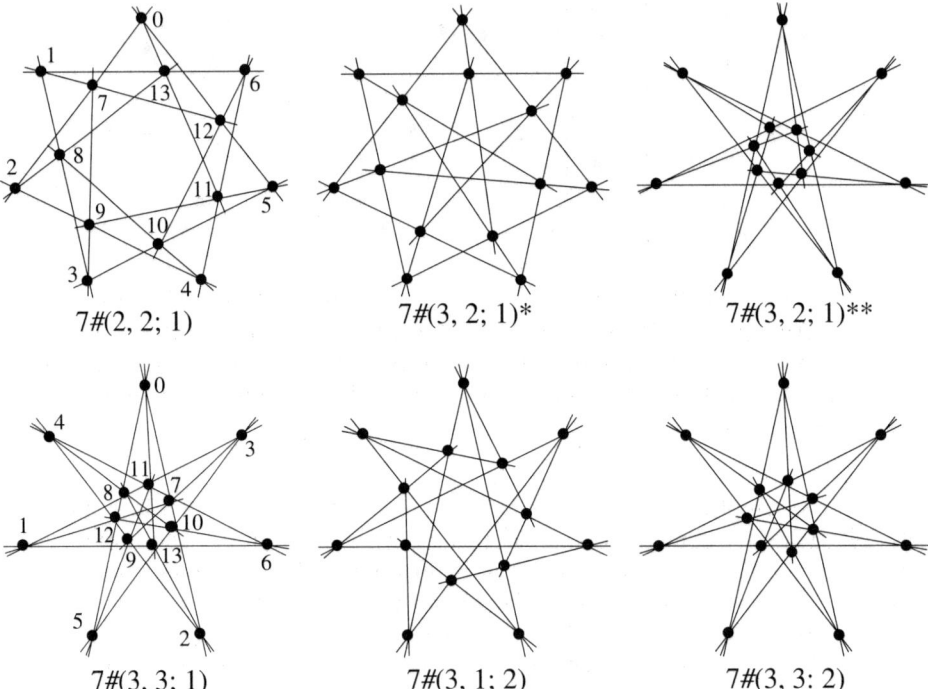

$7\#(2, 2; 1)$ $7\#(3, 2; 1)^*$ $7\#(3, 2; 1)^{**}$

$7\#(3, 3; 1)$ $7\#(3, 1; 2)$ $7\#(3, 3: 2)$

FIGURE 28. The six geometrically distinct 2-cyclic configurations (14_3), and their symbols. The four symbols without asterisks correspond to selfpolar configurations. The other two are polars of each other; since they have the same parameters, they are isomorphic as well. The two configurations in the left column are isomorphic; the labels indicate an isomorphism. Thus there are four non-isomorphic 2-cyclic configurations (14_3).

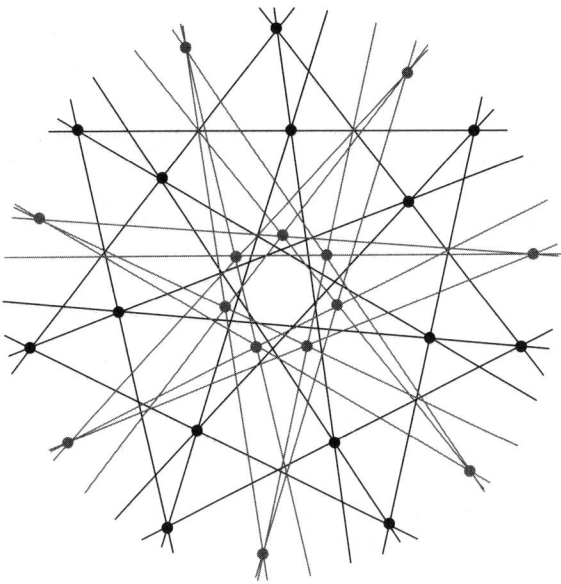

FIGURE 29. A pair of 2-astral configurations (14_3) polar to each other. Black: $7\#(3,2;1)^* = 7\#(3,2;1)$; gray: $7\#(3,2;1)^{**} = 7\#(2,3;4)$.

2- and 3-cyclic configurations (n_3); he comments on their star-shaped appearance, and mentions that other such configurations may be formed — but does not discuss the symmetries as such, nor any general methods of construction.

It is obvious that each line of a 2-cyclic configuration must contain two points from the same orbit, and one from another orbit. The notation used in these examples will be explained later; it is a special case of the notation

$$m\#(b_1, b_2, \ldots, b_h; b_0; \lambda_1, \lambda_2, \ldots, \lambda_{h-2})$$

— or in a shorter symbol $m\#(b_1, b_2, \ldots, b_h; b_0)$ — for h-cyclic configurations of this kind. Here $n = hm$ and we have $h - 2$ real parameters λ_j besides $h + 1$ discrete ones b_j. Together these parameters lead to a quadratic equation for an additional parameter. This equation can have 2, 1 or 0 real solutions — in the last case there are no corresponding configurations.

In particular, 2-cyclic configurations have no free parameters λ_j. The quadratic equation and the remaining real parameter are determined by the integer parameters $m\#(b_1, b_2; b_0)$. In case there are two distinct values for the real parameter, if necessary we append to the symbol $m\#(b_1, b_2; b_0)$ one or two asterisks * or **, to indicate whether the larger or the smaller value is used.

As pointed out in [**B8**], the dual of a configuration $m\#(b_1, b_2, \ldots, b_h; b_0)$ is the configuration $m\#(b_h, b_{h-1}, \ldots, b_1; b_1 + b_2 + \cdots + b_h - b_0)$. This is illustrated by the example in Figure 29. It should be noted that for 2-cyclic configurations, in the case the determining equation has two distinct real roots, the configurations $m\#(b, c; d)^*$ and $m\#(b, c; d)^{**}$ are duals of each other. In fact, as illustrated in Figure 29, they are related by a polarity.

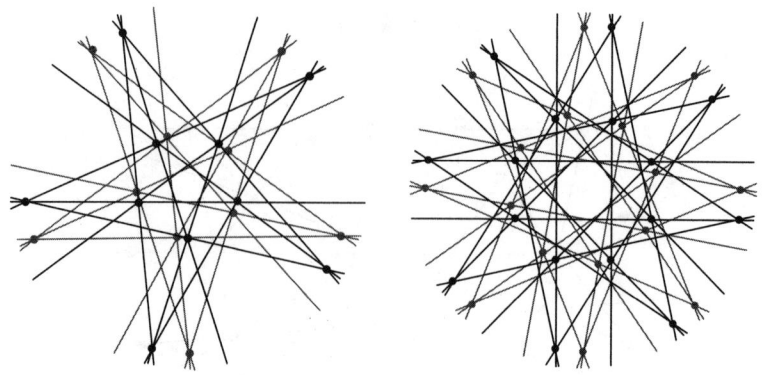

FIGURE 30. Negatively selfpolar configurations 5#(2,2;1) and 8#(3,3;2).

Extensive experimental evidence shows that symbols $m\#(b, c; d)$ of representable 2-cyclic configurations must satisfy:

$$0 < c \le b < m/2,$$
$$c - b + 1 < 2d \le b + c,$$
$$0 \ne d \ne c,$$
$$2\cos(b\pi/m)\cos(c\pi/m) \le 1 + \cos((b + c - 2d)\pi/m).$$

CONJECTURE 6.1. *The above conditions are necessary and sufficient for the existence of a 2-cyclic configuration $m\#(b, c; d)$.*

The 2-cyclic *selfpolar* configurations (n_3) are quite interesting; they come in three varieties. The first kind corresponds to symbols of the type $m\#(b, b; d)$, with $1 \le d < b$; they are *negatively selfpolar*. By this is meant that the polar of the configuration is a mirror image of the configuration itself. This is illustrated by the examples in Figure 30. The only other selfpolar configurations are the ones with symbol $m\#(b, c; d)$, where $b \ne c$ and $d = (b + c)/2$. They are *positively selfpolar*, that is, the polar is congruent to the original without reflection. But depending on the parity of b (and c), the polar either *coincides* with the original, or *differs* from it by a non-trivial rotation. These two possibilities are illustrated in Figures 31 and 32.

CONJECTURE 6.2. *There are no selfpolar 2-cyclic configurations (n_3) besides those described above.*

We turn now to explain the notation for h-cyclic configurations in which each line is incident with points of two orbits only; the remaining case will be described below. Our explanation is illustrated in Figure 33, using a 3-cyclic configuration (27_3) as an example.

As mentioned before, the symbol for an h-cyclic configuration (n_3), where $n = hm$, is of the form $m\#(b_1, b_2, \ldots, b_h; b_0; \lambda_1, \lambda_2, \ldots, \lambda_{h-2})$; the parameters are again determined by a *characteristic path*. The entries $b_1, \ldots, b_h$ are the spans of the diagonals in the different regular m-gons that are determined by the path; the diagonals are all oriented in the same way — clockwise or counterclockwise, and the real numbers $\lambda_1, \lambda_2, \ldots, \lambda_{h-2}$ denote the ratios in which each diagonal determined

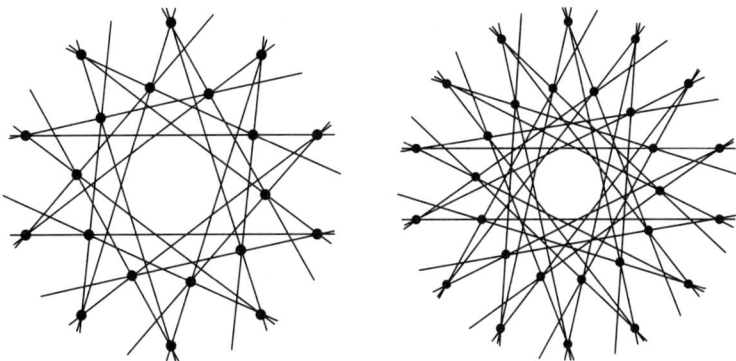

FIGURE 31. Positively selfpolar configurations $10\#(4,2;3)$ and $13\#(6,4;5)$, with b and c even.

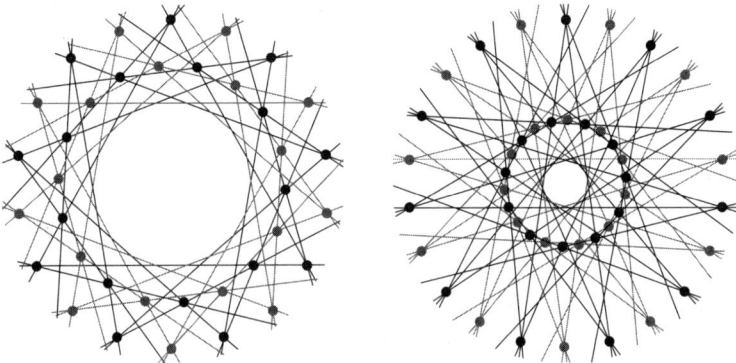

FIGURE 32. Positively selfpolar configurations $9\#(3,1;2)$ and $11\#(5,1;3)$, with b and c odd.

by a segment of the path is divided by the endpoint of the segment. The path returns to the starting polygon, but not necessarily to the starting point of the path. The parameter b_0 indicates the vertex of the first polygon at which the characteristic path ends. These data lead to a quadratic equation for the ratio λ_{h-1} on the next-to-last segment; the ratio applicable to the last segment is then completely determined. Thus there are either two, or one, or no real geometric configurations corresponding to a given symbol. There are also possibilities of unintended incidences similar to the ones we encountered earlier, hence we are in general talking about *representations* of the symbols, rather than *realizations*. In case the parameters $\lambda_1, \lambda_2, \ldots, \lambda_{h-2}$ in a symbol $m\#(b_1, b_2, \ldots, b_h; b_0; \lambda_1, \lambda_2, \ldots, \lambda_{h-2})$ are not relevant or not known, we abbreviate the symbol to $m\#(b_1, b_2, \ldots, b_h; b_0)$.

The example in Figure 33 presents a 3-cyclic configuration with symbol $9\#(2,3,2;6;.5)$. The points of the three orbits are denoted by B_j, C_j, D_j. The determination of the symbol is highlighted by the three-step characteristic path. Note that the ratio λ_1 can be chosen freely, and in the illustration it was taken as $\lambda_1 = 0.5$. Once the first $h - 2 = 1$ ratios λ_j are chosen, the last ratio λ_{h-1} (determining the position of the point of last orbit on the penultimate diagonal)

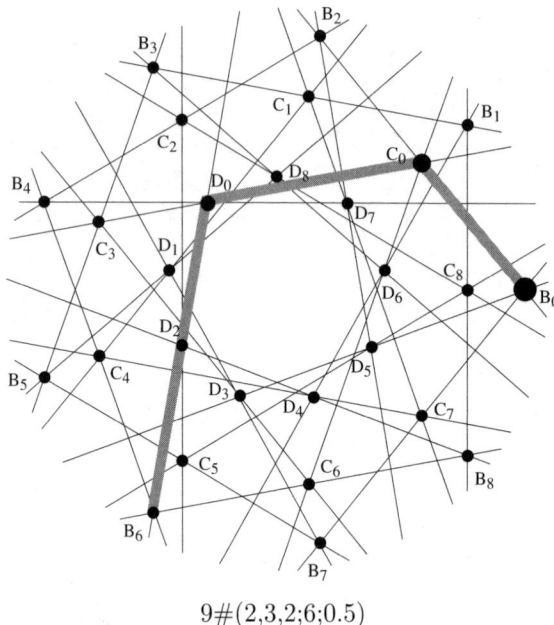

$$9\#(2,3,2;6;0.5)$$

FIGURE 33. The characteristic path $B_0C_0D_0B_6$ is represented in
the symbol by the first four parts; the first three integers are the
spans of the respective diagonals, the fourth indicates the point of
the first orbit at which the path ends. The last parameter indicates the ratio in which the segment B_0B_1 is divided by the point
C_0. This parameter can be freely chosen (with some limits); the
analogous parameters for the other two segments are then determined.

is determined by a quadratic equation. (For details see [**B8**].) In the illustration
$\lambda_{h-1} = \lambda_2$ is about 2/3. Naturally, the symbol is not unique since it depends,
besides the λ_j's for $h \geq 3$, on the orbit of the starting point, and on the orientation
chosen. The influence of the parameter λ_{h-2} is illustrated in Figure 34.

Using symbols like $u, v, w, \ldots$ for elements of the different orbits of points,
we can say that the h-cyclic configurations considered so far have lines of type
$\{u, u, v\}, \{v, v, w\}, \ldots$. But other possibilities exist in which the incidences of lines
with orbits of the points are different. For example, in case $h = 3$, it is possible
to have three orbits of lines, all three of the type $\{u, v, w\}$, or else, one of the
type $\{u, v, w\}$ and the other two of types $\{u, v, v\}$ and $\{u, w, w\}$. Three examples
of the former variety are shown in Figure 35, while examples of the second kind
are illustrated in Figure 36; the rightmost diagram in Figure 3 shows the $(9_3)_3$
configuration, which is of this kind. A notation for the configurations in Figure
35 is explained in the caption; no notation has been proposed for the kind of
configurations shown in Figure 36. No additional details about either of the kinds
are available as of this writing.

h-cyclic configurations in which lines of one or more orbits are incident with
points of three orbits have not been studied at all.

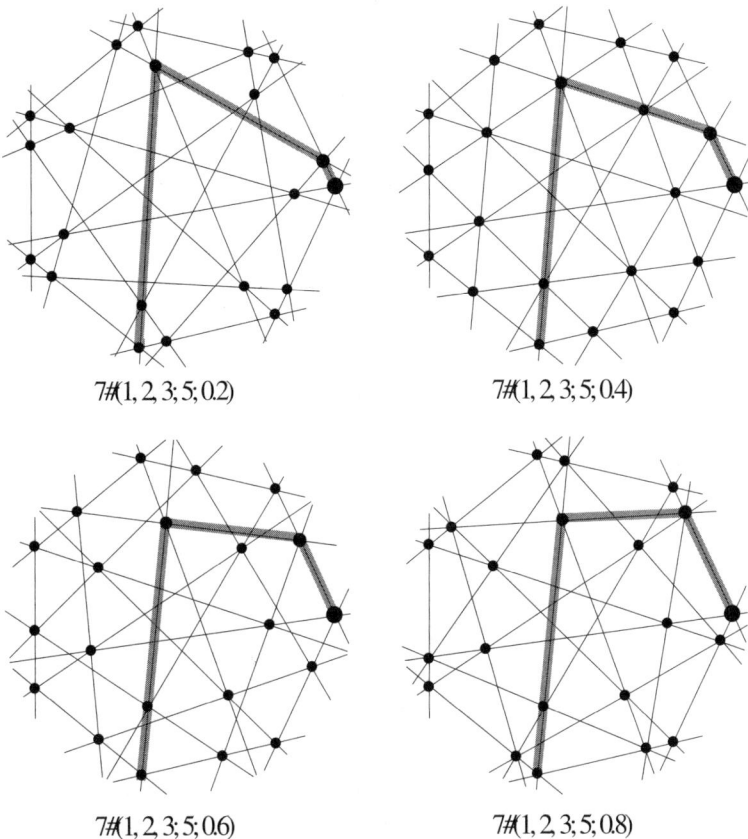

$7\#(1, 2, 3; 5; 0.2)$ $7\#(1, 2, 3; 5; 0.4)$

$7\#(1, 2, 3; 5; 0.6)$ $7\#(1, 2, 3; 5; 0.8)$

FIGURE 34. The dependence of the configuration $7\#(1,2,3;5;\lambda_1)$ on the parameter λ_1. For a value of λ_1 somewhat smaller than 0.8 the representation would not be a realization. The last diagram can be interpreted as nearly illustrating this case.

6.3. *h-dihedral configurations* are also largely uninvestigated. One difference in comparison with h-cyclic configurations is that even for $h = 2$, isomorphic 2-hedral configurations admit a real-valued parameter. This is illustrated in Figure 37 by a family of 2-hedral configurations (12_3); the configurations in each row are polars of each other. Examples of other kinds of 3-dihedral configurations are shown in Figures 38 and 39.

There are many open problems related to h-astral configurations. In most cases the information available at this time does not lead to any specific conjectures. Here are a few examples.

- Are non-Hamiltonian h-astral configurations (n_3)?
- Determine the number of different isomorphism classes of h-cyclic configurations (n_3), at least for $h = 2$. What relations exist among the parameters of isomorphic h-cyclic configurations?
- Are there combinatorial configurations (n_3) that can be represented by h-cyclic configurations for different values of h?
- Characterize self-dual polydihedral configurations.

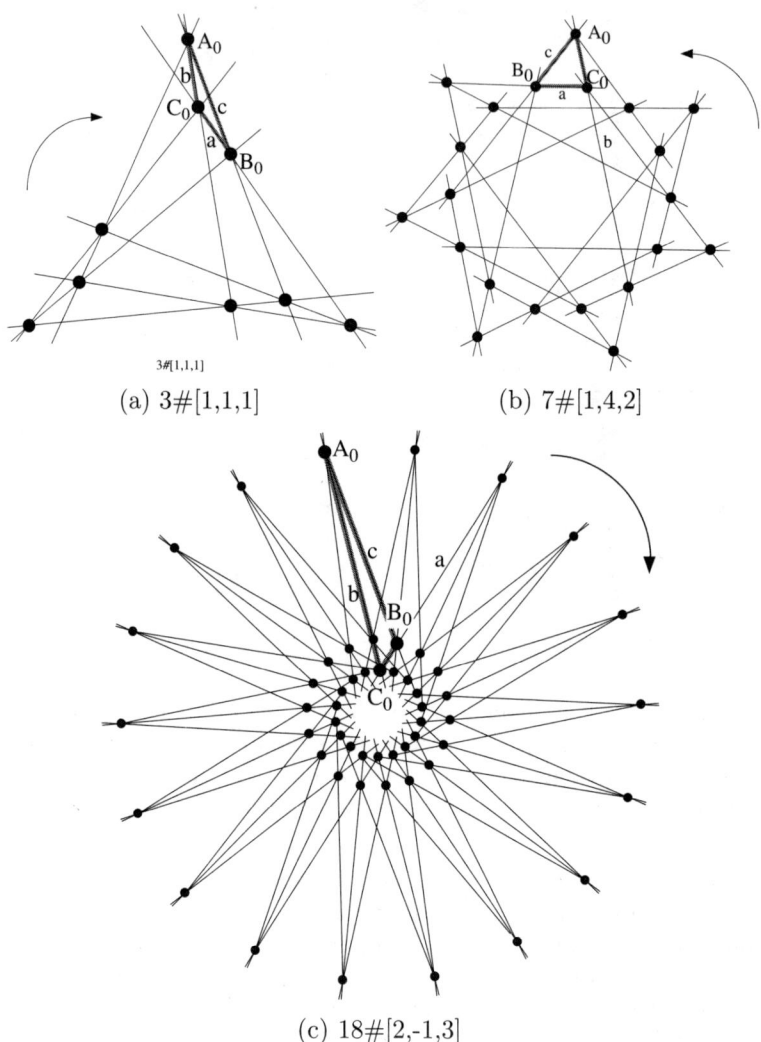

(a) 3#[1,1,1]   (b) 7#[1,4,2]

(c) 18#[2,-1,3]

FIGURE 35. Three examples of 3-cyclic configurations, in which
each line is incident with one point in each of the three point orbits.
The symbol near each can be used as a description. After choosing
a triangle (gray lines) with vertices in the three orbits, the numbers
in brackets indicate which of the points A_i, B_j, C_k is one of the lines
a, b, c, respectively. (a) is a realization of the Pappus configuration
$(9_3)_1$. The configuration in (b) has as its Levi graph the only cyclic
3-valent graph with 42 vertices; it appears in the Foster census
[F1].

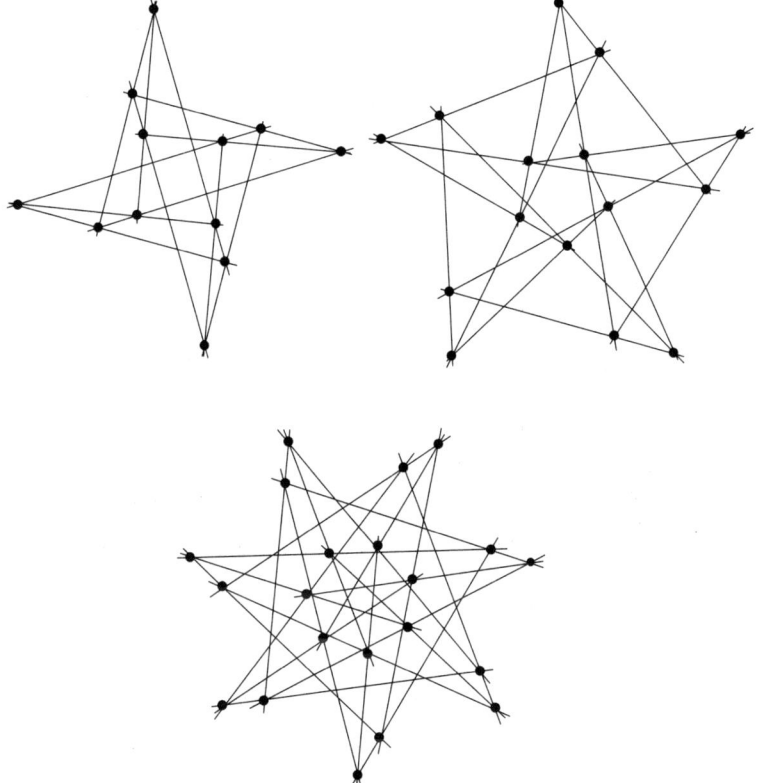

FIGURE 36. In these 3-cyclic configurations there is one orbit of lines each of which is incident with one point of each of the three point orbits. There are two orbits of lines each of which is incident with points from only two point orbits each.

- Does every astral configuration (n_4) contain an astral subconfiguration (n_3) ?
- Which 2-cyclic (or 2-dihedral) configurations (n_3) are only representations (not realizations) of the underlying combinatorial configurations?

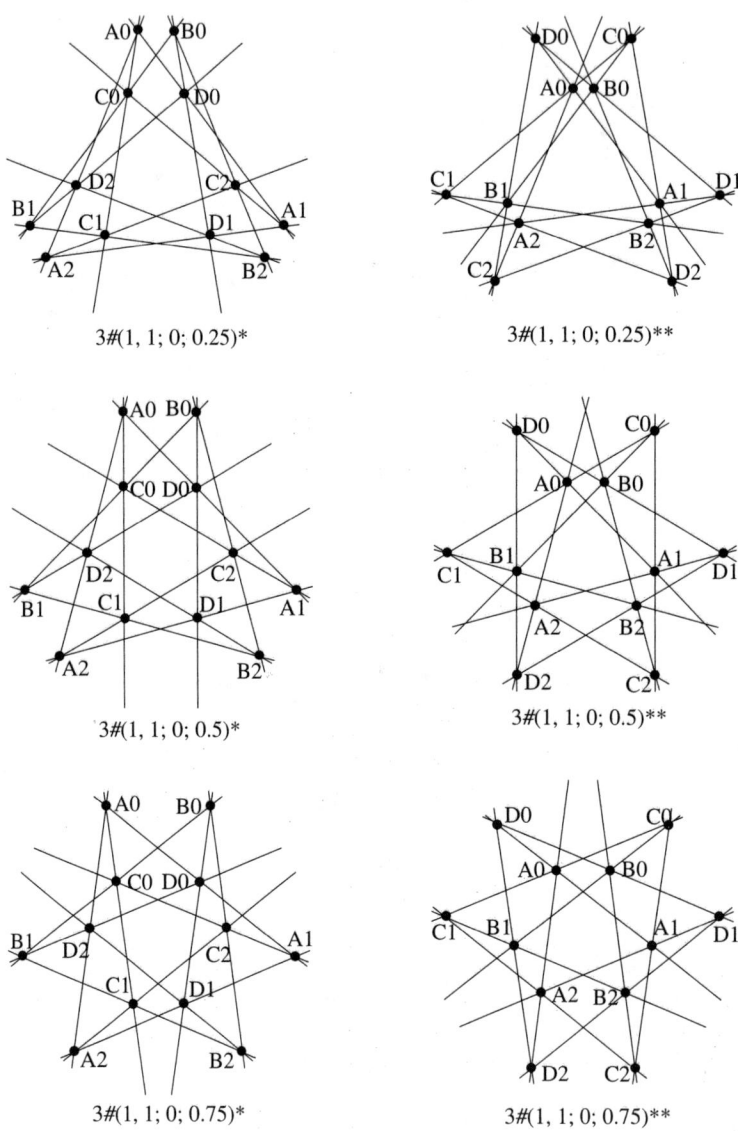

FIGURE 37. Pairs of mutually polar 2-dihedral configurations (12_3), all isomorphic, illustrating the dependence of the appearance of the realization on the value of a convenient parameter.

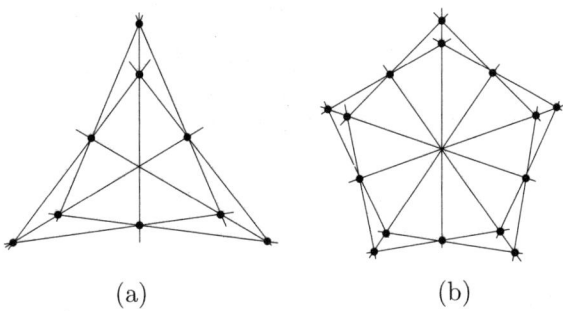

(a)   (b)

FIGURE 38. Two 3-dihedral configurations, (a) is another realization of the Pappus configuration (9_3); this is the smallest 3-dihedral
configuration. (b) A (15_3) configuration.

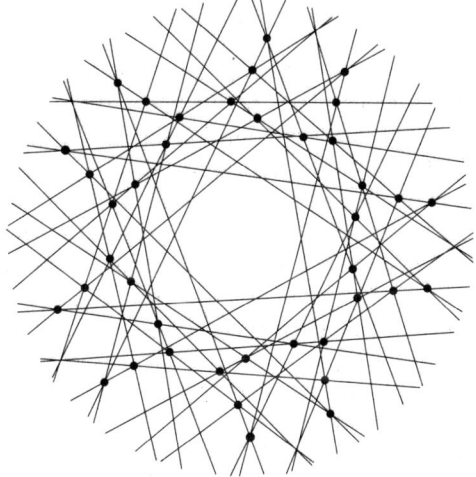

FIGURE 39. A 5-dihedral configuration (40_3). (Courtesy of L. Berman)

FIGURE 40. A 3-astral configuration $(48_5, 60_4)$. Adding the 12 points of infinity in the direction of the lines of that configuration yields a (60_5) configuration 3-astral in the extended Euclidean plane E^2.

7. Other astral configurations

It is easy to show that the only configurations (n_4) that are astral in the projective (extended Euclidean) plane E^{2+} are also astral in the Euclidean plane. The same goes for astral configurations (n_6) — but vacuously: there exist no astral configurations (n_6) (see Berman [**B2**]). However, the situation is different for astral configurations (n_5). Starting with two concentric copies of the astral configuration (24_4) shown in Figure 12, rotated by $\pi/24$ with respect to each other, and adding 12 diametral lines, an astral configuration $(48_5, 60_4)$ is obtained. By adding 12 points at infinity, a (60_5) configuration astral in the extended Euclidean plane E^{2+} results, see Figure 40. A similar construction works with all astral configurations (n_4). The astral configuration (50_5) indicated in Figure 41 is obtained by a variant of that construction, possible in this case. These diagrams are the first published depictions of any (n_5) configuration, as well as of any configuration of type $[5, 4]$. It may be conjectured that the (50_5) configuration in Figure 41 is the smallest (n_5) configuration astral in the extended Euclidean plane E^{2+}. However, we have:

CONJECTURE 7.1. *There exist no (n_5) configurations astral in the Euclidean plane E^2.*

Nothing seems to be known about the existence or nonexistence of configurations (n_5) or (n_6) that are h-astral in the Euclidean plane E^2 for $h \geq 4$.

Concerning geometric configurations of type $[q, k]$ that are not equinumerous (that is, for which $q \neq k$) very little information is available. Data on combinatorial configurations of this kind can be found in [**G4**] and [**G5**].

It is well known that for each integer $r \geq 3$ there exists a combinatorial configuration $((4r)_3, (3r)_4)$, and these are the only possible ones of type $[3, 4]$. There is no geometric configuration $(12_3, 9_4)$, but for every $r \geq 4$ there exist geometric configurations $((4r)_3, (3r)_4)$ astral in E^2. Typical examples are shown in Figure 42.

FIGURE 41. Without the lines through the center, this is the (40_4) configuration with symbol 10#(4,3,1,2,3,4,2,1). Including these lines gives a 3-astral configuration $(40_5, 50_4)$. With the addition of ten ideal points we obtain a configuration (50_5) that is 3-astral in the extended Euclidean plane E^{2+}.

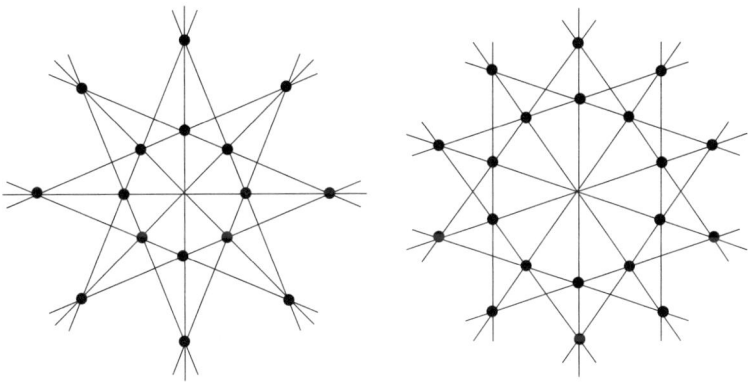

FIGURE 42. The smallest geometric configuration of type [3,4], astral in E^2.

Configurations $(12_4, 16_3)$, have been investigated (at least in special cases) for close to two centuries. Details and references are available in [**G3**], where it is stated that there are 574 combinatorial configurations $(12_4, 16_3)$. It is not known how many of them are geometric. Polars of the configurations in Figure 42 are configurations of type [4, 3] astral in the extended Euclidean plane E^{2+}. It is not known whether any are astral in E^2.

For each integer $r \geq 4$ there exists a combinatorial configuration $((5r)_3, (3r)_5)$; no other configurations of type [3, 5] are possible. There is no geometric configuration $(20_3, 12_5)$, but for every $r \geq 4$ there exist geometric configurations $((5r)_3, (3r)_5)$

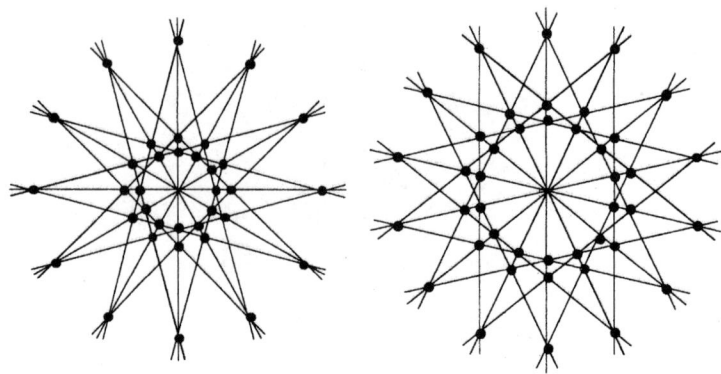

FIGURE 43. Configurations $(36_3, 18_6)$ and $(42_3, 21_6)$ astral in the Euclidean plane.

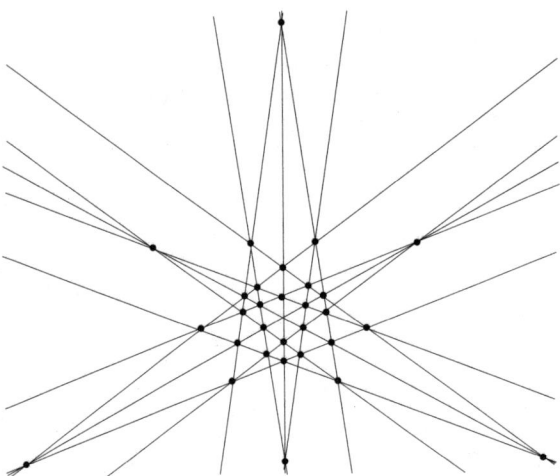

FIGURE 44. A configuration $(30_3, 15_6)$. Deleting the six outermost points and adding the point at the center yields a configuration $(25_3, 15_5)$.

astral in E^2; their construction is analogous to the one of configurations of type [3, 4] in Figure 42.

It can be shown that combinatorial configurations of type [3, 6] exist if and only if the parameters are $((2r)_3, r_6)$ with $r \geq 13$. For all $j \geq 6$ there exist geometric configurations $((6j)_3, (3j)_6)$ astral in the Euclidean plane; typical examples are shown in Figure 43. It can be shown that astral configurations $((2r)_3, r_6)$ do not exist if r is not a multiple of 2 or a multiple of 3; however, no example is known in which r is not a multiple of 3. It is also not known whether there exist any geometric configurations $(28_3, 14_6)$, but a non-astral configuration $(30_3, 15_6)$ exists and is indicated in Figure 44.

This essentially exhausts the current information available about astral and other geometric configurations. It clearly leaves many kinds of open problems, ranging from very elementary to highly technical. It is my hope that the present exposition will contribute to a renewed interest in geometric configurations, and in their generalizations (such as configurations of pseudolines).

References

[B1]   L. W. Berman, *A characterization of astral (n₄) configurations*. Discrete Comput. Geom. 26(2001), no. 4, 603–612.

[B2]   L. W. Berman, *Even astral configurations*. Electron. J. Combin. 11(2004), Research Paper 37, 23 pp. (electronic).

[B3]   A. Betten and D. Betten, *Tactical decompositions and some configurations v_4*. J.Geom. 66(1999), 27–41.

[B4]   A. Betten, G. Brinkmann and T. Pisanski, *Counting symmetric configurations v_3*. Discrete Appl. Math. 99(2000), 331–338.

[B5]   M. Boben, *Uporaba teorije grafov pri kombinatori/nih in geometri/nih konfiguracijah.* (In Slovenian) [The use of graph theory in the study of combinatorial and geometric configurations]. Ph.D. thesis, University of Ljubljana, November 2003.

[B6]   M. Boben, *Irreducible (v_3) configurations and graphs.* (To appear)

[B7]   M. Boben, B. Grünbaum, T. Pisanski and A. Žitnik, *Small triangle-free configurations of points and lines*. Dept. of Math, Univ. of Ljubljana, Preprint series 42(2004), #938.

[B8]   M. Boben and T. Pisanski, *Polycyclic configurations*. Europ. J. Combin. 24(2003), 431–457.

[B9]   J. Bokowski and B. Sturmfels, *Computational Synthetic Geometry*. Lecture notes in Mathematics #1355, Springer, New York 1989.

[B10] G. Bol, *Beantwoording van prijsvraag no. 17*. Nieuw Archief voor Wiskunde 18(1931), 14–66 (1933).

[B11] G. Brunel, *Polygones á autoinscription multiple*. Proc. Verb. Séances Soc. Sci. phys. nat. Bordeau, 1897/98, pp. 43–46.

[B12] W. Burnside, *On the Hessian configuration and its connection with the group of 360 plane collineations*. Proc. London Math. Soc. (2) 4(1907), 54–71.

[C1]   H. G. Carstens, T. Dinski and E. Steffen, *Reduction of symmetric configurations n_3*. Discrete Appl. Math. 99(2000), 401–411.

[C2]   H. S. M. Coxeter, *Configurations and maps*. Reports of a Math. Colloq. (2) 8(1949), 18–38.

[C3]   H. S. M. Coxeter, *Self-dual configurations and regular graphs*. Bull. Amer. Math. Soc. 56(1950), 413–455. (= Twelve Geometric Essays, Southern Illinois Univ. Press, Carbondale, Il 1968 = The Beauty of Geometry, Dover, Mineola, NY 1999, pp. 106–149.

[C4]   H. S. M. Coxeter, *Desargues configurations and their collineation groups*. Math. Proc. Cambridge Philos. Soc., 78(1975), 227–246.

[C5]   H. S. M. Coxeter, *The Pappus configuration and the self-inscribed octagon. I, II, III*. Nederl. Akad. Wetensch. Proc. Ser. A 80 = Indag. Math. 39(1977), pp. 256–269, 270–284, 285–300.

[C6]   H. S. M. Coxeter, *My graph*. Proc. London Math. Soc. (3) 46(1983), 117–136.

[D1]   R. Daublebsky von Sterneck, *Die Configurationen* 11₃. Monatshefte Math. Phys. 5(1894), 325–330 + 1 plate.

[D2]   R. Daublebsky von Sterneck, *Die Configurationen* 12₃. Monatshefte Math. Phys. 6(1895), 223–255 + 2 plates.

[D3]   R. Daublebsky von Sterneck, *Über die zu den Configurationen* 12₃ *zugehörigen Gruppen von Substitutionen*. Monatshefte Math. Phys. 14(1903), 254–260.

[D4]   H. L. Dorwart and B. Grünbaum, *Are these figures oxymora?* Mathematics Magazine 65(1992), 158–169.

[F1]   R. M. Foster, *The Foster Census*. Charles Babbage Research Centre, Winnipeg, MB, 1988, viii+240 pp.

[G1]   H. Gropp, *Il methodo di Martinetti (1887) or Configurations and Steiner systems S(2, 4, 25)*. Ars Combinatoria 24B(1987), 179–188.

[G2]   H. Gropp, *On the existence and nonexistence of configurations n_k*. J. Combinatorics, Information and System Science 15(1990), 34–48.

[G3]   H. Gropp, *The construction of all configurations (12₄, 16₃)*. Fourth Czechoslov. Symp. on Combinatorics, Graphs and Complexity, J. Nešetřil and M. Fiedler, eds. Elsevier 1992.

[G4]   H. Gropp, *Non-symmetric configurations with deficiencies 1 and 2.* In *"Combinatorics '90"*, A. Barlotti et al., eds. Elsevier 1992, pp. 227–239.

[G5]   H. Gropp, *Nonsymmetric configurations with natural index.* Discrete Math. 124(1994), 87–98.

[G6]   H. Gropp, *Configurations and their realizations.* Discrete Math. 174(1997), 137–151.

[G7]   H. Gropp, *On combinatorial papers of König and Steinitz.* Acta Applicandae Math. 52(1998), 271–276.

[G8]   H. Gropp, *Configurations between geometry and combinatorics.* Discrete Appl. Math. 138(2004), 79–88.

[G9]   B. Grünbaum, *Convex Polytopes.* Wiley, New York 1967. 2nd ed. Springer, New York 2003.

[G10]   B. Grünbaum, *Astral (n_k) configurations.* Geombinatorics 3(1993), 32–37.

[G11]   B. Grünbaum, *Astral (n_4) configurations.* Geombinatorics 9(2000), 127–134.

[G12]   B. Grünbaum, *Connected (n_4) configurations exist for almost all n – an update.* Geombinatorics 12(2002), 15–23.

[G13]   B. Grünbaum, *A 3-connected non-Hamiltonian configuration (50_3).* (In preparation.)

[G14]   B. Grünbaum and J. F. Rigby, *The real configuration (21_4).* J. London Math. Soc. (2) 41(1990), 336–346.

[H1]   D. Hilbert and S. Cohn-Vossen, *Anschauliche Geometrie.* Springer, Berlin 1932. English translation: *Geometry and the Imagination*, Chelsea, New York 1952.

[K1]   F. Klein, *Über die Transformationen siebenter Ordnung der elliptischen Funktionen.* Math. Ann. 14(1879), 428–471.

[K2]   S. Kantor, *Über eine Gattung von Configurationen in der Ebene und im Raume.* Wien. Ber. LXXX(1879), 227.

[K3]   S. Kantor, *Über die Configurationen (3,3) mit den Indices 8, 9 und ihren Zusammenhang mit den Curven dritter Ordnung.* Wien. Ber. LXXXIV(1881), 915–932.

[K4]   S. Kantor, *Die Configurationen $(3,3)_{10}$.* Wien. Ber. LXXXIV(1881), 1291–1314 + plate.

[K5]   D. König, *Über Graphen und ihre Anwendung auf Determinantentheorie und Mengenlehre.* Math. Ann. 77(1916), 453–465.

[L1]   R. Laufer, *Die nichkonstruierbare Konfiguration (10_3).* Math. Nachrichten 11(1954), 303–304.

[L2]   F. Levi, *Geometrische Konfigurationen.* Hirzel, Leipzig 1929.

[L3]   F. W. Levi, *Finite Geometrical Systems.* University of Calcutta, Calcutta, 1942.

[L4]   G. Loria and E. K. Lampe, Review of [M1]. Jahrbuch Fortschr. Math. 19(1887), 587–589.

[M1]   V. Martinetti, *Sulle configurazioni piane μ_3.* Annali di matematica pura ed applicata (2) 15(1887), 1–26.

[M2]   E. Merlin, *Sur les configurations planes n_4.* Bull. Cl. Sci. Acad. Roy. Belg. 1913, 647–660.

[M3]   A. F. Möbius, *Kann von zwei dreiseitigen Pyramiden eine jede in Bezug auf die andere um- und eingeschrieben zugleich heisen?* J. reine angew. Math. 3(1828), 273–278 = Gesammelte Werke 1(1885), 439–446.

[P1]   W. Page and H. L. Dorwart, *Numerical patterns and geometrical configurations.* Math. Magazine 57(1984), 82–92.

[P2]   T. Pisanski, Talk at the Ein Gev conference, 2000.

[P3]   B. Poonen and M. Rubinstein, *The number of intersection points made by the diagonals of a regular polygon.* SIAM J. Discrete Math. 11(1998), 135–156.

[R1]   T. Reye, *Geometrie der Lage. I.* 2nd ed. (1876).

[R2]   T. Reye, *Das Problem der Configurationen.* Acta Math. 1(1882), 93–96.

[R3]   J. F. Rigby, *Multiple intersections of diagonals of regular polygons, and related topics.* Geom. Dedicata 9(1980), 207–238.

[R4]   C. Rodenberg, Review of [K4]. Jahrbuch Fortschr. Math. 13(1881), 460.

[S1]   A. Schönflies, *Über einige ebene Configurationen und die zugehörigen Gruppen von Substitutionen.* Nachr. Ges. Wiss. Göttingen 1887, 410–417.

[S2]   A. Schönflies, *Über die regelmässigen Configurationen n_3.* Math. Ann. 31(1888), 43–69.

[S3]   H. Schroeter, *Über lineare Konstruktionen zur Herstellung der Konfigurationen n_3.* Nachr. Ges. Wiss. Göttingen 1888, 193–236.

[S4]   H. Schroeter, *Über die Bildungsweise und geometrische Construction der Configurationen 10_3.* Nachr. Ges. Wiss. Göttingen 1889, 239–253.

[S5]   H. Schubert, Review of [K3]. Jahrbuch Fortschr. Math. 13(1881), 460.

[S6]   E. Steinitz, *Über die Construction der Configurationen n_3.* Ph.D. Thesis, Breslau 1894.

[S7]   E. Steinitz, *Über die Unmöglichkeit, gewisse Configurationen n_3 in einem geschlossenen Zuge zu durchlaufen.* Monatshefte Math. Phys. 8(1897), 293–296.

[S8]   E. Steinitz, *Konfigurationen der projektiven Geometrie.* Encyklopädie der math. Wissenschaften, Vol. 3 (Geometrie), Part IIIAB5a(1910), pp. 481–516 .

[S9]   E. Steinitz, *Über Konfigurationen.* Archiv Math. Phys., 3rd Ser., 16(1910), 289–313.

[S10]   E. Steinitz and E. Merlin, *Configurations.* French translation of [S8], incomplete. Encyclopédie des Sciences Mathématiques, edition française. Tome III, Vol. 2(1913), pp. 144–160.

[S11]   R. Sternfeld, D. Koster, D. Kiel and R. Killgrove, *Self-dual confined configurations with ten points.* Ars Combinat. 67(2003), 37–63.

[S12]   B. Sturmfels and N. White, *Rational realizations of 11_3- and 12_3-configurations.* In *"Symbolic Computations in Geometry"*, by H. Crapo, T. F. Havel, B. Sturmfels, W. Whiteley and N. L. White, IMA Preprint Series #389, Univ. of Minnesota 1988, pp. 92–123.

[S13]   B. Sturmfels and N. White, *All 11_3- and 12_3-configurations are rational.* Aequat. Math. 39(1990), 254–260.

[T1]   E. Togliatti, Review of [Z2]. Zentralblatt Math. 43(1952), p. 358.

[V1]   E. Visconti, *Sulle configurazioni piane atrigone.* Giornale di Matematiche di Battaglini 54(1916), 27–41.

[Z1]   M. Zacharias, *Streifzüge im Reich der Konfigurationen: Eine Reyesche Konfiguration (15_3), Stern- und Kettenkonfigurationen.* Math. Nachrichten 5(1951), 329–345.

[Z2]   M. Zacharias, *Die ebenen Konfigurationen (10_3).* Math. Nachrichten 6(1951), 129–144.

[Z3]   M. Zacharias, *Bemerkung zu meiner Arbeit: Die ebenen Konfigurationen (10_3).* Math. Nachrichten 12(1954), p. 256.

DEPARTMENT OF MATHEMATICS, UNIVERSITY OF WASHINGTON, SEATTLE, WA 98195-0001 USA

E-mail address: grunbaum@math.washington.edu

Meditations on Ceva's Theorem

Jürgen Richter-Gebert

*This paper is dedicated to the unforgettable H. S. M. Coxeter,
who had a striking ability to relate visual thinking to formal notions*

ABSTRACT. This paper deals with the structure of incidence theorems in projective geometry. We will show that many of these incidence theorems can be interpreted as cyclic structures on a suitably chosen orientable manifold. Here the theorems of Ceva and Menelaus play the roles of basic building blocks for building larger theorems with greater complexity. In particular, we show how some other proofs for incidence theorems can be systematically translated into such "Ceva/Menelaus-proofs".

0 Introduction

*Classical projective geometry was a beautiful
field in mathematics. It died, in our opinion,
not because it ran out of theorems to prove,
but because it lacked organizing principles by
which to select theorems that were important.*

R. MacPherson, M. McConnell, 1988 [**17**]

A "proof" is something where many things come together and in the end everything closes up nicely to form a conclusion. In this paper we will see that this very naïve view of mathematical proofs is almost literally correct for certain classes of geometric incidence theorems. We will show that by a certain pasting process many non-trivial incidence theorems can be generated by gluing many copies of Ceva and Menelaus configurations. These small building blocks are arranged at the faces of a manifold and the final conclusion of the theorem corresponds to the fact that the manifold is topologically closed.

The article deals with three different aspects of this construction principle.

2000 *Mathematics Subject Classification.* Primary 51A20, 51A05; Secondary 05B30, 52C35.

- It will be shown how one can systematically generate incidence theorems by pasting together basic building blocks.
- We will see how a given incidence theorem can be analyzed and an underlying manifold structure can be revealed.
- It will be shown how to translate other proving techniques (in particular biquadratic final polynomials, as introduced in [5, 8, 18]) into Ceva/Menelaus proofs. For this translation process the theory of Tutte Groups, as introduced by Dress and Wenzel in a series of papers [9, 10, 11, 22] plays a decisive role.
- Finally, we will see how surgery on the manifolds, which underlie the proofs, can be used, even to generate *spaces of theorems*.

This article is meant as an introduction to these fascinating interrelations between geometry, topology, combinatorics and algebra. We will, whenever possible, present pictures and diagrams to explain the basic concepts, rather than presenting purely formal notions.

To get a feeling of what this article is about, we start by reproducing a whole page of the famous book "Geometry Revisited" by Coxeter and Greitzer which appeared in 1967 [6]. The page contains a proof of Pappus's Theorem by a nice symmetric way of combining six Menelaus configurations that are substructures in Pappus's configuration. The pattern of the proof is (as we will see) as simple, as it is powerful: Try to locate sub-configurations corresponding to hypotheses such that each of them implies a relation of the form: $\frac{[\ldots][\ldots][\ldots]}{[\ldots][\ldots][\ldots]} = 1$. Multiply all the expressions. Perform an "orgy of cancellation" (Coxeter's words) and interpret what is left after the cancellation as the conclusion.

Proofs of this kind may be found all over the literature of projective incidence theory (already dating back to very old papers, such as those of Poncelet and Chasles at the very early days of projective geometry). This is not very surprising, since these kind of proofs are often the only ones that make an elegant use of the underlying algebraic structures. The aim of this paper is to give a systematic approach to this class of proofs.

1. Binomial proofs

My personal starting point for the investigations in this article was a single nice proof for Pappus's Theorem and the desire to generalize the pattern of the proof to more general contexts. The original investigations were related to a systematic study of non-realizability proofs for arrangements of pseudolines (resp. oriented matroids) [5]. In the very last section of this article we will see an application of the more general theory in this context.

1.1. Pappus's Theorem. Let us start with Pappus's Theorem and the above mentioned proof of it. Pappus's Theorem is in a certain sense the smallest purely projective incidence theorem about points and lines. It states that, if one starts with two lines in the real projective plane and with three distinct points $1, 2, 3$ and $4, 5, 6$ on each line, then the three points $(1 \vee 5) \wedge (2 \vee 4), (1 \vee 6) \wedge (3 \vee 4), (2 \vee 6) \wedge (3 \vee 5)$ are automatically collinear as well. There are many different proofs for this fundamental theorem. We will consider a particularly well structured one. The theorem could be restated in the following way: *If in $\mathbb{RP}^2$ the triples of points $(1, 2, 3), (1, 5, 9),$*

another, and if the three lines AB, CD, EF meet DE, FA, BC, respectively, then the three points of intersection L, M, N are collinear.

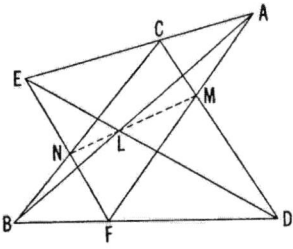

Figure 3.5A

The "projective" nature of this theorem is seen in the fact that it is a theorem of pure incidence, with no measurement of lengths or angles, and not even any reference to *order*: in each set of three collinear points it is immaterial which one lies between the other two. Figure 3.5A is one way of drawing the diagram, but Figure 3.5B is another, just as relevant. We can cyclically permute the letters *A, B, C, D, E, F,* provided we suitably re-name *L, M, N.* To avoid considering points at infinity, which would take us too far in the direction of projective geometry, let us assume that the three lines *AB, CD, EF* form a triangle *UVW,* as in Figure 3.5C. Applying Menelaus's theorem to the five triads of points

$$LDE, \quad AMF, \quad BCN, \quad ACE, \quad BDF$$

on the sides of this triangle *UVW,* we obtain

$$\frac{VL}{LW}\frac{WD}{DU}\frac{UE}{EV} = -1, \quad \frac{VA}{AW}\frac{WM}{MU}\frac{UF}{FV} = -1, \quad \frac{VB}{BW}\frac{WC}{CU}\frac{UN}{NV} = -1,$$

$$\frac{VA}{AW}\frac{WC}{CU}\frac{UE}{EV} = -1, \quad \frac{VB}{BW}\frac{WD}{DU}\frac{UF}{FV} = -1.$$

Dividing the product of the first three expressions by the product of the last two, and indulging in a veritable orgy of cancellation, we obtain

$$\frac{VL}{LW}\frac{WM}{MU}\frac{UN}{NV} = -1,$$

whence *L, M, N* are collinear, as desired. [**17**, p. 237.]

Coxeter/Greitzer's proof of Pappus's Theorem.

$(1,6,8), (2,4,9), (2,6,7), (3,4,8), (3,5,7), (4,5,6)$ *are collinear, then* $(7,8,9)$ *is collinear as well.*

We give a proof of the theorem under the additional non-degeneracy assumptions that no two lines of the picture coincide and that furthermore $(1,4,7)$ is not collinear.

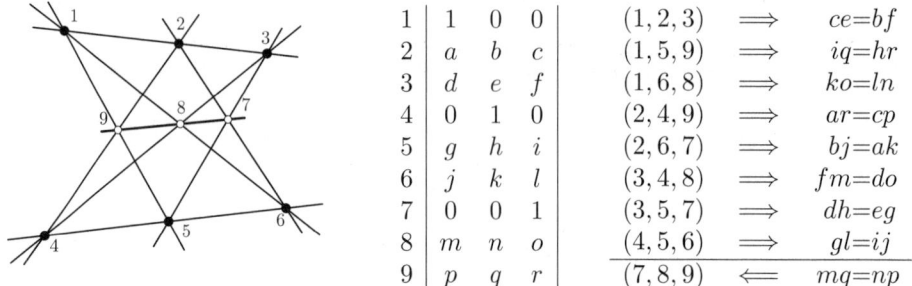

1	1	0	0	$(1,2,3)$	$\implies$	$ce=bf$
2	a	b	c	$(1,5,9)$	$\implies$	$iq=hr$
3	d	e	f	$(1,6,8)$	$\implies$	$ko=ln$
4	0	1	0	$(2,4,9)$	$\implies$	$ar=cp$
5	g	h	i	$(2,6,7)$	$\implies$	$bj=ak$
6	j	k	l	$(3,4,8)$	$\implies$	$fm=do$
7	0	0	1	$(3,5,7)$	$\implies$	$dh=eg$
8	m	n	o	$(4,5,6)$	$\implies$	$gl=ij$
9	p	q	r	$(7,8,9)$	$\impliedby$	$mq=np$

Pappus's Theorem, the smallest projective theorem.

PROOF. Assume that the points are given by homogeneous coordinates as indicated by the matrix above. By the assumption that $(1,4,7)$ is not collinear we can in particular assume that (if necessary, after a suitable projective transformation) the points 1, 4 and 7 are represented by the three unit vectors. With this special choice of the coordinates, each of the eight collinearities of the hypotheses can be expressed as the vanishing of a certain 2×2 sub-determinant of the coordinate matrix. If we write down all these equations (compare picture above), multiply all left sides and multiply all right sides we are left with another equation $mq = np$ which translates back to the collinearity of $(7,8,9)$. By our non-degeneracy assumptions all variables involved in the proof will be non-zero, therefore the cancellation process is feasible.   □

The non-degeneracy assumptions made for the proof may at first sight seem a bit unnatural. In fact, Pappus's Theorem stays valid also in the absence of these assumptions. However, for most more complicated theorems additional non-degeneracy assumptions are unavoidable (compare [18]). If not otherwise stated, for the rest of the paper, we will always assume that different lines of a configuration that share a common configuration point will not coincide. This ensures in general that during the cancellation process in our proofs we will never divide by "0".

1.2. The general pattern. If one tries to generalize the proof structure of the previous section, one encounters unexpected difficulties. The proof heavily relies on the very special choice of the basis. This choice made it possible to express every collinearity as a 2×2 determinant, since every collinear triple contained at least one basis point. Already for simple theorems like Desargues's Theorem (which consists of 10 points and 10 lines) such a special choice is no longer possible. No matter how we choose a basis, there will always be one collinear triple that does not contain a basis point and therefore must be expressed by the vanishing of a 3×3 determinant.

The key observation that helps to bypass this dilemma is the fact that the equations of the form, $XY = WZ$ can be interpreted as a certain fragment of a suitably chosen Grassmann Plücker relation. Let $x_1, x_2, \ldots, x_n \in \mathbb{R}^3$ be the homogeneous coordinates of points in $\mathbb{RP}^2$. We abbreviate $[a,b,c] := \det(x_a, x_b, x_c)$. For sake of simplicity we will identify a point with its index.

It is a well-known fact that for arbitrary points $a, b, c, d, e \in \mathbb{RP}^2$ we have

$$[a,b,c][a,d,e] - [a,b,d][a,c,e] + [a,b,e][a,c,d] = 0.$$

If in addition (a, b, c) is collinear, then we have automatically $[a, b, c] = 0$ and this, together with the Grassmann Plücker relation, implies the binomial equation

$$[a, b, d][a, c, e] = [a, b, e][a, c, d].$$

Conversely, if we know that this equation holds, we can conclude that either (a, b, c) or (a, d, e) is collinear since the first term of the Grassmann Plücker relation then has to vanish.

In principle, the equation $[a, b, d][a, c, e] = [a, b, e][a, c, d]$ may be considered as a kind of coordinate free version of a 2×2 determinant. We can use this interpretation directly to generate very well structured proofs for projective incidence theorems. We will demonstrate this technique (so called *binomial proofs*) by two examples.

First consider Desargues's Theorem (shown in the next picture). If (in $\mathbb{RP}^2$) all collinearities except $(7, 6, 8)$ are present as in the picture and no two of the lines coincide then $(7, 6, 8)$ is also automatically collinear.

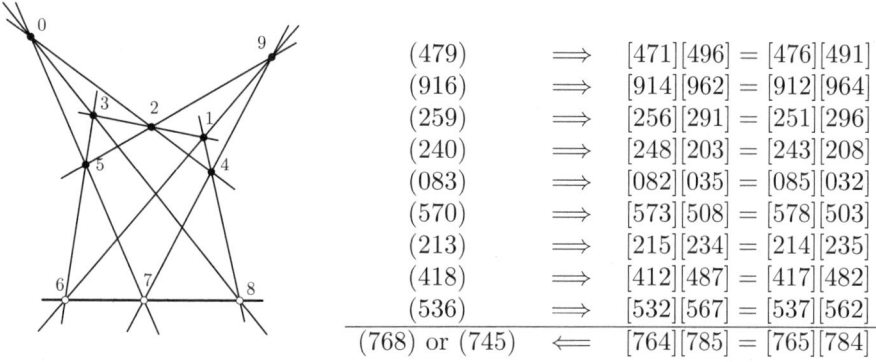

(479)	$\Longrightarrow$	$[471][496] = [476][491]$
(916)	$\Longrightarrow$	$[914][962] = [912][964]$
(259)	$\Longrightarrow$	$[256][291] = [251][296]$
(240)	$\Longrightarrow$	$[248][203] = [243][208]$
(083)	$\Longrightarrow$	$[082][035] = [085][032]$
(570)	$\Longrightarrow$	$[573][508] = [578][503]$
(213)	$\Longrightarrow$	$[215][234] = [214][235]$
(418)	$\Longrightarrow$	$[412][487] = [417][482]$
(536)	$\Longrightarrow$	$[532][567] = [537][562]$
(768) or (745)	$\Longleftarrow$	$[764][785] = [765][784]$

Desargues's configuration . . .   . . . and its proof

PROOF. For the proof consider the equations to the right of the picture. The first 9 equations are consequences of the 9 hypotheses of the theorem. If we multiply all left sides and all right sides and cancel terms that occur on both sides, we are left with the last equation (the cancellation process is possible since all determinants that occur in the equations are non-zero by our non-degeneracy assumptions). By the Grassmann Plücker argument the last equation implies that either $(7, 6, 8)$ or $(7, 4, 5)$ is collinear. Since the non-collinearity of $(7, 4, 5)$ is among our non-degeneracy assumptions, we can conclude that the triple $(6, 7, 8)$ has to be collinear.   □

As a final example of this kind let us consider the configuration shown in the next picture. It shows a certain 10_3-configuration (10 points, 10 lines, and three points on each line) that has the property that it is geometrically not realizable without additional degeneracies (observe that the line $(5, 6, 0)$ is slightly bent). The equations on the right of the picture demonstrate that, if all 10 collinearities are satisfied as indicated in the picture, then (by the usual cancellation argument) we can conclude that also $[578][670] = [570][678]$ holds. This however implies that either $(5, 1, 2)$ or $(5, 7, 0)$ is collinear. Both cases force a massive degeneration of the configuration.

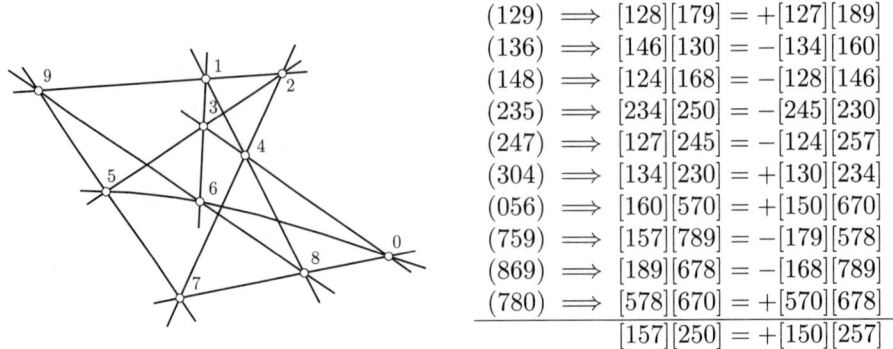

$$
\begin{aligned}
(129) &\implies [128][179] = +[127][189] \\
(136) &\implies [146][130] = -[134][160] \\
(148) &\implies [124][168] = -[128][146] \\
(235) &\implies [234][250] = -[245][230] \\
(247) &\implies [127][245] = -[124][257] \\
(304) &\implies [134][230] = +[130][234] \\
(056) &\implies [160][570] = +[150][670] \\
(759) &\implies [157][789] = -[179][578] \\
(869) &\implies [189][678] = -[168][789] \\
(780) &\implies [578][670] = +[570][678] \\
\hline
& \quad\;\; [157][250] = +[150][257]
\end{aligned}
$$

A non-realizable 10_3 configuration.

1.3. Automatic proving. The above proving technique was successfully applied in [**8, 18**] to create algorithms that prove many theorems in projective geometry automatically. The algorithm has the particularly nice feature that, if it finds a proof, the proof admits a clear readable structure, such that its correctness can be checked by hand easily. This is an interesting contrast to automatic proving techniques that are based on methods of commutative algebra (like Gröbner Basis or Ritt's algebraic decomposition method) that produce proofs consisting of large polynomials of generally high degree.

The basic idea of such a binomial-based proving algorithm is simple. First one generates all binomial relations that are consequences of the hypotheses of the theorem. Then one creates binomial expressions that imply the conclusion and finally tests whether one of the conclusion binomials can be generated as a suitable combination of the hypotheses binomials. In fact, this last step can be carried out, in principle, by a linear equation solver, since one is interested in linear combinations of the exponent vectors. In this approach the determinants themselves are treated as formal symbols (variables) and one has never to go down to the concrete level of coordinates.

Although this method has the potential to find nice and well structured proofs of this kind, if they exist, it has a great disadvantage. When in the third step the linear equation solver searches for a suitable dependence it has "forgotten" all structural information about the theorem. In essence it searches "blindly" for a linear dependence in a space in which the variables correspond to the $\binom{n}{3}$ determinants, and where each collinearity corresponds to many binomial equations. Since the calculations all have to be carried out in exact arithmetic, one may easily run into space or time problems. It would be much more desirable to have some insight in the possible structures of such proofs to rule out many unreasonable cancellation patterns in advance. This is what the rest of this paper is about.

2. Theorems on manifolds

In this section we will give a different view on cancellation patterns that can be used to prove incidence theorems. The cancellation pattern used now will have the additional feature that it can be interpreted directly in a topological way.

2.1. The theorems of Ceva and Menelaus. We will sketch a remarkable relation between incidence theorems and cycles on manifolds. At first sight the presented approach to incidence theorems seems to be very special but indeed we will demonstrate in Section 6 that this approach is as expressive as the binomial proofs described in the previous section.

Our main protagonists are the theorems of Ceva and of Menelaus. Ceva's Theorem states that if in a triangle the sides are cut by three concurrent lines that pass through the corresponding opposite vertex, then the product of the three (oriented) length ratios along each side equals 1. Menelaus's Theorem states that this product is -1 if the cuts along the sides come from a single line.

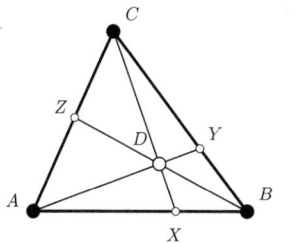   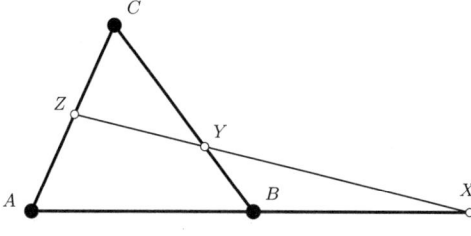

Ceva's Thm: $\frac{|AX|}{|XB|} \cdot \frac{|BY|}{|YC|} \cdot \frac{|CZ|}{|ZA|} = 1$ Menelaus's Thm: $\frac{|AX|}{|XB|} \cdot \frac{|BY|}{|YC|} \cdot \frac{|CZ|}{|ZA|} = -1$

In fact, these theorems are almost trivial if one views the length ratios as ratios of certain triangle areas. For this observe that, if the line (A, B) is cut by the line (C, D) at a point X, then we have

$$(*) \qquad \frac{|AX|}{|XB|} = -\frac{\Delta(C, D, A)}{\Delta(C, D, B)},$$

where $\Delta(A, B, C)$ denotes the oriented triangle area.

In order to prove Ceva's Theorem we consider the obvious identity:

$$\frac{\Delta(CDA)}{\Delta(CDB)} \cdot \frac{\Delta(ADB)}{\Delta(ADC)} \cdot \frac{\Delta(BDC)}{\Delta(BDA)} = -1,$$

(note that the triangle area Δ is an alternating function and that each triangle in the denominator occurs as well in the numerator). Applying the above identity $(*)$ we immediately get Ceva's Theorem. Similarly, a proof of Menelaus's Theorem is derived. For this consider the special line as being generated by two points D and E. We have

$$\frac{\Delta(DEA)}{\Delta(DEB)} \cdot \frac{\Delta(DEB)}{\Delta(DEC)} \cdot \frac{\Delta(DEC)}{\Delta(DEA)} = 1,$$

Applying the identity $(*)$ yields Menelaus's Theorem. Observe that the expressions of Ceva and Menelaus carry an orientation information. If in the future we talk about the "Ceva-expression" (or "Menelaus-expression") for the triangle A, B, C the letters A, B, and C are assumed to be ordered as in the expression above. A, C, B would generate the reciprocal expression. Furthermore, we will call the points X, Y, and Z in the above drawing the *edge points* of the configuration. The points A, B, and C will be called the *vertices* of the configuration. Point D in Ceva's configuration will be called the *Ceva point* and the cutting line in Menelaus's configuration is the *Menelaus line*.

2.2. A homotopy argument. Now, consider the situation where two triangles that are equipped with a Ceva configuration share an edge and the corresponding edge point on this edge (see the picture below). The triangle A, B, C yields a relation $\frac{|AZ|}{|ZB|} \cdot \frac{|BX|}{|XC|} \cdot \frac{|CY|}{|YA|} = 1$ while the triangle C, B, D yields $\frac{|CX|}{|XB|} \cdot \frac{|BV|}{|VD|} \cdot \frac{|DW|}{|YW|} = 1$. The quotient $\frac{|BX|}{|XC|}$ occurs in the first expression and its reciprocal occurs in the second expression. If we multiply both expressions, this quotient cancels and we are left only with terms that live on the boundary of the figure. We obtain

$$\frac{|AZ|}{|ZB|} \cdot \frac{|CY|}{|YA|} \cdot \frac{|BV|}{|VD|} \cdot \frac{|DW|}{|YW|} = 1.$$

We now consider a triangulated topological disc. All triangles of the triangulation should be equipped with Ceva configurations that have the additional property that points on interior edges are the shared edge points of the two adjacent triangles. We consider the product of all corresponding Ceva-expressions.

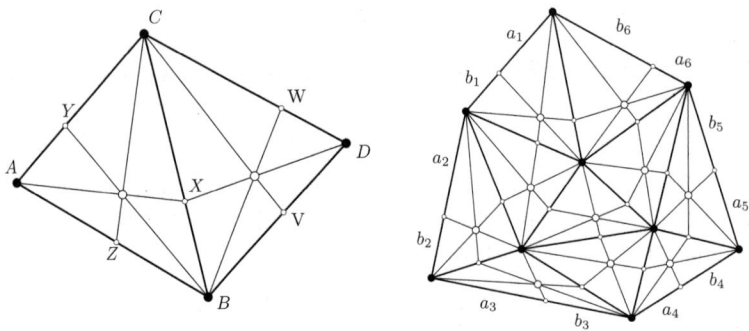

Gluing two Ceva Configutarions   Gluing many Ceva Configutarions

If the triangles are oriented consistently (adjacent triangles use the common edge in opposite directions), all quotients related to inner edges will cancel. We are left with an expression that only depends on the position of the boundary points (including the edge points along the boundary edges). If in the last picture on the right the letters $a_1, b_1, , \ldots, a_6, b_6$ correspond to the oriented lengths around the boundary we can conclude immediately that we must have

$$\frac{a_1}{b_1} \cdot \frac{a_2}{b_2} \cdot \frac{a_3}{b_3} \cdot \frac{a_4}{b_4} \cdot \frac{a_5}{b_5} \cdot \frac{a_6}{b_6} = 1.$$

Now, consider any triangulated manifold that forms an oriented 2-cycle. This cycle serves as a kind of *frame* for the construction of an incidence theorem. It is important to mention in what category we understand the term "triangulated manifold". We consider compact, orientable 2 manifolds without boundary and subdivisions by CW-complexes whose faces are triangles. So in principle, already a subdivision of a 2-sphere by two topological triangles, which are identified along the edges, would be a feasible object for our considerations.

Consider such a cycle as being realized by flat triangles (it does not matter if these triangles intersect, coincide or are coplanar as long as they represent the combinatorial structure of the cycle). By the above argument the presence of Ceva configurations on all but one of the faces will imply automatically the existence of a Ceva configuration on the final face. Thus at the final face the three lines connecting the edge points and the vertices will meet automatically, and we have an incidence theorem. In what follows we will study many concrete examples of this amazingly rich construction technique.

As a first example take the projection of a tetrahedron $(ABCD)$ to $\mathbb{R}^2$. Now, choose Points U, V, W, X, Y, Z one on each of the edges of the tetrahedron. Assume that for three of the faces these points form a Ceva configuration. Then they automatically form a Ceva configuration on the last face — an incidence theorem.

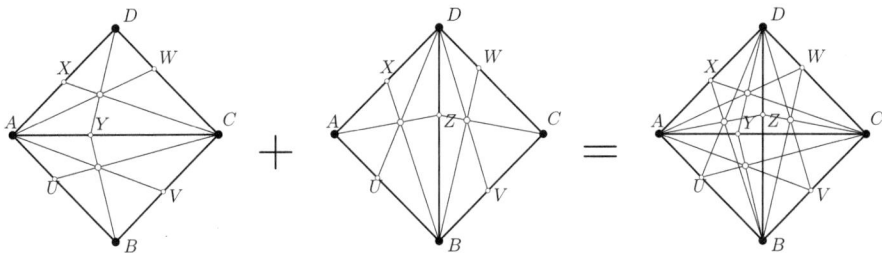

Although the proof of this incidence theorem is already evident by our above homotopy argument, we still want to present the algebraic cancellation pattern in detail. Consider the following formula

$$\left(\frac{|AU|}{|UB|} \cdot \frac{|BV|}{|VC|} \cdot \frac{|CY|}{|YA|}\right) \cdot \left(\frac{|CW|}{|WD|} \cdot \frac{|DX|}{|XA|} \cdot \frac{|AY|}{|YC|}\right) \cdot$$
$$\left(\frac{|AX|}{|XD|} \cdot \frac{|DZ|}{|ZB|} \cdot \frac{|BU|}{|UA|}\right) \cdot \left(\frac{|BZ|}{|ZD|} \cdot \frac{|DW|}{|WC|} \cdot \frac{|CV|}{|VB|}\right) \quad = \quad 1.$$

This formula is obviously true, since all lengths of the numerator occur in the denominator as well and vice versa (this property is inherited from the cyclic structure). On the other hand, each of the factors in brackets being 1 states the Ceva condition for one of the faces. Thus three of these conditions imply the last one. The essential fact that makes this proof work is that whenever two faces meet in an edge the two corresponding ratios cancel. In general we obtain:

> For any triangulated oriented 2-CW-cycle choose a point on each edge such that for every face either a Ceva or a Menelaus condition is generated. If altogether an even number of Menelaus configurations is involved, then the conditions on all but one of the triangles automatically imply the condition on the last triangle.

We need an even number of Menelaus configuration since each Menelaus configuration accounts for a factor of -1 in the product. We will call such a cycle equipped with Ceva/Menelaus configurations a Ceva/Menelaus-cycle. Instead of drawing the whole incidence structure we often simply draw a schematic diagram, in which we indicate the combinatorial structure of the cycle and attach a label C or M to each of the faces. We will usually draw these schemes as a planar net of triangles for which we specify which vertices and which edges have to be identified.

3. A census of incidence theorems

At first instance the method described in the last section is very useful for producing geometric incidence theorems by pasting together triangles that carry Ceva or Menelaus configurations. In this chapter we want to elaborate on this aspect. We will at least for small numbers of triangles list several examples that can be produced by this philosophy. In any case we need a configuration of triangles that forms a closed orientable CW-cycle. It may happen that two triangles are identified along more than one edge. However, to avoid trivial cases (in which the two configurations of these triangles together with the edge points simply coincide), we have to assume that in such a case one triangle is equipped with a Ceva configuration and the other with a Menelaus configuration. Later on such a sub-configuration consisting of two triangles that coincide along two edges will be called a "pocket". Let us now start with a census of small incidence theorems. Observe that the number of triangles involved in a Ceva/Menelaus-cycle must be even.

3.1. Two triangles. For two triangles there is only one possibility of forming a Ceva/Menalaus-cycle. We can only identify the two triangles along their edges. Then we have to assign to one of the triangles a Ceva configuration and to the other one a Menelaus configuration. In the real projective plane this configuration is not realizable at all, since one triangle forces the product of ratios to be -1 the other forces the product of ratios do be 1. However, if we consider a field of characteristic two, then this configuration is an incidence theorem. In fact, this configuration is nothing else but the well-known Fano plane. In the triangle scheme below the labels at the vertices of the triangles indicate which vertices have to be identified.

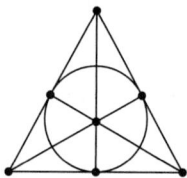
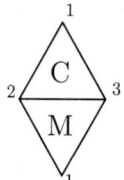

The Fano configuration . . .   . . . and its triangle scheme

3.2. Four triangles. The structure becomes considerably richer if we consider four instead of only two triangles. First of all there are two combinatorially different ways of creating a manifold involving four triangles: either they could form a tetrahedron, or they could form a manifold, in which two *pockets* formed by two triangles are pasted. It is important to observe that for the last case one has not only to make clear which vertices have to be identified, but also which edges (and with them the edge points) have to be identified.

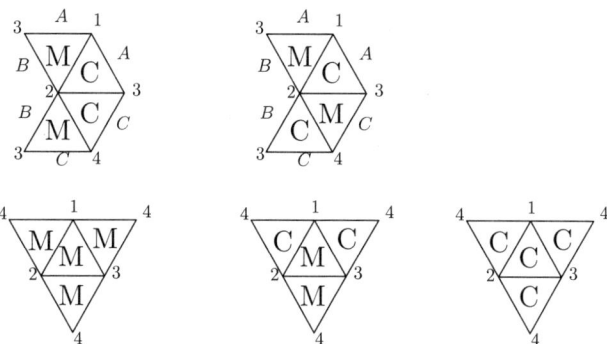

All five possibilities to form an incidence theorem on four triangles.

The first row shows the two possible cases for the "pocket" version. The second row shows the three possibilities of assigning Ceva and Menelaus configurations to a tetrahedron.

Let us now analyze the geometric interpretation of these cases as incidence theorems. The situation for the first example of the first row is shown in the next picture. In principle, two adjacent triangles that carry a Ceva configuration are overlaid with two adjacent triangles that carry Menelaus configurations. The final coincidence of the resulting configuration is satisfied automatically. The resulting image is the well-known configuration that shows that for three given points on a line one can construct a harmonic point by erecting a drawing of a tetrahedron over the line.

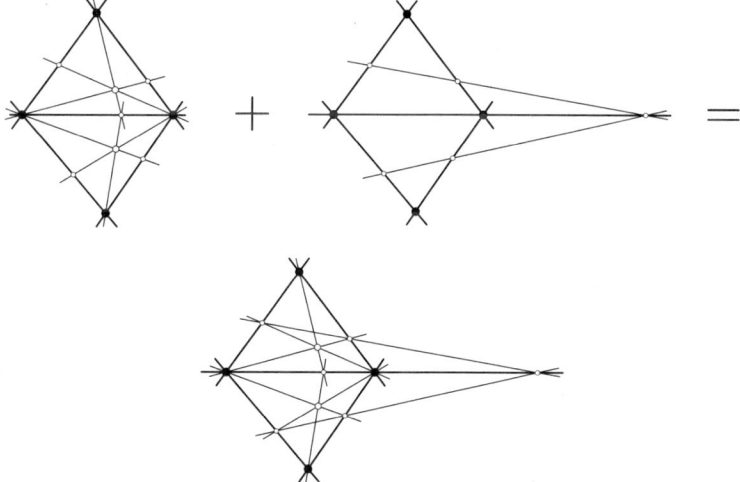

Although the second picture in the first row is combinatorially different from the first one, the resulting incidence configuration is exactly the same. The patient reader is invited to check this.

The first scheme in the second row is nothing else but a description of Desargues's Theorem, which we already encountered in Section 1. The following picture shows the decomposition of the tetrahedron in a front and a back part.

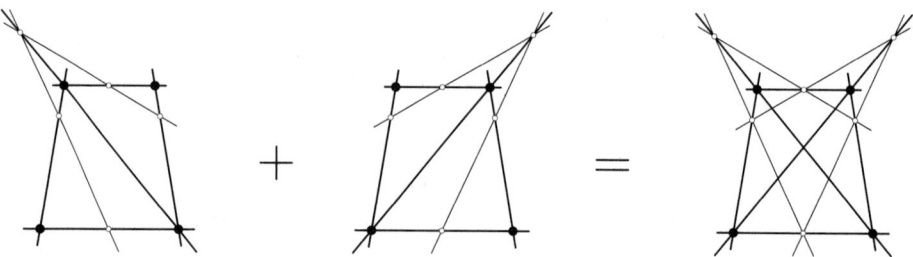

The last two theorems in the second row correspond to the incidence configurations shown in the next picture. In fact, the last example is the one we already have studied when we introduced the proving technique three pages earlier. This configuration has a few remarkable special cases in which one or more points do coincide. However, we will not analyze them here in detail.

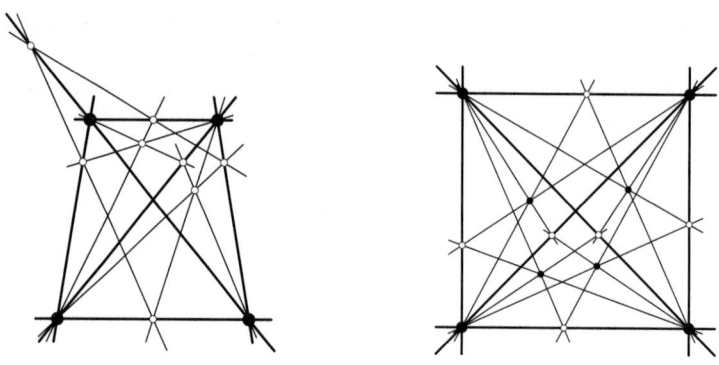

3.3. Six triangles. Already with six triangles the situation becomes quite elaborate. There are altogether six combinatorially different types of cycles, and 21 different types of underlying incidence configurations. We will not list all of them here. Some of the manifolds involve "pocket" sub-configurations and will be considered in a more general setting later on in Section 4. The only cycles that do not contain pockets are shown in the next picture. The cycle on the left is a double pyramid over a triangle. There are 8 non-isomorphic ways to equip it with Ceva and an even number of Menelaus configurations. We will not study these cases here. The second cycle is topologically more interesting. It is the smallest CW-decomposition of a torus into triangles. It has only three vertices and nine edges. There are exactly four ways of assigning a Ceva/Menelaus-cycle to it: all triangles Ceva, exactly two adjacent triangles Menelaus, exactly two adjacent triangles Ceva, all triangles Menelaus. The next subsection will be dedicated to a detailed discussion of the first and the last one of these situations.

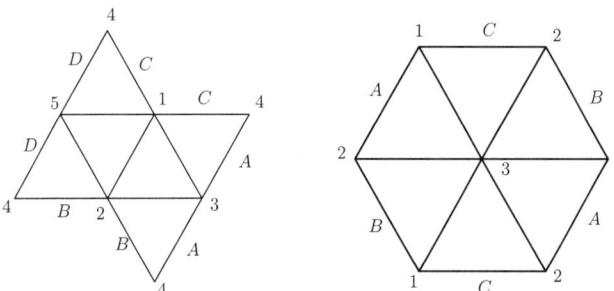

3.4. Pappus revisited. We now will deal exclusively with the situation of six triangles that form a topological torus. First observe, that if we realize this cycle by flat triangles, then the only way of doing this is by making all triangles coincident. Assume the triangles are realized in this way and furthermore assume that we assign a Ceva configuration to each of these triangles. The left drawing of the next picture shows the unfolded configuration, while the middle drawing shows the overlay. Observe that for each of the triangles we get one Ceva point. Lines of two adjacent triangles that share the corresponding edge point become identified. So, all together we have the three original points of the triangle, and six Ceva points. These nine points together with the nine interior lines have exactly the combinatorics of Pappus's Theorem. Thus we have obtained a "Ceva-proof" for Pappus's Theorem. It is an amazing fact that by the pairwise identifications of the interior lines the edge points and also the original edges of the triangles do not play any role in the theorem we just proved. The edge points are just intersections of two lines. After deleting these points, the edges of the triangles are just joins of two points and can be also deleted. The picture on the right shows the situation after the deletion. It is just a nice and symmetric drawing of Pappus's Theorem.

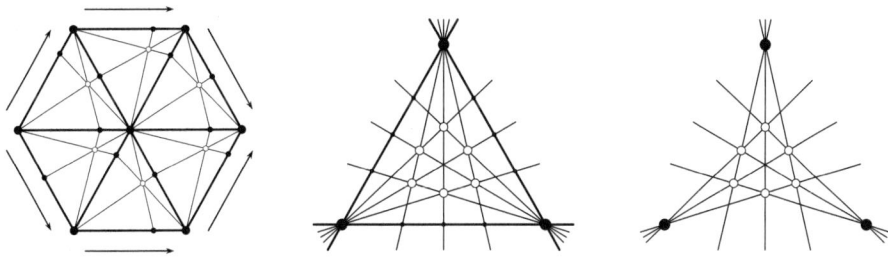

A Ceva-proof for Pappus's Theorem

It is a really surprising fact that also in the case in which we assign to all six triangles a Menelaus configuration we get a drawing of Pappus's Theorem. The next picture shows a drawing of this configuration. The original triangle of the manifold is the dark triangle in the right of the drawing. The additional six lines in the drawing are the six Menelaus lines that cut all three edges of the triangles. The nine points on the left of the configuration correspond to the nine edges of the triangles in our torus. Similar to the fact that in the last proof the edges of the triangles could be neglected, this time, the three vertices are superfluous. This way of decomposing Pappus's Theorem into six Menelaus configurations was exactly the method we encountered in the original copy of Coxeter's page shown in Section 1.

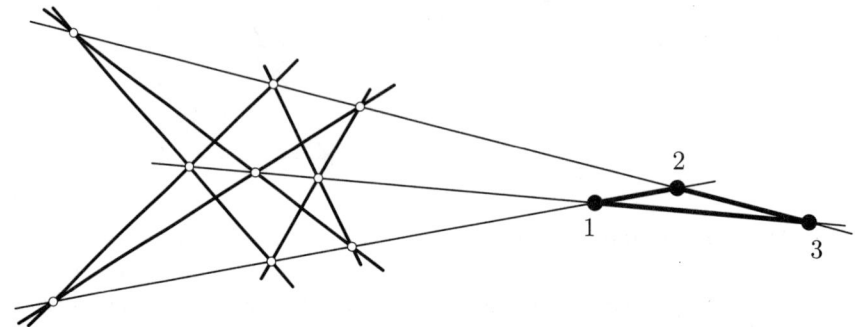

A Menelaus-proof for Pappus's Theorem

4. Basic building blocks

In this section we want to review a few basic substructures that are very often present in Ceva/Menelaus-proofs. In general, these basic substructures can be directly assigned to elementary principles of projective geometry – like cross ratios, harmonic points, or perspectives.

4.1. Pockets, ... In the last section we already encountered the situation in which two triangles were joined along *two* edges. In this case it was necessary to equip one of the triangles with a Ceva configuration and the other with a Menelaus configuration (otherwise the third edge-point would coincide as well). In Section 3.1 we already saw that we can identify the two triangles along the third line only in fields of characteristic 2. Here we study the case, in which we do not identify the edge points on the third line. In this case we will have exactly four points on this line: two original triangle edges, one Ceva-edge-point and one Menelaus-edge-point. The next picture (left) shows the corresponding incidence configuration. Topologically we can consider the pocket as a disk that is bounded by two edges $1 - C - 2$ and $2 - M - 1$. If we multiply the Ceva and the Menelaus condition we immediately get

$$\frac{|1C|}{|C2|} \cdot \frac{|2M|}{|M1|} = -1.$$

In other words $1, 2, C$ and M are in harmonic position.

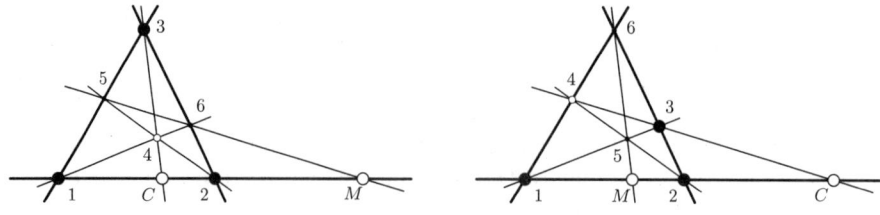

Accidentally, we can interpret exactly the same incidence configuration also the other way around. If we interchange the role of the points $3, \ldots, 6$ according to $3 \leftrightarrow 6$, $4 \leftrightarrow 5$, $5 \leftrightarrow 4$, $6 \leftrightarrow 3$, we see that for the (old) triangle $1, 2, 6$ we obtain an interchanged role of the Ceva and the Menelaus edge point. This shows that if we have a pocket in a Ceva/Menelaus-cycle, we can freely interchange the role of the "C" and the "M" on the pocket without changing the incidence configuration.

4.2. ... tunnels, ... A very common situation in projective incidence theorems is that "information" about a certain substructure is transfered from one part of the configuration to another by projection. For instance, if four points on a line have a certain cross ratio, any projection of these points onto another line will result in four points with the same cross ratio. We can perfectly model this situation by suitable substructures of Ceva/Menelaus-cycles. Assume that we have a CW-triangulation of a manifold that is homeomorphic to $S^1 \times [0, 1]$ (a sphere with two holes). The boundary of this manifold consists of two disjoint circles. If each triangle is equipped with a Ceva or Menelaus configuration (such that the number of Menelaus configuration is even), then the product of the length ratios along the first cycle must be identical to the product of the length ratios on the second (or its inverse, depending on the orientation). Thus in a sense the configuration transfers information from the first boundary cycle to the second one.

The smallest possible situation is shown in the next picture. It consists of 4 triangles that form a simple cycle. Combinatorially the situation is a tetrahedron, with two cuts along two opposite edges. The cycles consist just of two edges, each.

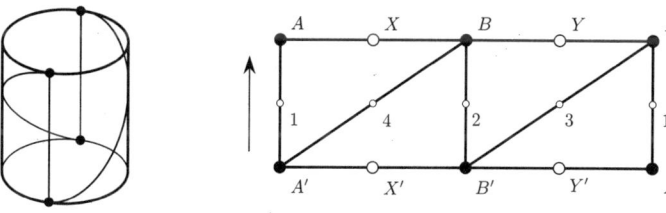

If the situation is as in the picture, then for the corresponding products of length ratios we get:

$$\frac{|AX|}{|XB|} \cdot \frac{|BY|}{|YA|} \;=\; \frac{|A'X'|}{|X'B'|} \cdot \frac{|B'Y'|}{|Y'A'|}.$$

In other words, the configuration transfers the cross ration of A, B, X, Y to the points A', B', X', Y'. If all triangles are equipped with Menelaus's configurations, the corresponding incidence configuration is shown in the following picture (left).

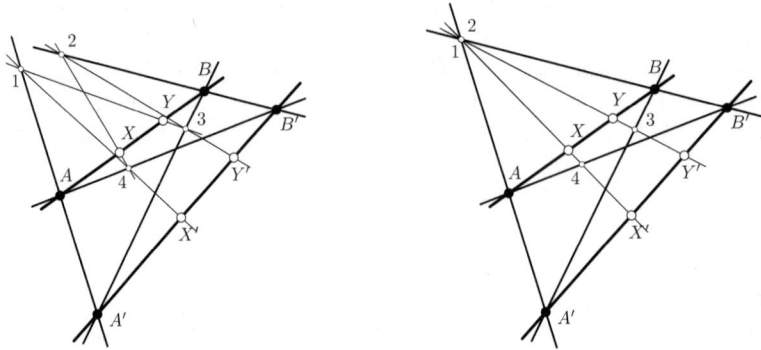

The right picture shows a special instance of this configuration, in which the points 1 and 2 (and several lines) coincide. In this picture the points A, B, X, Y and the points A', B', X', Y' are connected to each other by a perspectivity. So we obtained a Ceva/Menelaus-proof for the fact that the cross ratios are preserved by perspectivities.

4.3. ... and toothpicks. In this subsection we will study the semantics of degenerate Ceva triangles and degenerate Menelaus triangles. Usually, a Ceva configuration produces a relation $\frac{|AX|}{|XB|} \cdot \frac{|BY|}{|YC|} \cdot \frac{|CZ|}{|ZA|} = 1$. This relation remains also valid if the points A, B, C are collinear. In this case, however, the points X, Y, Z will coincide with the vertices. And we will get:

$$\frac{|AC|}{|CB|} \cdot \frac{|BA|}{|AC|} \cdot \frac{|CB|}{|BA|} = 1.$$

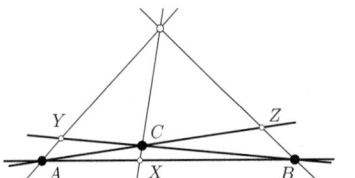
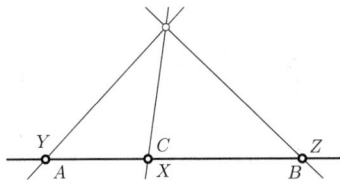

Similarly, applying Menelaus's Theorem to a degenerate triangle A, B, C we see that the Menelaus line intersects the line A, B, C at a unique point X, since all edge points will coincide. In this case we get the identity

$$\frac{|AX|}{|XB|} \cdot \frac{|BX|}{|XC|} \cdot \frac{|CX|}{|XA|} = -1.$$

Here X can be any point on the line through A, B, C.

One might wonder whether such degenerate configurations may ever occur in a real proof of an incidence theorem. From the point of view of Ceva/Menelaus proofs the length ratios $\frac{|AC|}{|CB|}$ must be considered as indecomposable units. The degenerate subconfiguration gives us the possibility to replace $\frac{|AC|}{|CB|}$ by the product $\frac{|CA|}{|AB|} \cdot \frac{|AB|}{|BC|}$. In particular, this gives us the possibility to exchange the role of edge points and vertices. In fact, allowing degenerate triangles considerably broadens the applicability of our proving methods. We want to discuss briefly how to derive a trivial (but useful) identity in a Ceva/Menelaus setup. Assume that A, B, X, Y are four points on a line. Then we obviously have:

$$\frac{|AX|}{|XB|} \cdot \frac{|BY|}{|YA|} = \frac{|BY|}{|XB|} \cdot \frac{|AX|}{|YA|}.$$

However, if we consider the four length ratios as unbreakable symbols the identity is far from being trivial. We can establish this identity in our setup by pasting two degenerate Ceva configurations (triangles A, B, X and A, B, Y) and two degenerate Menelaus configurations (triangles X, Y, A and X, Y, B). The following calculation proofs the identity:

$$\left(\frac{|AX|}{|XB|} \cdot \frac{|BA|}{|AX|} \cdot \frac{|XB|}{|BA|}\right) \cdot \left(\frac{|BY|}{|YA|} \cdot \frac{|AB|}{|BY|} \cdot \frac{|YA|}{|AB|}\right) \cdot$$
$$\left(\frac{|XB|}{|BY|} \cdot \frac{|YB|}{|BA|} \cdot \frac{|AB|}{|BX|}\right) \cdot \left(\frac{|YA|}{|AX|} \cdot \frac{|XA|}{|AB|} \cdot \frac{|BA|}{|AY|}\right) \quad = \quad 1.$$

Ratios that do not cancel are underlined. They yield exactly the desired expression. The combinatorics of this expression is again a kind of tunnel: A tetrahedron in which two holes are cut along opposite edges. If we close these two slots of the tetrahedron by two pockets, we immediately get a proof for the following incidence theorem:

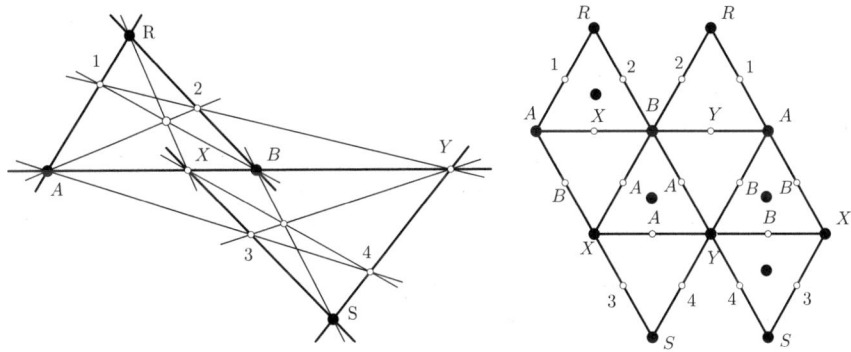

The picture on the right indicates the manifold structure with complete labels for the vertices and for the edge points. The Menelaus triangles are marked by a black dots.

5. Other primitives

So far we used only Ceva configurations and Menelaus configurations as basic entities. They both "lived" on a triangle and had exactly one edge point per edge. In this section we want to give a glimpse of what happens if we allow more than one edge point or if we consider polygons other than triangles.

5.1. Conics. There is a beautiful theorem of Carnot, which can be considered to be a generalization of Ceva's Theorem.

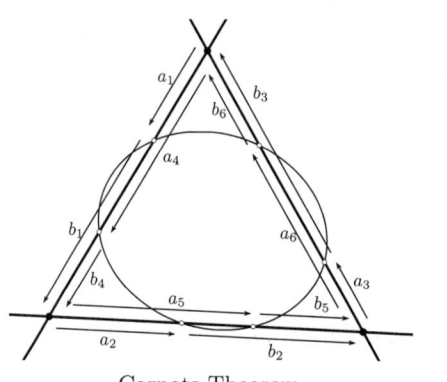
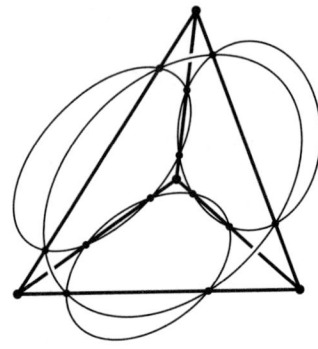

Carnots Theorem   Gluing many Carnot configurations

We consider a triangle with exactly two (distinct) edge points per edge. We assume the edge points are labeled $1, \ldots, 6$ and the corresponding length ratios are a_i/b_i. Carnot's Theorem states that we have

$$\frac{a_1}{b_1} \cdot \frac{a_2}{b_2} \cdot \frac{a_3}{b_3} \cdot \frac{a_4}{b_4} \cdot \frac{a_5}{b_5} \cdot \frac{a_6}{b_6} \;=\; 1$$

if and only if the six edge points lie on a common conic. We can immediately use Carnot's configuration as a "primitive" to build theorems that also involve conics. In the above picture on the right we present a small theorem that only involves Carnot-faces: If one has a tetrahedron with two distinct points on each edge and if the six edge points of three faces are co-conical, then they will be co-conical for the last face automatically. In fact, there is nothing special about the tetrahedron. Any oriented triangulated 2-manifold would serve as a frame as well.

It is also easily possible to prove, for instance, Pascal's Theorem with this method. For this one combines one Carnot configuration with four Menelaus configurations. Since each edge of the Carnot configuration has two edge points, one has to count them with multiplicity two and each of these edges has to be glued to two Menelaus configurations – one for each edge point.

5.2. n-gons. If we consider n-gons instead of triangles we immediately can produce several nice generalizations of Ceva's and Menelaus's Theorem. Many of them have been studied in the literature. Most of them are of the form that a certain (combinatorially symmetric) incidence configuration forces a certain product of length ratios (often with cyclic symmetry) to be 1 or -1. For an extensive treatment of this topic see the article series [**12, 13, 14, 15, 21**]. Here we only want to present three of these theorems, without proofs.

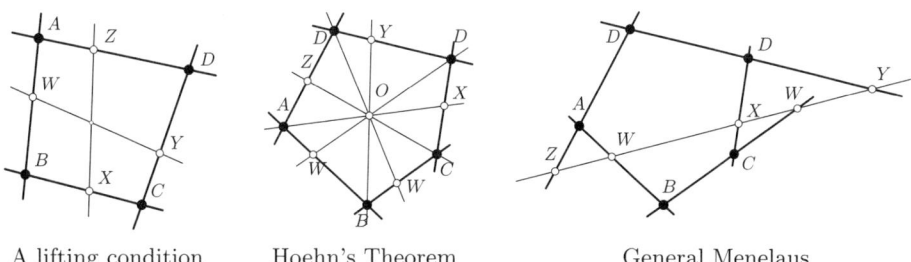

| A lifting condition | Hoehn's Theorem | General Menelaus |

If we consider a quadrangle A, B, C, D (first picture above) with edge points W, X, Y, Z, then the condition

$$\frac{|AW|}{|WB|} \cdot \frac{|BX|}{|XC|} \cdot \frac{|CY|}{|YD|} \cdot \frac{|DZ|}{|ZA|} = 1$$

holds if and only if the six lines of the picture are tangent to a common conic. Another equivalent condition for this is that there is a proper lifting of these lines to three-space that preserves all nine incidences.

The second drawing shows a theorem which is known under the name *Hoehn's Theorem*. Consider an n-gon with odd n and an additional point O. If we construct edge points by intersecting each edge with the line that connects O to the opposite vertex, then the cyclic product of the length ratios will be 1. This theorem is a direct generalization of Ceva's Theorem where we have $n = 3$. It has to be mentioned that the converse of Hoehn's Theorem is not true for $n > 3$, since the cyclic product being 1 does not necessarily imply that all central lines meet in a point (which can be seen by a simple degree-of-freedom count). One can prove this theorem by exactly the same idea we used in Section 2.1 to prove Ceva's Theorem.

The last picture shows a generalization of Menelaus's Theorem. Consider an n-gon where edge points are generated by cutting the edges with a single line (not necessarily in the interior) then the cyclic product of the length ratios equals $(-1)^n$. The theorem can be proved easily by triangulating the n-gon, applying the usual Menelaus Theorem to all triangles and canceling all interior edges.

As one application for using Hoehn's Theorem as a basic building block we consider a theorem of Saam, whose incidence configuration is given in the left picture below (compare [**19**, **20**]). One starts with a central point and an odd number of lines passing through it. On each of these lines one chooses a projection point P_i. Then one starts with a point A_1 (compare to the picture) and projects this point through P_1 onto the next line. If one proceeds projecting, then after cycling around twice one again reaches the initial point A_1. The picture on the right shows a way to prove this theorem by a cycle construction. One forms an n-gon by the points A_i with odd i and forms a pyramid over this n-gon. On all triangular faces one imposes a Ceva configuration and on the n-gon one imposes a Hoehn configuration. The usual cancellation procedure proves the theorem. It also becomes obvious that Saam's theorem is only a special case of a more general theorem that has the same underlying manifold proof, but in which the apex of the pyramid does not coincide with the center-point of Hoehn's configuration.

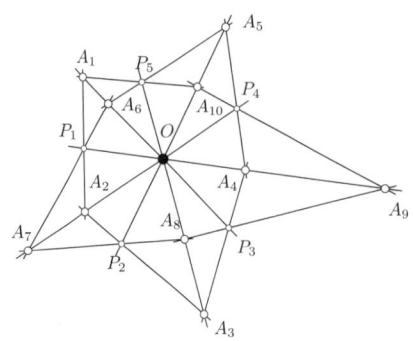
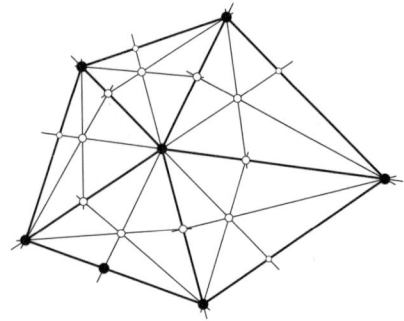

Saam's Theorem   The underlying manifold

6. The connection of Ceva/Menelaus proofs and binomial proofs

In this section we will discuss the connection of Ceva/Menalaos proofs with the binomial proving techniques presented in Section 1. At first sight it is not clear at all whether and how the two proving concepts are related. In particular, binomial proofs seem to carry much less structure (they "just" admit a cancellation pattern but no obvious topological structure) than Ceva/Menelaus proofs. Therefore it would not be surprising if the class of theorems provable by binomial methods would be considerably richer than the class of theorems provable by Ceva/Menelaus constructions.

Amazingly, this is not the case. There is even an algorithmic way of systematically translating one structure into the other. Space restrictions do not allow us to give a full formal proof of this result here. We only want to sketch the basic ideas and demonstrate the relation by a concrete example.

6.1. Symbols and relations. One problem of the formal treatment of our proving techniques is that we treat the same object on very different levels. First of all there is the combinatorial structure of the incidence theorem that serves as a basis for all other levels. Secondly there are concrete realizations of the hypotheses and non-degeneracy requirements of the theorem, in which we have concrete coordinates and can calculate concrete values for determinants and length ratios. On the other hand, if we consider binomial proofs, the determinants play merely the role of formal symbols. We will not assign specific values to them, but the incidence structure of the theorem determines rules how these formal symbols are related to each other. In the case of binomial proofs the decisive relations on the symbols are given by the identities of the form $[\ldots][\ldots] = [\ldots][\ldots]$ that are used in our proofs. If we consider Ceva/Menelaus proofs instead, the role of formal symbols is played by the ratios of oriented lengths. The existence of Ceva or Menelaus sub-configurations is translated into the presence of relations among these length ratios.

The essence of our proofs is that the purely formal treatment of the symbols (formal determinants or formal length ratios) allows to conclude additional relations that have to be present in every realization of the hypotheses configuration. The idea of treating the abstract properties of an incidence configuration on a formal level is not at all new. In his article *The bracket ring of combinatorial geometry*

[23] N. White discusses extensively the ring structure of formal determinants that is imposed by an underlying incidence structure. Similarly, in a series if articles on the *Tutte Group* of a matroid [9, 10, 11, 22] A. Dress and W. Wenzel analyze the consequences of incidence relations on the abstract multiplicative group on several formal kinds of symbols. The abstract symbols that are studied there are *abstract determinants*, *ratios of (adjacent) abstract determinants*, and *abstract scalar products*. For a fixed underlying matroid each of these setups produces a group and the three groups are very closely related. Up to isomorphism they just differ by a factor of $\mathbb{Z}^k$. The isomorphism is not at all trivial and relies on homotopy arguments on so called Maurer graphs [10, 16].

It turns out that the essence of our proving techniques can be nicely expressed in terms of Tutte groups and it turns out that the equivalence of binomial proofs and Ceva/Menelaus proofs relies on the equivalence of the different setups for Tutte groups. For reasons of space limitations we will not present the formal setup here, since it involves quite a lot of technical machinery. Rather than that we will explain the basic concepts and describe the translation process by a concrete example.

6.2. Binomial proofs vs. Ceva/Menelaus proofs. For our explanations we will make a few simplifying assumptions. We will assume that we study a concrete incidence configuration with a concrete matroid underlying it. We restrict the considerations to the rank 3 case only. The hypotheses (and the conclusion) of our theorem will be expressed by certain non-bases of the matroid. The non-degeneracies will result in the presence of certain bases of the matroid. In general an incidence theorem will not determine the underlying matroid uniquely. Only a partial structure of it will be fixed by the hypotheses non-degeneracies. We will neglect this technical difficulty and assume that we have a fixed underlying matroid **M** with set of bases **B**.

We consider the graph $\mathcal{G}_\mathbf{M}$ whose vertices correspond to the bases **B** and whose edges are those pairs of bases that differ exactly by one element. We will identify each determinant that occurs in a binomial proof (and therefore will be a basis) with the corresponding vertex of $\mathcal{G}_\mathbf{M}$. Assume that $(1,2,3)$ are collinear and that $[1,2,x][1,3,y] = [1,2,y][1,3,x]$ is a corresponding binomial equation. Bases involved in such a binomial equation correspond to a 4-cycle in $\mathcal{G}_\mathbf{M}$ without diagonal edges. We will call such a quadrangle a *binomial quadrangle*. A binomial quadrangle with vertices $(1,2,x), (1,3,y), (1,2,y), (1,3,x)$ is called *degenerate* if at least one of the triples $(1,2,3)$ or $(1,x,y)$ is not a basis. If both triples are non-bases it is called *pure*.

We will bicolor degenerate quadrangles, which correspond to binomial equations, such that the bases on the left of the equations are colored white and the bases on the right are colored black. A binomial proof now corresponds to a collection of (bicolored) binomial degenerate quadrangles, such that in this collection each vertex is as often colored black, as it is colored white.

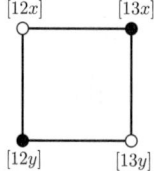

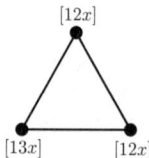

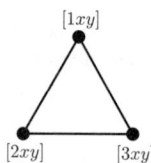

a binomial quadrangle   a triangle of first kind   a triangle of second kind

In the picture above the first drawing shows such a quadrangle with labeled bases for the binomial relation $[1, 2, x][1, 3, y] = [1, 2, y][1, 3, x]$. In the graph $\mathcal{G}_{\mathbf{M}}$ there can be exactly two combinatorially different types of triangles. They are represented by the two labeled triangles in the picture above. In the context of Tutte groups these triangles are called *triangle of first kind* and *triangle of second kind*, respectively. One of the fundamental properties of such bases graphs of matroids is that every cycle in the graph can be generated as a sum of triangles and of pure quadrangles (here sum is meant in the sense that if coinciding edges are added with opposite directions, then they cancel each other). If the rank three matroid does not contain loops or parallel elements, each cycle can even be considered as sum of triangles only. This result is the technical kernel of the fact that the different Tutte groups are isomorphic up to a $\mathbb{Z}^k$-factor. It will also be our main tool for the translation of binomial proofs into Ceva/Menelaus proofs.

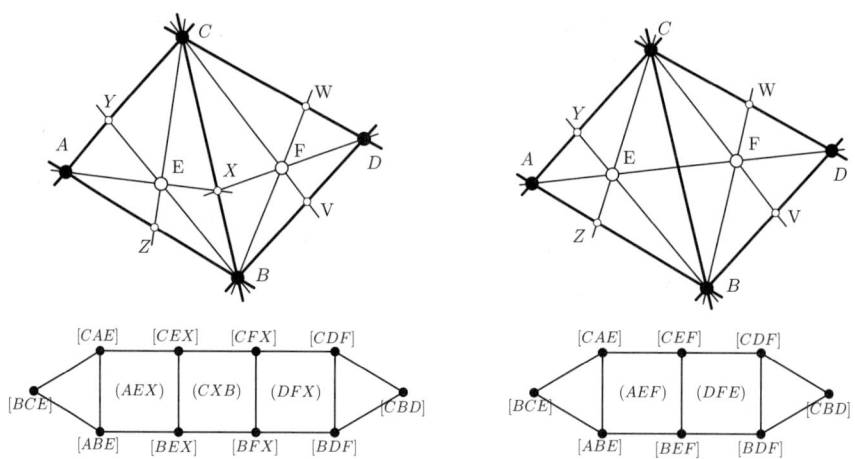

Bases graphs of glued Ceva configurations.

There is a close connection of triangles of first and second kind to Ceva's and Menelaus's configuration. For this we consider the bases as non-zero triangle areas in our configuration. An oriented edge can be interpreted as a quotient of two such triangle areas. As explained in Section 2.1. such a quotient can also be considered as a ratio of lengths on a line. The three ratios encoded by each of the two triangles are exactly those ratios that we used in our original proofs of Ceva's and Menelaus's Theorem in Section 2.1.

Now, let us study the situation where in a Ceva/Menelaus proof two triangles are glued along an edge. What does the corresponding situation in the bases graphs look like? We will only elaborate on two situations; the rest is essentially analogous. Let us consider the situations shown in the picture on the preceding page, where two Ceva configurations are joined. In each of the corresponding cases a substructure of the basis graph is shown. The quadrangles that are visible in the graphs are actually degenerate binomial quadrangles (the triple inside the quadrangle indicates the non-basis that is responsible for the degeneracy). Recall that the quadrangles represent a relation of the form $\frac{[1,a,x]}{[1,a,y]} = \frac{[1,b,x]}{[1,b,y]}$. Both sides of the equation can be interpreted as representing the same length ratio, however represented by different area quotients. The chains of quadrangles that occurs in the examples above serve as a kind of translator between the two Ceva triangles. They make sure that the area ratios in one triangle really represent the same length ratio as the area ratio of the other triangle. If we consider the bases graph underlying a configuration that admits a Ceva/Menalaus proof, we will find the following substructure. Each Ceva or Menelaus configuration of the proof corresponds to a suitable triangle in the graph. All edges of these triangles are paired by chains of degenerate binomial quadrangles (this chain may have length zero). The collection of binomial equations corresponding to exactly these quadrangles forms a binomial proof of the theorem. This gives the translation of Ceva/Menelaus proofs into binomial proofs.

Slightly more complicated is the translation of a binomial proof into a Ceva/Menelaus proof (this is not very surprising, since the second kind of proof carries much more structural information). We will demonstrate the basic procedure by an example that shows essentially all important features. We revisit the theorem below (recall Section 4.3) and consider a concrete proof that was generated by an implementation of an algorithm that produces binomial proofs.

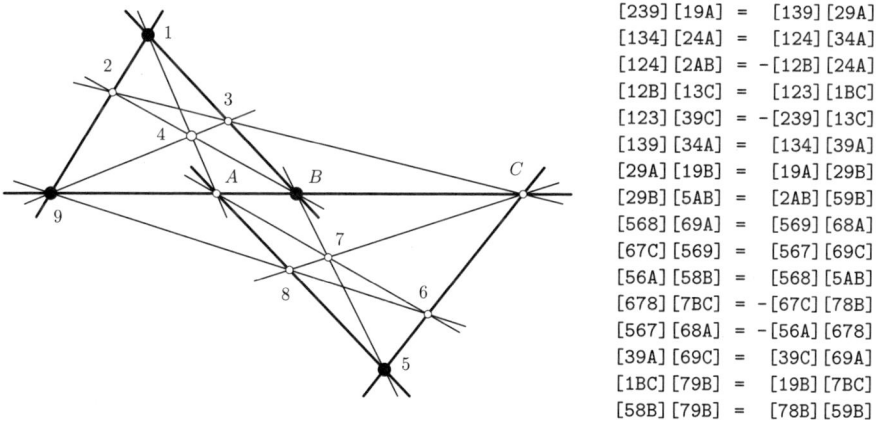

$$\begin{aligned}
[239][19A] &= [139][29A] \\
[134][24A] &= [124][34A] \\
[124][2AB] &= -[12B][24A] \\
[12B][13C] &= [123][1BC] \\
[123][39C] &= -[239][13C] \\
[139][34A] &= [134][39A] \\
[29A][19B] &= [19A][29B] \\
[29B][5AB] &= [2AB][59B] \\
[568][69A] &= [569][68A] \\
[67C][569] &= [567][69C] \\
[56A][58B] &= [568][5AB] \\
[678][7BC] &= -[67C][78B] \\
[567][68A] &= -[56A][678] \\
[39A][69C] &= [39C][69A] \\
[1BC][79B] &= [19B][7BC] \\
[58B][79B] &= [78B][59B]
\end{aligned}$$

How can we derive a Ceva/Menelaus proof from such a given binomial proof? How can we reveal and reconstruct the manifold structure? Here we will give a kind of cooking recipe without a formal explanation why this recipe works. As a first step we reconstruct a suitable substructure of the underlying basis graph. For each binomial equation of the proof we form a quadrangle with bicolored vertices and label the vertices by the corresponding bases. The bases on the left sides of the binomial equations will be colored white, the others will be colored black. We then try to stick the quadrangles together such that each white vertex meets its black counterpart (see picture below – capital letters on an exterior edge mean that this edge has to be identified with the corresponding edge labeled by the identical letter). This will always be possible since it was the defining property of a binomial proof. We will now interpret the quadrangles as substructures that are used to link Ceva/Menelaus configurations. For each quadrangle we have to make a decision which pair of opposite sides is considered as carrying the information of the length ratio. In our picture we have drawn these edges slightly darkened (observe that there is a lot of freedom in this choice). If two quadrangles are attached to each other, such that the two dark edges coincide we can neglect these two edges for the further considerations. We can also neglect the non-darkened edges. We are left with a collection of darkened edges that will form several edge-disjoint cycles. In our example we have seven such cycles: five triangles, one quadrangle and a pentagon. The choice which pair of edges will be darkened has a great influence on the number and size of cycles we will get. In our example we made this choice in a way that a maximum number of dark edges became identified and such that we get many small cycles.

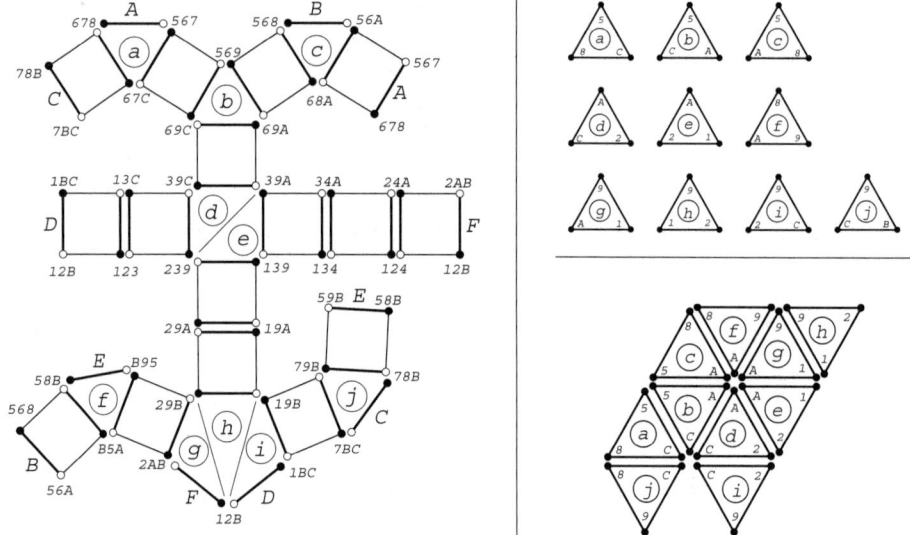

Now, Maurer's homotopy theorem tells us, that each cycle can be decomposed into triangles (perhaps for this we have to use new bases that are not already present in the substructure constructed so far. In our example this is not the case). In our example the quadrangle can be decomposed into two triangles and the pentagon is decomposed into three triangles. The collection of all these triangles forms the support for all our Ceva/Menelaus configurations that are needed for a Ceva/Menelaus proof. From the bases labels at the vertices of the graph we can easily read off whether a triangle has to be a considered as a Ceva or as a Menelaus configuration. In our example only the triangles labeled "c" and "h" carry Ceva configurations (in fact they are toothpicks). All other triangles are Menelaus configurations. We can also easily read off the vertices for each triangle. Finally, we can take all these triangles with vertices labelled by the corresponding points and can construct a closed and oriented manifold from them. In fact, the basic gluing structure is already given by the chaining quadrangles. However, we have to be a little careful to assign the correct orientation to the Ceva configurations (in the Ceva/Menelaus manifold this will be opposite to the picture of the quadrangle cancellations.). In our example the resulting manifold (right lower picture) turns out to be a sphere. This finishes the process of translating a binomial proof into a Ceva/Menelaus proof.

7. Spaces of theorems

In this section we want to give a glimpse on another aspect of Ceva/Menelaus proofs — namely how different proofs can be combined to form new proofs of more complicated theorems. We will study this in the simplest possible case, for which indeed even a complete classification is possible.

7.1. Grid theorems. For this we now come back to a structure that we have already met in Section 3.4, when we proved Pappus's Theorem. We will study those incidence theorems for which the Ceva/Menelaus proof requires only Ceva triangles and for which furthermore all Ceva triangles have the same vertices $1, 2, 3$. Pappus's Theorem was the smallest such example. In such a theorem we have three bundles of lines each bundle passing through one of the three vertices. Each Ceva point is met by one line of each bundle. Since the position of each line is uniquely determined by the length ratio of the corresponding Ceva point, there must be a manifold proof if the underlying incidence structure is a theorem. If we have a manifold proof for such a theorem, then adjacent triangles of the manifold must have opposite orientation with respect to the triangle $(1, 2, 3)$. In our configuration we color the Ceva points that use one orientation white and the others black. The crucial condition for us to have a theorem is that along each line we have as many black points as we have white points. The picture below shows the situation for Pappus's Theorem. The drawing in the middle is the situation in which the vertices of the triangle were moved to the line at infinity. We can associate a matrix to the situation that has an entry "+1" for every white point, a "-1" for every black point and zero otherwise. Our theorem property now translates into the fact, that such a matrix has column sums, row sums and sums in the north-east diagonal direction all zero.

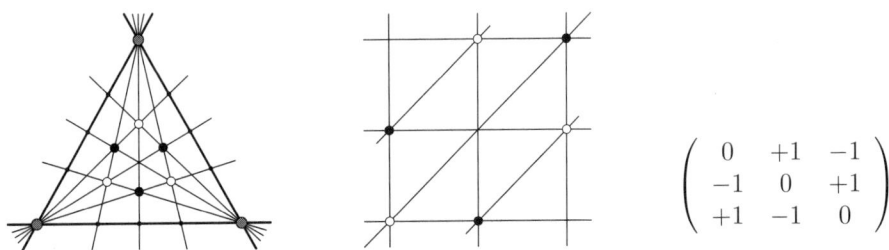

$$\begin{pmatrix} 0 & +1 & -1 \\ -1 & 0 & +1 \\ +1 & -1 & 0 \end{pmatrix}$$

Pappus's Theorem and its underlying matrix

7.2. Spaces of incidence theorems. Now, conversely consider an $n \times n$ matrix with integer entries and column sums, row sums, north-east diagonal sums being zero. Without loss of generality one can assume that at least one entry of the matrix is odd. If not, divide the matrix by a suitable power of two. The non-zero entry of the matrix will be called its *support*. Take a triangle $(1, 2, 3)$ as frame for an incidence configuration. Each row that has a support entry will correspond to a Ceva line through point 1, each column that contains a support entry will correspond to a Ceva line through point 2 and similarly each diagonal to a Ceva point through point 3. The support entries of the matrix correspond to the Ceva points, in which the corresponding row, column and diagonal lines meet. By this construction each support entry corresponds to a Ceva triangle that has to be counted with a multiplicity induced by the value of the matrix entry (the sign indicates the orientation). The cancellation pattern on the manifold is directly induced by the sum= 0 property of the matrix. We can conclude that the Ceva configuration at the odd support entry with value s must satisfy a relation $\left(\frac{|AX|}{|XB|} \cdot \frac{|BY|}{|YC|} \cdot \frac{|CZ|}{|ZA|} \right)^{s} = 1$ and hence (at least over the real numbers) it is the conclusion of an incidence theorem.

Observe that the column sum, row sum, diagonal sum conditions are just linear conditions over the vector space of $n \times n$ matrices. A careful count shows that exactly $4n - 4$ of them are linearly independent. Thus the space of all admissible matrices is generated by a suitable basis of $(n - 2)^2$ configurations (which have to be incidence theorems as well). It turns out that a nice basis is given by the Pappus configurations that are supported by adjacent rows and columns (there are exactly $(n - 2)^2$ of them). The picture below shows an example in which an incidence theorem on a 4×4 grid is generated as the sum of two Pappus configurations. Topologically this process of forming a sum corresponds to a surgery on the two Pappus tori. The tori are identified along two triangles, and after this the triangles are removed from the manifold.

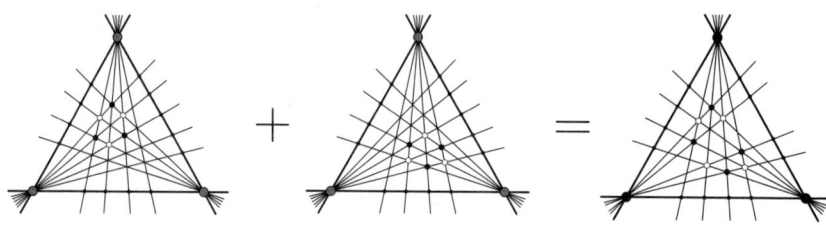

8. Conclusion

We just scratched the surface. In a sense this article is the outline of a whole program: How can cycle structures on manifolds be used to get a deeper understanding of theorems in geometry? There are many things that still have to be investigated. We want to mention a few of them here.

- *What happens if we allow fields other than* $\mathbb{R}$? Over the complex numbers the simple distinction of $+1$ and -1 evolves to an entire spectrum of possible units — new aspects of torsion enter the theory. Already simpler configurations result in non-trivial binomial relations. For instance the fact that A, B, C, D are cocircular in $\mathbb{CP}^1$ results in the fact that the cross ratio of these points is real. It is an easy task to take such cross ratio quadrangles and form manifolds from them that prove incidence theorems (see for instance [**2, 21**]).

- *What about second and third order syzygies?* In this article we studied relations among sub-configurations of an incidence theorem. In the last section we saw that there are cases in which spaces of such theorems emerge. What can we say about relations between these relations, or even about relations on the relations between the relations. At least in the context of the bracket ring such problems have been partially studied [7].

- *Back to non-realizability proofs.* The origin of the whole investigations were non-realizability proofs for arrangements of pseudolines. How can the results in this paper, be applied to get classification results with respect to realizability and non-realizability in this context. In particular, realizability of pseudoline arrangements that consist of three bundles of pseudolines can be characterized completely by the methods in Section 7. This implies also a characterization of liftable rhombic tilings with three directions.

- *Where is the complexity?* Proving theorems is probably hard. So the question arises about the limits of the proving techniques described here. What do theorems look like, that cannot be proved by Ceva/Menelaus proofs? Can the theory be extended to cover even these theorems?

- *Relations to integrability theory.* There is a whole community that works on a setup for a theory of discrete analogues for integrable and differentiable structures (see for instance [1, 3, 4]). In these setups smooth structures are replaced by discrete samples of points that still carry fundamental properties of integrability. One of these fundamental properties is that a homology theoretical discrete generalization of Green's and Stokes's Theorems hold. The structures that appear there are very similar to the cancellation patterns on manifolds. What exactly is the relation?

- *Make good implementations.* Finally, one should take advantage of the knowledge of underlying topological structures to implement automatic geometry provers that do not "blindly" test all cancellation patterns. The topological information should be used to rule out at least the most stupid dead ends of the search tree.

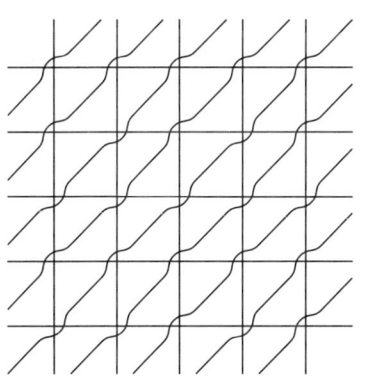 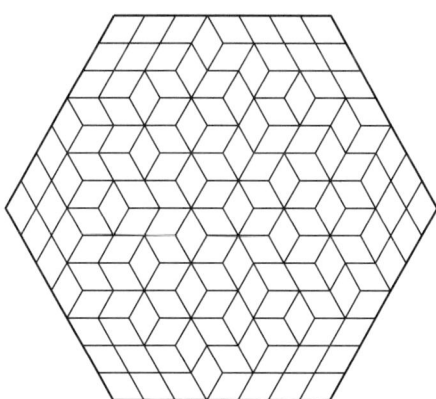

A perturbed arrangement of pseudolines, and a rhombic tiling

References

[1] Adler, V.E., Bobenko, A.I. & Suris, Yu.B., *Geometry of Yang-Baxter Maps: pencils of conics and quadrilateral mappings*, Comm. in Analysis and Geometry, **12** (2004), 967–1007.

[2] Below, A., Krummeck, V. & Richter-Gebert, J., *Matroids with complex coefficients – phirotopes and their realizations in rank 2*, in Discrete and Computational Geometry – The

Goodman-Pollack Festschrift B. Aronov, S. Basu, J. Pach, M. Sharir (eds), Algorithms and Combinatorics **25**, Springer Verlag, Berlin 2003, pp. 205-235.

[3] Bobenko, A.I., Hoffmann, T. & Suris, Yu.B., *Hexagonal Circle Patterns and Integrable Systems: Patterns with Multi-Ratio Property and Lax Equations on the Regular Triangular Lattice*, International Math. Res. Notes, (2002), No 2, 112–164.

[4] Bobenko, A.I., & Suris, Yu.B., *Integrable Systems on Quad-Graphs*, International Math. Res. Notes, (2002), No 11, 574-611.

[5] Bokowski J., & Richter, J., *On the finding of final polynomials*, Europ. J. Combinatorics, **11** (1990), 21–34.

[6] Coxeter, H.S.M. & Greitzer S.L., *Geometry Revisited*, Mathematical Association of America, Washington, DC, 1967.

[7] Crapo, H., *Invariant-Theoretic Methods in Scene Analysis and Structural Mechanics*, J. Symb. Comput., **11**, (1991), 523–548.

[8] Crapo, H. & Richter-Gebert, J. *Automatic proving of geometric theorems*, in: "Invariant Methods in Discrete and Computational Geometry", Neil White ed., Kluwer Academic Publishers, Dodrecht, (1995), 107–139.

[9] Dress, A.W.M., Wenzel, W., *Endliche Matroide mit Koeffizienten*, Bayreuth. Math. Scr., **24** (1978), 94–123.

[10] Dress, A.W.M., Wenzel, W., *Geometric Algebra for Combinatorial Geometries*, Adv. in Math. **77** (1989), 1–36.

[11] Dress, A.W.M., Wenzel, W., *Grassmann-Plücker Relations and Matroids with Coefficients*, Adv. in Math. **86** (1991), 68–110.

[12] Grünbaum B., Shephard, G.C., *Ceva, Menelaus, and the Area Principle*, Mathematics Magazine, **68** (1995), 254–268.

[13] Grünbaum B., Shephard, G.C., *A new Ceva-type theorem*, Math. Gazette **80** (1996), 492–500.

[14] Grünbaum B., Shephard, G.C., *Ceva, Menelaus, and Selftransversality*, Geometriae Dedicata, **65** (1997), 179–192.

[15] Grünbaum B., Shephard, G.C., *Some New Transversality Properties*, Geometriae Dedicata, **71** (1998), 179–208.

[16] Maurer, S.B., *Matroid basis graphs I*, J. Combin. Theory B, **26** (1979), 159–173.

[17] MacPherson, R. & McConnell M., *Classical projective geometry and modular varieties*, in Algebraic Analysis, Geometry and Number Theory: Proceedings of the JAMI Inaugural Conference, ed. Jun-Ichi Igusa, John Hopkins U. Press, (1989), 237–290.

[18] Richter-Gebert, J., *Mechanical theorem proving in projective geometry*, Annals of Mathematics and Artificial Intelligence **13** (1995), 139–172.

[19] Saam, A., *Ein neuer Schließungssatz für die projektive Ebene*, Journal of Geometry **29** (1987), 36–42.

[20] Saam, A., *Schließungssätze als Eigenschaften von Projektivitäten*, Journal of Geometry **32** (1988), 86–130.

[21] Shephard, G.C., *Cyclic Product Theorems for Polygons (I) Constructions using Circles*, Discrete Comput. Geom., **24** (2000), 551-571.

[22] Wenzel, W., *A Group-Theoretic Interpretation of Tutte's Homotopy Theory*, Adv. in Math. **77** (1989), 27–75.

[23] White, N., *The Bracket Ring of Combinatorial Geoemtry I*, Transactions AMS **202** (1975), 79–95.

TECHNICAL UNIVERSITY MUNICH, ZENTRUM MATHEMATIK, BOLTZMANNSTR. 3, D-85748 GARCHING, GERMANY

E-mail address: `richter@ma.tum.de`

Coxeter and the Artists: Two-way Inspiration

Doris Schattschneider

ABSTRACT. H.S.M. Coxeter's delight in geometric form was the catalyst for fruitful two-way interactions with artists, sculptors, and model-makers, many having no formal mathematical training. His articles, books, and personal encouragement inspired them to create breathtaking expressions of the power and beauty of geometry. In many cases, the inspiration was reciprocal: the artists' expressions, often guided by their unique intuition, sparked Coxeter's geometric curiosity, and in their work, he often found new mathematical understanding.

Introduction

"For Coxeter, mathematics is an art that extends infinitely into all the arts, and hence, into life itself." [**H79**] His delight in the beauty and intricacy of geometric form (in all dimensions and spaces) was the catalyst for a lifetime of fruitful two-way interactions with artists, sculptors, and model-makers. Sometimes an artist began with an idea from a letter, publication, or suggestion by Coxeter; more often, intuition was the artist's only guide. Coxeter took great joy in analyzing the work of the artists and artisans, and appreciated their often startling insights and discoveries arrived at by unorthodox means. He enthusiastically shared his mathematical insights gleaned from their work (often baffling the artists with his explanations). Through his countless lectures all over the globe and in his many publications, he brought their work to the attention of a wide audience.

In this article, by telling the stories of several of these interactions, I attempt to reveal a part of Coxeter's life and influence that is little known in the mathematical world. Model-makers Alicia Boole Stott, Michael Longuet-Higgins, Paul Donchian, Magnus Wenninger, and Marc Pelletier, and artists M.C. Escher, George Odom, Peter McMullen, A.G. (Tony) Bomford, John Robinson, George Hart, and Rinus Roelofs are known recipients of Coxeter's wisdom and encouragement. His articles, books, and personal encouragement inspired them to create breathtaking expressions of the power and beauty of geometry. In many cases, the inspiration was reciprocal: the artists' expressions sparked Coxeter's geometric curiosity, providing new problems to solve as well as new ways to understand.

2000 *Mathematics Subject Classification.* Primary 01A60, 01A80, 51M20, 52B15.

The stories of Coxeter's interactions with Alicia Boole Stott, Michael Longuet-Higgins, Peter McMullen, and A. G. Bomford are only briefly summarized here; I give details in [S05].

Alicia Boole Stott (1860-1940), the third daughter (of five) of the prominent mathematician George Boole, was 70 years old when introduced to Coxeter by her nephew Sir Geoffrey I. Taylor. Until her death 10 years later, she shared with Coxeter her friendship and her insights into 4-dimensional polytopes through models; he fondly called her "Aunt Alice." [C87], [C49, pp. 258–259] **Michael Longuet-Higgins** (b. 1925), an avid model-maker from his school days (along with his older brother Christopher), is a physical oceanographer. Between 1940 and 1942, independently of Coxeter and J.C.P. Miller (who had done their work between 1930 and 1932), he and his brother enumerated all but one of the uniform polytopes, for which he fashioned wire edge-models. In 1952, Freeman Dyson (who had been at school with the two brothers) told Coxeter about the models; Coxeter and Miller were then in the final stages of writing up their monumental paper *Uniform Polyhedra* [CHM54]. The models made such an impression that Coxeter and Miller asked Longuet-Higgins to construct the missing model ("Miller's monster") and invited him to become a co-author [H04]. **Peter McMullen** (b. 1942), Professor of Mathematics at University College, London, was a first-year undergraduate at Cambridge when he took it upon himself to work out and make drawings of the projections of all the regular 4-polytopes into corresponding projections of their Petrie polygons. When Coxeter saw these in 1968 he was impressed; he included several of these and other McMullen drawings in his *Regular Complex Polytopes* [C74]. **A.G. Bomford** (1927-2003) was a surveyor for the Royal Corps of Engineers (England) and from 1961-1982, senior surveyor and then director for the Australian Division of National Mapping. An adventurer, he retired at age 55 to travel to remote places to hike, climb, kayak, and explore. At home, he was a patient and precise maker of woolen hooked rugs and polyhedron models of wood. Through a 20-year correspondence with Coxeter, he worked out technical details of hyperbolic designs for the rugs and dimensions for crafting the polyhedra, which included the "golden" zonohedra. [D04], [BC]

Although the line is often blurred between model-makers and artists, we will first discuss those we have cast as model-makers, and then discuss the artists.

The Model-makers

Paul S. Donchian In the summer of 1933, a remarkable set of hand-made models of polytopes of cardboard and of wire was displayed at the Century of Progress Exposition in Chicago, which had the theme of progress through science (Figure 1). During the 1932-33 academic year, Donald Coxeter had been a Rockefeller Fellow at Princeton and at the end of his stay, visited the Chicago exposition with his father, who was visiting for a month. The model-maker was Paul S. Donchian (1895-1967), from Hartford, Connecticut, who for forty years operated a carpet business, inherited from his father. Of Armenian descent, several of his ancestors were jewelers and handicraftsmen; his own craftsmanship might have stemmed from this background. He had no background in higher mathematics, yet at the age of 30, in response to several strange dreams, he determined to make a thorough study of the geometry of hyperspace, aiming to reduce it to simplest terms so that anyone without formal mathematical background could understand it. In pursuit of this goal, he devoted years to making beautiful and precise mathematical models.

FIGURE 1. Donchian models on display at the Century of Progress Exposition. Chicago, 1933. Photo courtesy of Marc Pelletier.

Coxeter was impressed not only by Donchian's models, but also by his unusual insight, and began to make Donchian's work known in the mathematical world. The first occurrence was a joint presentation with Donchian and Patrick Du Val at the American Association of Science meeting in Pittsburgh in January, 1935 [**H35**]. He also saw to it that Donchian's result on an n-dimensional extension of the Pythagorean Theorem was published in the *Mathematical Gazette* (1935), listing Donchian as first author [**DC35**]. In the years that followed, Coxeter displayed and wrote about Donchian's work in several publications. In revising the 11th edition of Rouse Ball's *Mathematical Recreations and Essays* (1938), Coxeter added a new chapter on polyhedra, using photographs of Donchian's cardboard models to illustrate the Platonic and Archimedean solids, equilateral zonohedra, compounds of regular polyhedra, and the Kepler-Poinsot solids. In describing the zonohedra, Coxeter wrote that they "were first investigated by E.S. Fedorov. Their interest has been enhanced by P.S. Donchian's observation that they may be regarded as three-dimensional projections of n-dimensional *hyper-cubes* (or *measure-polytopes*)." [**B38**, p. 141] He then described how to construct an equilateral zonohedron, crediting Donchian with the method.

In his 1947 book *Regular Polytopes*, the culmination of almost 25 years' work, Coxeter used eight full-page photos of Donchian's models. The first three show cardboard models with highly reflective (painted gilt and silver) faces, numbered for ready reference: regular, quasi-regular and rhombic solids (Plate I), some equilateral zonohedra (Plate II), and regular star-polyhedra and compounds (Plate III). In Chapter 13, after proving that every zonohedron that has octahedral or icosahedral symmetry is an orthogonal shadow of a measure polytope γ_n, Coxeter credits Donchian's insight: "The idea of projecting measure polytopes into zonohedra is due to P. Donchian... By an intuitive feeling for the fitness of things, he suspected, many years ago, that his 'spherically symmetrical' zonohedra would be orthogonal shadows of measure polytopes. That conjecture is here, at last, justified." [**C49**, p. 262] The remaining photographs were of edge models of projections

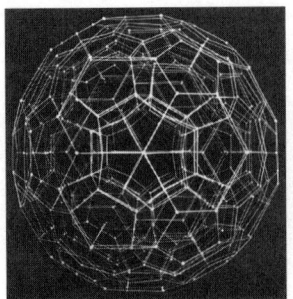
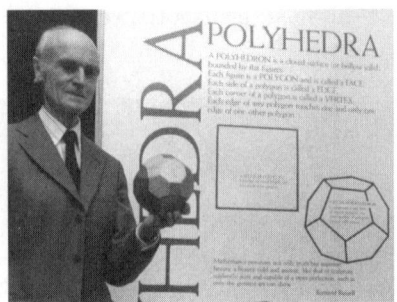

FIGURE 2. Left: Donchian's wire model of $\{5,3,3\}$ from Plate VIII of *Regular Polytopes*. Reprinted with permission of Dover Publications.

FIGURE 3. Right: H.S.M. Coxeter at the opening of the exhibit of Donchian models at the Franklin Institute. Philadelphia, November 1977. Photo courtesy of Marc Pelletier.

of four-dimensional polytopes, fashioned out of "straight pieces of wire with globules of solder for vertices." [**C49**, p. 242] Plate VI showed different projections of $\{4,3,3\}, \{3,3,3\}, \{3,3,4\}$, and $\{3,4,3\}$ (which Coxeter termed "some of the simpler hyper-solids"), IV and VII had different views of the "vertex first" and "cell first" projections of $\{3,3,5\}$, and V and VIII had different views of the "cell first" projection of the 120-cell $\{5,3,3\}$ (Figure 2). Donchian made the models by taking "the central section as an 'exterior shell', but for the rest, he made use of various plane projections published by Schoute and Van Oss, regarding them as plan and elevation, like an architect." [**C49**, p. 260] Coxeter praised the wire models, noting that "parts that would fall into coincidence [by projection] have been artificially separated by slightly altering the direction of projection, or introducing a trace of perspective... [They are] more like portraits, involving a minimum of distortion. As [Donchian] says, '... Each component part is distinctly visible and tangible, in practically its true position and relationship'." [**C49**, p. 243] Photographs of Donchian's wire model of the 120-cell appear in several other Coxeter publications, including [**C57**], and [**C61**].

Ten years after Donchian's death, his brother Dick donated a large collection of his models to the Franklin Institute in Philadelphia, and Coxeter was present at the opening exhibition (Figure 3). The models still reside there, but most are not on display.

Fr. Magnus Wenninger, O.S.B. In 1958, Magnus Wenninger (b. 1919), a Benedictine monk who then headed the mathematics department and taught at St. Augustine's College in Nassau, Bahamas, enrolled in a graduate summer course at Columbia Teachers College in New York City (he completed his Master's degree in Mathematics Education in 1961). It was not the course, but a display of polyhedra models made by Col. Robert S. Beard, housed in a display case in the corridor at Columbia, that piqued his interest in model-making. During the following year, he made his first set of polyhedron models, using Coxeter's Chapter

5 in *Mathematical Recreations and Essays* as his main source. Once started, he was hooked; between 1959 and 1961, he made all the polyhedra in Cundy & Rollett's *Mathematical Models* (in which the authors, in their preface, give credit to Coxeter for "some of the drawings and for his friendly interest and encouragement.") [**CR51**, p. 8]. By now quite proficient, Wenninger then worked out his own nets and built all the stellations in *The Fifty-nine Icosahedra* by Coxeter, du Val, Flather, and Petrie [**CDFP38**]. "The set graced the back wall of my classroom, growing as I completed each one between 1961 and 1963. The average working time spent on each was about eight hours, plus three or four hours each to discover suitable nets." [**W71**, p. ix]

Wenninger's success in this project gave him both an idea and the courage to write to Coxeter, asking about the article *Uniform Polyhedra* [**CHM54**], which had been cited by Cundy and Rollet. This was his first contact with Coxeter, and began a warm relationship that would last until Coxeter's death.

> [Coxeter] sent me a complimentary copy, one of three he still had in his possession. ...[For] the purposes of making the models I inspected the drawings, done by J.C.P. Miller ... to discover the facial planes from which I derived the parts. ...
>
> My working time on the non-convex uniform polyhedron models varied greatly. The simpler ones took three or four hours each, the average would be near eight or ten hours each; a few of the complex models took twenty or thirty hours work. Two of the non-convex snubs required more than 100 hours work each. [**W71**, p. ix–x]

By the time Wenninger and Coxeter first met (in 1968, at an NCTM meeting in Philadelphia), Wenninger had already published a booklet, *Polyhedron Models for the Classroom* [**W66**], and several articles. With Coxeter's encouragement, he produced his now-classic *Polyhedron Models* (1971) for which Coxeter wrote the foreword. "In his infectiously enthusiastic style, the author gives clear instructions for making models of many kinds of polyhedra ... illustrated by photographs of his own collection, including what is almost certainly the only complete set ever made of the known uniform polyhedra. ... The most complicated 'snub' solids are not only extremely difficult to make, but also highly decorative: a perfect instance of the connection of truth and beauty." [**W71**, p. xii] (Figure 4) In 1973, in the third edition of *Regular Polytopes*, Coxeter noted that despite the disinterest of most mathematicians, the subject of polyhedra remained alive, witnessed by the success of L. Fejes Toth's *Regular Figures*, B. Grünbaum's *Convex Polytopes*, and M.J. Wenninger's *Polyhedron Models*.

Wenninger and Coxeter exchanged letters frequently over the years, and often Coxeter would send reprints of recently published articles. It was *Virus macromolecules and geodesic domes* [**C71**] that propelled Wenninger into the production of spherical models from circular bands, culminating in his book *Spherical Models* (1979), which he dedicated "to Professor H.S.M. Coxeter, without whose inspiration this book could never have been written." [**W79**] Coxeter was delighted with the book, and wrote to Wenninger, "It surpasses my highest expectations. Its style and choice of subjects are just right. And I particularly admire your treatment of Geodesic Domes. You have avoided the clumsy work of Pugh (and Bucky himself)!" [**WC**] (Figure 5) Four years later, Wenninger produced his third volume,

FIGURE 4. Left: Magnus Wenninger. Great inverted retrosnub icosidodecahedron, 1970. Paper, diam. approx. 30 in. Model #117 in [**W71**]. Reprinted with permission of Cambridge University Press.

FIGURE 5. Right: Magnus Wenninger. The 8-frequency geodesic dome $\{3, 5+\}_{8,0}$, 1978. Paper, diam. approx. 15 in. From [**W79**, p. 98]. Reprinted with permission of Cambridge University Press.

Dual Models (1983) which complements and completes the others. His preface ends with words that echo Coxeter's, "Only when you handle a model yourself will you see the wonders that lie hidden in this world of geometrical beauty and symmetry." [**W83**, p. xii]

Throughout their 30-year relationship, Wenninger and Coxeter often met at professional meetings and at conferences all over the globe, on geometry, polyhedra, and symmetry. Most often, Coxeter would give an enthusiastic presentation about the wonders of a special geometric object or property, while Wenninger would set up a small exhibit of models and quietly work in his polyhedra-building "shop" at a table in a spot frequented by participants, witnessing in this way to his own enthusiasm and skill in model-making. There were also some personal visits, notably, to Coxeter's home in 1981, and in 1993, when Wenninger was at Dartmouth. Their last time together was at a fitting event: September 2000 at an international symposium on symmetry in Stockholm [**HL00**].

Marc Pelletier On February 15, 2002, in a memorable ceremony, the Fields Institute gave recognition to a very frail Donald Coxeter. The sparkle in his piercing eyes was as keen as ever, and his smile was broad as he viewed, hanging high in the atrium of the Institute, a beautiful steel sculpture of the "3-D shadow of the 4-D dodecahedron (Figures 7, 8). The cell-first projection of $\{5, 3, 3\}$ was made by Marc Pelletier (b. 1959), who had suggested the 95th birthday gift. Pelletier grew up in New Hampshire, and as a young teenager, became interested in polyhedra and geodesic domes through the books of Cundy and Rollett, Buckminster Fuller, and Coxeter's *Regular Polytopes* (where he was struck by the photo of Donchian's wire model shadow of the 120-cell $\{5, 3, 3\}$). When he read Baer's *Zome Primer* [**B70**] he was convinced that he could make a model of the 120-cell using Baer's system, which

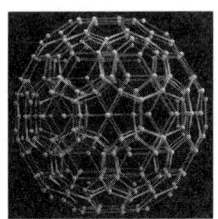

FIGURE 6. Left: Model of $\{5,3,3\}$ made with *Zometool*.

FIGURE 7. Right: Mobile sculpture of $\{5,3,3\}$ by Marc Pelletier, at the Fields Institute, 2002. Arc-welded stainless steel, diam. 5 ft. Photo by Amina Buhler-Allen.

was based on icosahedral symmetry. At the age of 18, he hitchhiked to Albuquerque to meet Baer and find out more about the '31-zone system.' ("Zome" is a melding of "zonohedron" and "dome".) He built a crude first model that confirmed his belief, and over the next ten years, working with Paul Hildebrandt, developed a ball joint that would become the unique feature of today's *Zometool*, an injection-molded plastic ball-and-stick system based on sticks in golden ratio, and joints that can join them in any of the 31 directions of the rotation axes of an icosahedron. When the first balls and sticks were produced in June 1992, Pelletier took them to the Art and Mathematics conference held at SUNY, Albany and there, to the amazement of participants, built a waist-high projection of the 120-cell (Figure 6).

Pelletier was a devoted admirer of Coxeter and made his first contact with a telephone call to introduce himself and tell him about Baer's system. Baer writes

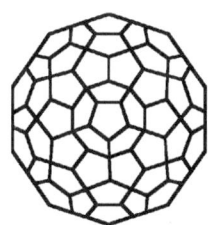

IN APPRECIATION OF
PROF. H.S.M. (DONALD) COXETER
for his lifelong contributions to Geometry

FIGURE 8. Left: Donald Coxeter gazing at Pelletier's sculpture (see Figure 7). Photo by Amina Buhler-Allen.

FIGURE 9. Right: Icon and partial inscription on the plaque presented to Coxeter Feb. 15, 2002.

that Marc was an evangelist for the system in the 1980s, "For years Pelletier wandered the country extending knowledge of the system and sharing it with all who were interested." [B92, p. 209]. Later he made a visit to Coxeter's home in Toronto. When Coxeter learned that Pelletier was amassing all he could find about Paul Donchian, he gave him his Donchian file of clippings and some correspondence. He also wrote an endorsement for *Zometool*, "Zome System considerably simplifies the procedure of construction and unifies the study of space frame structures into one coherent system, of great educational value in the teaching of solid geometry, science, art, engineering and architecture." [Z]

Pelletier's acquaintance with another model-maker was serendipitous: drawings in the margins of a used book [H71] he owned were done by Vincent Roark, a prolific artist with a passion for 3-and 4-dimensional polytopes. Pelletier traveled to Kansas City to meet the artist, and in the early 1990s, the two collaborated on the design and production of a few large arc-welded stainless steel models of the 120-cell, for public venues and private collections. The gift for Coxeter's birthday was an edition of this sculpture. It should be noted that Pelletier's sculpture differs from Donchian's wire model in that it has only half as many edges; those that would be directly behind a visible edge in the cell-first projection are not included. At the celebration, John H. Conway spoke on new ideas on polytopes and Pelletier gave a presentation on Donchian's work; the Fields web site [F02s] contains slides and audio of these. A plaque was presented to Coxeter, engraved with an icon representing the sculpture (Figure 9), and expressing appreciation to "Prof. H.S.M. (Donald) Coxeter, for his lifelong contributions to Geometry."

The Artists

M.C. Escher In 1954, Coxeter attended the International Congress of Mathematicians in Amsterdam. For this occasion, a special exhibition of the work of the Dutch graphic artist M.C. Escher (1898-1972) was organized at the Stedelijk Museum. N.G. de Brujn wrote in the catalog, "In view of the fact that Mr. Escher's work may be said to be a point of contact between art and mathematics, the Organizing Committee ... took the initiative in inaugurating this exhibition. Probably mathematicians will not only be interested in the geometrical motifs; the same playfulness which constantly appears in mathematics in general and which, to a great many mathematicians is the peculiar charm of their subject, will be a more important element." [dB54] This exhibition introduced Congress participants to many of Escher's fascinating prints and carved wooden balls, as well as some of the artist's hand-drawn and colored drawings of periodic tessellations of reptiles, birds, and other interlocked creatures. Coxeter had a special introduction to the artist through his wife, Hendrina (Rien). He related that "being Dutch, [she] naturally talked to [Escher] when he was exhibiting his art to the mathematicians. So she got to know him, and that was very helpful; we kept up correspondence." [H96]

Three years later, Coxeter's Presidential address to the Royal Society of Canada was devoted to the topic of symmetry in the Euclidean and hyperbolic planes, and on a sphere's surface [C57]. He remembered Escher's charming drawings of tessellations, in which the fundamental region for a symmetry group was a half-bug or a whole horseman (rather than a mathematician's parallelogram or triangle), and wrote to Escher, requesting the use of two drawings to illustrate planar symmetry. Escher happily agreed, and Coxeter's article became the first mathematical

FIGURE 10. Left: Escher's circles and lines penciled on the reprint sent by Coxeter. Photo and computer overlay by Doris Schattschneider.

FIGURE 11. Right: M.C. Escher. *Circle Limit I*, November 1958. Woodcut, diam. 16.5 in. All M.C. Escher works ©2004 The M.C. Escher Company, The Netherlands. All rights reserved. Used by permission.

publication to contain an Escher tessellation. As a courtesy, Coxeter sent Escher a reprint of the article, which, as he learned six months later, provided an unintended jolt to the artist. Escher wrote to Coxeter (December 5, 1958),

> Did I ever thank you for sending me ... "A Symposium on Symmetry"? I was so pleased with this booklet and proud of the two reproductions of my plane patterns!
>
> Though the text of your article ... is much too learned for a simple, self-made plane pattern-man like me, some of the text-illustrations and especially figure 7, page 11, gave me quite a shock. ...
>
> I tried to find out how this figure was geometrically constructed, but I succeeded only in finding the centres and the radii of the largest inner-circles (see enclosure). If you could give me a simple explanation how to construct the following circles, whose centres approach gradually from the outside till they reach the limit I should be immensely pleased and very thankful to you! Are there other systems besides this one to reach a circle-limit? [**C79**, p. 19], [**EC**]

Escher's enclosure was a large drawing on tracing paper of intersecting circles that showed he had figured out much of the basic scheme of Coxeter's Figure 7, which was a hyperbolic tessellation of triangles. He had first worked over the reprint's figure with a compass, trying to decipher the pattern of circle centers. Figure 10 shows Escher's penciled circles and lines (enhanced by computer overlay) that he drew on the reprint he received from Coxeter. Escher also sent Coxeter, by sea post, two prints: his first *Circle Limit* print of rather primitive black and white fish, which made clear how much he had understood (Figure 11), and a "line limit" print of reptiles [**Es57**].

Coxeter replied on December 29, before the arrival of the prints,

I am glad you like my "Fig. 7", and interested that you succeeded in reconstructing so much of the surrounding "skeleton" which serves to locate the centres of the circles. This can be continued in the same manner. For instance, the point that I have marked on your drawing (with a red o on the back of the page) lies on three of your circles with centres 1, 4, 5. These centres therefore lie on a straight line (which I have drawn faintly in red) and the fourth circle through the red point must have its centre on this same red line.

In answer to your question "Are there other systems besides this one to reach a circle limit?" I say yes, infinitely many! This particular pattern is denoted by [4,6] because there are 4 white and 4 shaded triangles coming together at some points, 6 and 6 at others. But such patterns $[p, q]$ exist for all greater values of p and q and also for $p = 3$ and $q = 7, 8, 9, \ldots$. See Coxeter and Moser, "Generators and Relations", Ergebnisse der Math., Heft. **14** (Springer, Berlin 1957) bottom of p. 53. A different but related pattern, called $\{{}^p_q\}$, is obtained by drawing new circles through the "right angle" points, where just 2 white and 2 shaded triangles come together. I enclose a spare copy of $\{{}^3_7\}$ .... If you like this pattern with its alternate triangles and heptagons, you can easily derive from [4,6] the analogous $\{{}^4_6\}$, which consists of squares and hexagons. [**EC**]

While Coxeter no doubt thought that his letter answered Escher's question and provided help to the artist, Escher, having no formal mathematical training beyond high school, was sorely disappointed. Coxeter had pointed out the location of exactly one more circle center on Escher's drawing, explained some notation, and said there were an infinite number of such hyperbolic tessellations. Yet, as was to happen in encounters with other artists, Coxeter's "help" only increased the artist's self-reliance on figuring things out for himself. Escher wrote to his son George (Feb. 15, 1959),

Coxeter's letters show that ... an infinite number of higher values can be used as a basis [for hyperbolic tessellations]. He encloses an example of using the values three and seven, of all things! However, this odd seven is no use to me at all; I long for two and four (or four and eight) because I can use these to fill a plane in such a way that all the animal figures whose body axes lie on the same circle also have the same "colour", whereas in the other example [Circle Limit I] two white ones and two black ones constantly alternate. My great enthusiasm for this sort of picture and my tenacity in pursuing the study will perhaps lead to a satisfactory solution in the end. Although Coxeter could probably help me by saying just one word, I prefer to find it myself for the time being, also because I'm so often at cross-purposes with those theoretical mathematicians, on a variety of points. In addition, it seems to be very difficult for Coxeter to write intelligibly to a layman. Finally, no matter how difficult it is, I feel all the more satisfaction from solving a problem like this in my own bumbling fashion. [**B82**, p. 92]; [**C85**, p. 64]

After his first attempt with *Circle Limit I*, much of Escher's energy in 1959-1960 was spent on working out (in his own way) the geometric underpinnings of other hyperbolic tessellations and creating three new circle limit prints that "captured infinity" within the confines of a circle. When writing to George about doing this work, he called it "Coxetering." [**B82**, p. 94]. (In the mathematical world, Coxeter's name is an adjective for several mathematical structures, but for Escher, it was a verb.) Escher was especially pleased with his complex and colorful *Circle Limit III* (Figure 12), based on the hyperbolic tessellation $\{8, 3\}$, which overcame what he saw as shortcomings of *Circle Limit I*.

On May 1, 1960 he sent Coxeter the print, explaining, "The whole area is filled up with series of theoretically an endless number of fishes, swimming head-to-tail in the same colour. The white curved lines through their bodies accentuate the continuity of every series. A minimum of four wood blocks, one for [each of the four colors] and a fifth for the black lines was needed. Every block has roughly the form of a segment of 90°. This implicates [sic] that the complete print is composed of $4 \times 5 = 20$ printings." [**B82**, p. 100], [**EC**] In that same letter, Escher mentioned that he would present a slide lecture at the 1960 summer meeting of the International Union of Crystallography, in Cambridge, England, and offered to give the lecture in Toronto in the Fall, when he would be visiting his son in Canada. Coxeter immediately wrote to Dr. Lee Ritcey, asking for the sponsorship of the Canadian Mathematical Congress, "[Escher's] latest woodcut (in 4 colors and black, of the Miraculous Draught of Fishes) is quite remarkable and has interesting applications to the theory of groups... . This looks like a fine opportunity for Canadian mathematicians and artists to see something unusual and to learn to appreciate each other in a new way." [**EC**]

In his letter of thanks to Escher (May 16, 1960), Coxeter told the artist the print was "very successful; both interesting and beautiful." He couldn't resist explaining the print's rich mathematical content (with several detailed references to his *Regular Polytopes* and *Generators and Relations for Discrete Groups*), pointing out the generators of its symmetry group in hyperbolic geometry, the placement of various vertices of the $\{8, 3\}$ tessellation, centers of rotational symmetry, the implications for coloring seen in the "compound $\{3, 8\}[6\{8, 8\}]\{8, 3\}$ of six $\{8, 8\}$'s inscribed in a $\{3, 8\}$," and an idea for a "handsome piece of modern sculpture" obtained by identifying one octagonal face of $\{8, 8\}$ with its pairs of opposite sides to obtain a surface of genus two, like "the surface of a solid figure-of-eight." [**EC**] Again, Escher was overwhelmed by Coxeter's explanations, and wrote to George, "I had an enthusiastic letter from Coxeter about my coloured fish, which I sent him. Three pages of explanation of what I actually did... It is a pity that I understand nothing, absolutely nothing of it... ." [**B82**, p. 100–101]

In the Fall of 1960 Escher gave two invited lectures at the University of Toronto and was hosted overnight at the Coxeter's home. He expressed his appreciation with a gift of the just-published book *Grafiek en Tekeningen M.C. Escher* [**Es60**] which he inscribed "To Professor and Mrs. H.S.M. Coxeter, with many thanks for his helpful hand with 'Circular Limits', which enabled me to make the woodcut no. 10 of this book [*Circle Limit I*, in Dutch, *Kleiner en Kleiner II*]. Toronto, November 9th, 1960."

FIGURE 12. Left: M.C. Escher. *Circle Limit III*, December 1959. Woodcut in four colors and black, diam. 16.375 in.

FIGURE 13. Right: M.C. Esher *Square Limit*, April 1964. Woodcut in two colors, 13.375 in. × 13.375 in.

Indirectly, Coxeter was instrumental in introducing Escher's work to a wide public audience in North America. During the post-Sputnik period, the U.S. government generously funded mathematical training, and Coxeter gave courses of lectures at summer institutes for school teachers as well as public lectures in New York City to the Friends of Scripta Mathematica. His lectures on geometry were wide-ranging, far beyond the usual high school fare, and included an emphasis on transformations. These lectures became the basis for his innovative but now-classic text *Introduction to Geometry*, published in 1961. Chapter 4, "Two-dimensional crystallography" illustrated the principles of planar symmetry groups with the two Escher tessellations shown in his 1957 article on symmetry. Although today it is commonplace for Escher prints or tessellations to be found in mathematical texts, Coxeter was the first to use Escher's work in this way. The unusual text caught Martin Gardner's attention, who devoted his entire "Mathematical Games" column in the April 1961 *Scientific American* to an enthusiastic review. Roughly one-quarter of the review discussed Escher's work, and reproduced an Escher tessellation from Coxeter's text; in addition, the cover of that issue displayed an Escher tessellation of birds ("colorized" by an editor). [**G61**]

In 1964, Escher again wrote to Coxeter and sent him a print, *Square Limit* (Figure 13). In his usual self-effacing way, Escher explained to "Professor Coxeter" the geometric underpinnings of the print, "I fear that the subject won't be very interesting seen from your mathematical point of view, because it's really simple as a flat filling. None the less it was a headaching job to find an adequate method to realise the subject in the simplest possible way. After many other tentatives I chosed [sic] the scheme shown in this figure." [**EC**] The scheme, drawn on squared graph paper, showed the dissection of a square into four isosceles right triangles, and in each of these, a self-similar series of ever-diminishing right triangles that followed an algorithm of dissection (rather than diminishing by dilatation), having as limit the edge of the square. After successfully completing four hyperbolic circle limit prints, Escher had devised his own Euclidean method of creating a square limit, extending to four directions the limiting process of his earlier "line limit" print sent to Coxeter in 1958.

Coxeter wrote more articles about Escher's work than that of any other artist. His first was an invited essay, *The Mathematical Implications of Escher's Prints* for the catalog of the 1968 Escher retrospective exhibition at the Gemeentemuseum in the Hague, The Netherlands [**C68**]. The essay gives a mathematical tour of Escher's work, pointing out properties and some history of Möbius bands, poly-hedra, transformations, symmetry groups, color symmetry, loxodromes, hyperbolic tessellations and their duals, and impossible constructions (which Coxeter called "whimsical works"). After discussing some of the mathematical properties of *Circle Limit III*, Coxeter remarked, "In my opinion, this woodcut would have been still more beautiful without the white arcs, which artificially divide each fish into two unequal parts and have no mathematical significance." When three years later, his essay was included in the book *The World of M.C. Escher* [**C71a**, p. 51], Coxeter had evidently changed his mind about the remark and omitted the last five words.

Escher died in 1972, so never saw the two articles devoted to *Circle Limit III* that Coxeter later wrote, celebrating those very white arcs. (In fact, Coxeter once expressed to me his regret that he was never able to tell Escher personally how perfect his work was.) In *The Non-Euclidean Symmetry of Escher's Picture 'Circle Limit III'* [**C79**] he wrote,

> Of all Escher's pictures with a mathematical background, the most sophisticated is his 1959 woodcut, *Circle Limit III* which uses four colours in addition to black and white. Queues of fishes of each colour are swimming along white arcs that cut the pe-ripheral circle at a certain angle. ... [We] shall see why *all* the white arcs 'ought' to cut the circumference at the same angle, namely 80^0 (which they do, with remarkable accuracy). Thus Escher's work, based on his intuition, without any computation, is perfect, even though his poetic description of it... 'perpendic-ularly from the boundary' was only approximate. [pp. 19–20]

What Coxeter had discovered after careful study (including measuring angles in the print) was that the arcs were not hyperbolic lines, as he (and others) first supposed, but rather, branches of equidistant curves that cut through corresponding vertices of the octagons of the underlying tessellation $\{8, 3\}$. He used hyperbolic trigonometry to prove the asserted angle measure. (Notably, the article's Figure 1 was a version of Coxeter's earlier infamous Figure 7, but now with the full scaffolding that Escher had hoped to be shown 21 years earlier.) In 1996, Coxeter revisited the same problem (of computing the angles the arcs made with the bounding circle) in *The Trigonometry of Escher's Woodcut "Circle Limit III"*, this time giving a simplified argument using "the elements of trigonometry and the arithmetic of the biquadratic field $Q(\sqrt{2} + \sqrt{3})$: subjects of which [Escher] steadfastly claimed to be entirely ignorant." [**C96**, p. 42]

Coxeter wrote three other lengthy articles on symmetry, each richly illustrated with Escher's work. In *Angels and Devils*, he began, "About forty years ago, Abra-ham Sinkov and I wrote twin papers on the subject of groups determined by the periods of two generators S, T and of their commutator $S^{-1}T^{-1}ST$, never dreaming that twenty years later M.C. Escher would be using such groups (unconsciously) as symmetry groups for a carved ball and four other works of art." [**C81**, p. 197] He then explains symmetry groups in the plane, on the surface of a sphere, and in the hyperbolic plane, as exemplified by Escher's renditions of interlocked angels and devils on those three surfaces. After discussing the hyperbolic symmetry group $[4^+, 6]$ of *Circle Limit IV*, he writes,

> Escher used the analogous group $[3^+, 8]$ for his *Circle Limit II*, which is just as interesting mathematically although Bruno Ernst [in [**E76**]] dismissed it with a little joke. The underlying tessellation $\{3, 8\}$ has for vertices the centers of all the crosses. The points where the three colors come together, being the centers of the triangular faces of $\{3, 8\}$ are the vertices of the dual tessellation $\{8, 3\}$. On the other hand, the centers of the crosses of one color are the vertices of a tessellation of quadrangles, $\{4, 8\}$. ... In this sense, Escher may be said to have anticipated my discovery of the regular compound tessellation $\{3, 8\}[3\{4, 8\}]2\{8, 3\}$ ... which consists of 3 superposed tessellations $\{4, 8\}$ whose vertices belong to a single $\{3, 8\}$ while their faces have the same centers as the faces of the dual $\{8, 3\}$, each used twice. [**C81**, pp. 207–208]

In 1985, in his lengthy book review of *M.C. Escher: His Life and Complete Graphic Work* [**C85**], Coxeter revisited much of his earlier symmetry analyses of the circle limit prints, but also included new information on the hyperbolic geometry of *Circle Limit I*, and an analysis of Escher's compound of three octahedra in his print *Stars*. In *Coloured Symmetry*, Coxeter addressed the subject of color symmetry groups, introducing a definition of a "coloured symmetry group with N colours to be a set of three groups related in a special way, namely, a space group G and normal subgroup G_1 of finite index, along with a representation of the quotient group $\Gamma \cong G/G_1$ as a permutation group of degree N." [**C86**, p. 21] Eleven of Escher's tessellations are analyzed to illustrate. The last two sections of the article describe in detail how one could cover a torus with a cluster of 18 of Escher's butterflies, or cover a double torus with 24 of Escher's fish from *Circle Limit III* (as he had described to Escher in his May 1960 letter). In 1985, at Coxeter's urging, Douglas Dunham carried out the second project [**D86**]. In *Escher's Lizards*, Coxeter continued the discussion of color symmetry, admitting that "One of [Escher's] patterns of lizards, in which 4 colours are permuted by the 'octic group' D_4 is so subtle that I described it incorrectly in my talk on *Coloured Symmetry*. The correction of this error (which nobody noticed) provides an opportunity for a brief survey of the theory, and a comparison of the two notations $G/G_1 \cong \Gamma$ and $[G : H] = N$." [**C88**, p. 23]

All the Escher works mentioned here can be found in [**B82**] and [**S90**]; Dunham's article [**D03**] displays the circle limit prints with overlays and explanations of their underlying hyperbolic tessellations of polygons.

George Odom In 1983, the *American Mathematical Monthly* published problem E 3007, proposed by George Odom, Poughkeepsie, NY:

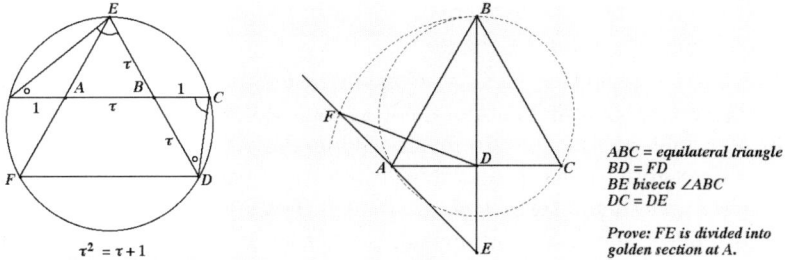

FIGURE 14. Left: Van de Craats' solution to *Monthly* problem E 3007.

FIGURE 15. Right: Another construction of the golden section by Odom.

> Let A and B be the midpoints of the sides EF and ED of an equilateral triangle DEF. Extend AB to meet the circumcircle (of DEF) at C. Show that B divides AC according to the golden section. [**O83**]

Although the discovery of this unexpected and beautifully simple method of constructing the golden section had indeed been Odom's, the person who formulated it as a problem and sent it to the *Monthly* was H.S.M. Coxeter. George Phillips Odom, Jr. (b. 1941) began his correspondence with Coxeter in the mid 1970's because he felt his work had "some mathematical interest." Over the years, their correspondence continued; they discussed not only Odom's mathematical insights, but psychology, philosophy, religion, and world affairs. Odom had no formal mathematical background, but had the ability to "see" mathematical relationships, and made drawings and models of his discoveries, communicating them to Coxeter. Over the years, Odom suffered intense periods of terrible depression, and since the early 1980s, he has been a resident at a psychiatric center. Coxeter's recognition and encouragement of Odom's mathematical insights, as well as his continuing friendship, were especially meaningful to this isolated man.

The solution to Odom's problem by Jan van de Craats was published in the *Monthly* three years after it was posed, and it was as beautiful as the relationship itself: no words accompanied it and none were necessary (Figure 14) [**vdC86**]. Odom returned to the golden section again and again, recasting the result several times in 3-dimensional versions that exploited the properties of regular polyhedra, and communicated these to Coxeter and also to Magnus Wenninger, who in turn communicated them in lectures, articles, and web-based forums on polyhedra. Three versions come from nesting the regular octahedron and tetrahedron in a cube:

(1) A line segment that joins midpoints A and B of adjacent edges of a regular tetrahedron and is extended from B to meet the circumsphere of the tetrahedron, is divided into golden section by the point B.

(2) A line segment that joins the centers A, B of adjacent faces of a cube and is extended from B to meet the circumsphere of the cube, is divided into golden section by the point B.

(3) If a regular octahedron is inscribed in a cube (with its vertices at face centers of the cube), and one edge AB of the octahedron is extended from

B to meet the circumsphere of the cube, this extended segment is divided into golden section by the point B.

A still different construction (sent to Coxeter in February, 1997) comes from the fact that the vertices of a regular icosahedron coincide with those of three mutually perpendicular golden rectangles; from this, Odom derived the drawing in Figure 15. Coxeter at first wrote to Odom that this "is not even a close approximation" to the golden section, but then later recanted, writing "I owe you a profound apology for my careless remark about [your drawing]. As Father Magnus kindly pointed out, I was mistaken: this new construction for the golden section is quite correct." The letter was signed "Your friend, HSM Coxeter." [OC]

A three-dimensional cardboard sculpture of Odom's (done in 1984) intrigued Coxeter: this consisted of four brightly painted "hollow" equilateral triangles that were interlocked, forming a beautifully rigid, symmetric form (Figure 16). Although Coxeter may have had the sculpture for some time, it was another sculpture of three interlocked hollow triangles, by John Robinson, done quite independently in time and place, that triggered Coxeter to write an article [C94] in which he closely analyzed each sculpture. (Robinson's sculpture is discussed in our next section.) In Odom's sculpture, Coxeter found an unusual and aesthetically satisfying form whose symmetry group is S_4, the group of permutations of four objects (in this case, the 4! rotations that permute the four interlocked triangles.) Each component triangle has the edges of its inner (cut out) triangle half as long that of the outer triangle, and in the assembled sculpture, the midpoints of the outer edges meet the corners of the inner triangles. In the article, Coxeter points out that a cuboctahedron may be coordinatized in 4-space with vertices the 12 permutations of $(1, 0, 0, -1)$, and by using this embedding and that of a concentric cuboctahedron obtained by doubling the coordinates of the vertices, one can join the edges of the two homothetic figures in such a way as to obtain the four interlocked triangles of Odom's sculpture.[1] Figure 17 shows how one hollow triangle is situated in this construction.

When Magnus Wenninger notified Odom of Coxeter's death in 2003, his reply was poignant, "I don't know what I would have done without you, Socantes (his doctor), and Coxeter. The three of you have been my only contact with the human race." [OW]

John Robinson In 1992, Coxeter received an unexpected gift of a book, *Symbolic Sculpture* [Ro92], sent by the author at the urging of his friend, mathematics professor Ronald (Ronnie) Brown at the University of Wales, Bangor. The author was sculptor John Robinson (b. 1935), whose work was known in England, but relatively unknown elsewhere. Robinson grew up in England, departed at age 17 for Australia, eventually established a sheep farm there, and then in 1969, sold the farm and returned with his wife and sons to England, intending to stay two years. With no formal art training, but a love of making sculpture, first awakened in boarding school and then reawakened once the sheep farm was successful, John

---

[1]Odom and Coxeter were both unaware of the earlier discovery of the four interlocked equilateral triangular rings by Alan Holden, who pictures the sculpture made of wooden dowels in [H71, p. 182], indicating that it can arise from what he terms a "nolid", a polyhedron whose regular polygon faces interpenetrate and has zero volume. He writes "A curious nolid arises from using the twelve corners of a cuboctahedron as the vertices of four equilateral triangles. It has all the axes of rotation symmetry of a cube, but it lacks planes of reflection symmetry and therefore comes in left- and right-handed forms." [H71, p. 125] In a later paper [H80], he uses the term "regular polylink" for the sculpture of four linked hollow triangles.

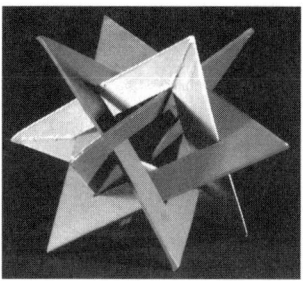
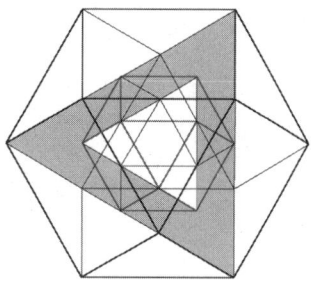

FIGURE 16. Left: George Odom. Four interwoven triangles, 1984. Painted cardboard. Photo by Alfred Thomas for [**C94**]. With kind permission of Springer Science and Business Media.

FIGURE 17. Right: The vertices of the triangles in Odom's sculpture coincide with the vertices of a cuboctahedron.

established a studio and sculpted full time. At first his figures were of children, athletes, and other human forms, which sold well, then gradually he turned to abstract forms, calling them "symbolic sculpture." He conceives these graceful forms intuitively, beautifully executing ovoids, knots, tori, borromean rings, cones, pyramids, spirals, helices, and other symbolic portrayals of his "interpretation of the cosmos, earth, animals, man and woman, birth, death, religion, civilisation, and the future." [**Ro92**, p. 106]

Coxeter was fascinated by what he saw in Robinson's book and thanked him with a copy of his *Introduction to Geometry*, pointing out several figures that might be of interest. He also said he would put Robinson's book on display at the ICME in Quebec that summer, and mention Robinson (and Escher) in his talk in the session on Mathematics and Art. One particular sculpture, *Intuition* (Figure 18), formed by interwoven hollow triangles, especially caught Coxeter's attention. On his web page, Robinson writes, "For me this sculpture represents a knotted core of stability within the centre of knowledge, from which comes sparks of originality and invention, often for no apparent reason. We call these sparks INTUITION. The sparks shoot in all directions, but come from the core of experience." [**RoI**] Coxeter wrote to Robinson that he planned to write an article about the sculpture, comparing it with Odom's. "Your title 'Intuition' was well-chosen because, although I am quite sure that George never saw any work of yours, there is something uncanny in their several points of resemblance: he and you had a similar 'intuition'." [**RoC**] Working only from the photo, Coxeter had assumed, and told Robinson in a letter, that the outer edges of his hollow triangles were twice the length of the inner edges, and that an inner vertex of any one of the triangles coincided with the midpoint of an outer edge of another. He also noted Odom's and Robinson's sculptures had an essential difference: while every two triangles in Odom's sculpture were interlocked, Robinson's three triangles were interlocked in the manner of Borromean rings. He closed his letter with "P.S. One important question: Is your structure inherently rigid, or would it collapse if placed on a slippery floor?" Robinson replied

FIGURE 18. Left: John Robinson. *Intuition*, 1992. Polished stainless steel, height 4 feet. Installed at the Fields Institute, Toronto. Photo courtesy of the Fields Institute.

FIGURE 19. Right: Equilateral triangles interlocked as in *Intuition* collapse when the ratio of outer to inner edge is 2:1.

immediately, " 'Intuition' is extremely rigid. ... It would make a self-supporting roof for a building." [**RoC**]

The next summer (1993), Coxeter brought a cardboard model of Robinson's sculpture to a workshop at the Regional Geometry Institute at Smith College, and showed participants how it could collapse (see Figure 19). He asked them to determine the point at which, when lifting the triangles to form the three-dimensional form (maintaining trigonal symmetry), it became rigid. (They were surprised when he said he didn't know the answer.) Two articles report the outcome. Coxeter reported [**C94**] that under an additional assumption (that seemed correct in eyeing the cardboard model), the ratio of outer to inner edge of each triangle was not 2:1, but 1.9663265, and the sought-after point at which the triangles locked was not the midpoint of the edge, as he had first assumed. (Even then, he maintained in a letter to Robinson, the sculpture would collapse of its own weight. Robinson steadfastly countered that it was rigid, and sent a sketch of a building whose roof was supported by the sculpture.) That same issue of the *Intelligencer* featured some of Robinson's sculpture at the University of Wales, Bangor, where he is an honorary Fellow [**Br94**]. When Robinson received Coxeter's article, his reply was reminiscent of Escher's reaction to Coxeter's analyses:

> I must confess that I don't understand the mathematics of your essay, but I do get immense satisfaction in looking at the equations and knowing that they relate directly to something that has 'popped' into my brain, while looking at the Borromeo Rings [sic]. The act of 'poppin' is why I called the sculpture INTUITION. [**RoC**]

Three participants continued working on the problem after the workshop and later reported their more general result, that the point of rigidity for the model was a function of the ratio of outer to inner edge of each triangle, which could vary

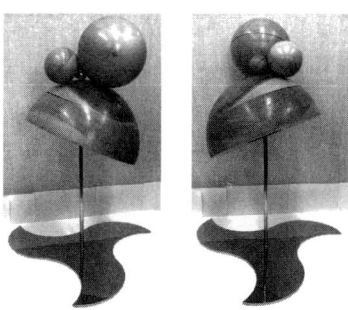

Four circles to the kissing come. ...
The sum of the squares of all four bends
Is half the square of their sum.
To spy out spherical affairs
An oscular surveyor
Might find the task laborious,
The sphere is much the gayer,
And now besides the pair of pairs
A fifth sphere in the kissing shares.
Yet, signs and zero as before,
For each to kiss the other four
The square of the sum of all five bends
Is thrice the sum of their squares. --F. Soddy

FIGURE 20. John Robinson. *Firmament*, 1997. Mutually tangent spheres, smallest of stainless steel, others of wood, diam. 1.5 cm, 2.8 cm, 5.3 cm, 10 cm, 18.8 cm. Base stainless steel, in the form of an 'Andalusian Tile.' Photos by Walter Whiteley.

between 1 and 2, and the model would lie flat when the ratio was 2:1 [**BFG96**].[2] Coxeter quickly noted their results in a postscript to his earlier article [**C95**].

Early in 1996, Robinson wrote to Coxeter after the installation of *Intuition* at the Isaac Newton Institute in Cambridge, mentioning that his friend Frederick Helson of Toronto proposed the idea of celebrating Donald's 90th birthday with an edition of *Intuition* in Toronto. John Chadam, head of the Fields Institute, embraced the idea; subsequently Robinson's patrons made financing possible, and permissions from the city were obtained. That June, while visiting Toronto to plan for the sculpture, Robinson lunched with Coxeter, who offered the artist an idea for a new work of five mutually tangent spheres having radii in geometric progression that would use Soddy's remarkable theorem: *The square of the sum of all five bends is thrice the sum of their squares* [**S36**]. (Here, the "bend" is the reciprocal of the radius.) Returning to England, Robinson wrote, "I was intrigued about your Five Spheres of increasing size fitting perfectly around one of the spheres, and that this process would go on ad infinitum, inwards and outwards. Can I please ask you to give me the magic 5 radii. I can't get the image out of my mind, and would love to experiment with the idea." [**RoC**] Coxeter's reply explained the "easily solved equation $x^5 - x^4 - x^3 - x^2 - x + 1 = 0$, gave its root x between 1 and 2 for which the radii are $\frac{1}{x^2}, \frac{1}{x}, 1, x, x^2$, noted the fact that the sequence could be continued indefinitely (as a loxodromic sequence in which any five consecutive spheres are mutually tangent), and finally ended by giving a numerical sequence of 7 radii, beginning with 1.5 cm and ending with 66.8 cm.

Robinson produced the challenging sculpture as a birthday gift to Coxeter. (Figure 20) The smallest ball is steel, and the four others are wooden, turned on a lathe, with the largest only a hemisphere. Coxeter's letter to Robinson is inside the

---

[2]No one seems to have asked Robinson about the measurements. The triangles in the sculpture at the Fields Institute measure 48 inches (outer edge) and 25.5 inches (inner edge), giving a ratio of 1.882352941. The points at which an inner (cutout) triangle catch the outer edges of a triangle thrusting through the hole measure 19.5 inches and 28.5 inches from the triangle's vertex, in good agreement with the result in [**BFG96**], which predicts those points to be 28.97514221 inches and 19.02473029 inches from the vertex.

top sphere, which is in two pieces. Robinson was awed at the physical realization of those equations:

> Only when I put the jigsaw puzzle together was I able to see the miracle that Donald had perceived through his mathematical vision. ... I called the sculpture FIRMAMENT because it reminds me of the marvellous 19th century working models of the Solar System that fascinated me as a child in the London Science Museum. I didn't understand what I was looking at then, just as I don't understand Donald Coxeter's mathematics now. What I do understand is that the Universe is a Miracle, Man is a Miracle, and that this kind of Mathematics is part of the Miracle. [**RoF**]

On Feb. 9, 1997, *Intuition* was unveiled by Frederick Helson at a gala event at the Fields Institute, and *Firmament* was presented to Coxeter by Ronnie Brown. On that occasion, Brown made a presentation on Robinson's sculptures, and Coxeter spoke about the work of Robinson, Odom, and Escher [**F97**]. Four months later, Donald and Rien showed Robinson his sculpture prominently displayed in their living room. In 2000, despite a mending fractured pelvis, Coxeter insisted on exhibiting *Firmament* and lecturing at a conference in honor of his (now retired) student Donald W. Crowe, so his daughter Susan gamely drove him, along with the sculpture, from Toronto to Madison, Wisconsin.

Tangent circles and tangent spheres were favorite subjects of Coxeter, appearing in *Introduction to Geometry* and many subsequent publications. Among the last articles he wrote were *Numerical Distances Among the Spheres in a Loxodromic Sequence* [**C97**], which discusses several mathematical properties of a sequence of spheres, and uses *Firmament* as an illustration; *Five Spheres in Mutual Contact* [**C00**], which focuses on the consequences of inversion in a circle or sphere; and *An Absolute Property of Four Mutually Tangent Circles* [**C05**], the opening lecture at the Bolyai Bicentennial conference in Budapest in July 2002. The last was his final paper; until two days before his death, he "persevered in putting the final touches on [that article]. He could not quite believe it when no further errors or typos could be found—he always took great pleasure in seeking out the mistakes in his papers and books, to be corrected in subsequent printings, of which there were always many." [**RW04**]

George Hart and **Rinus Roelofs** There are doubtless many artists who never had a personal relationship with Coxeter, yet for whom his publications (especially those on polyhedra) have been profoundly influential and a rich source of inspiration. George Hart (b. 1955) and Rinus Roelofs (b. 1954) are two contemporary sculptors who represent this group. Although they live on different continents, only became aware of each other's work via the internet around 1999, and met for the first time in Granada, Spain, in 2003 at a conference on arts, architecture, and mathematics [**MA03**], their life trajectories have some striking similarities.

Rinus Roelofs lives in the Netherlands and was introduced to Coxeter's work around 1974 while studying for his degree in Applied Mathematics at the Technical University of Enschede. Interested in group theory, he still uses his copy of *Generators and Relations for Discrete Groups* [**CM57**]. There he first learned about symmetry groups, and was led to other articles by Coxeter whose pictures inspired him to study polyhedra and the concept of duality (Figure 21). When he pursued

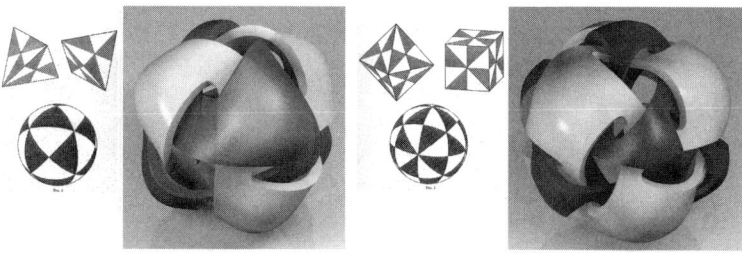

FIGURE 21. Rinus Roelofs. *Duality*, 2003. These digital sculptures are "translations" of Coxeter's figures depicting duality of regular polyhedra.

a second degree at the Enschede Art Academy with a specialization in sculpture, he says, "this all came back." Currently he is active as a full-time sculptor, and also teaches NURBS modeling and programming with *Rhinoceros*. George Hart lives in New York, and started making polyhedra at age 13, inspired by Coxeter's Chapter 5 in *Mathematical Recreations and Essays*, and Cundy and Rollet's *Mathematical Models*. When, as a freshman at MIT, he discovered Coxeter's *Regular Polytopes*, he "began to get serious." He holds a Ph.D. in Electrical Engineering and Computer Science from MIT, and currently leads a dual life as a sculptor and as research professor of computer science at Stony Brook University, New York.

The work of both artists is deeply rooted in the symmetry and duality properties of polyhedra, exploring and exploiting structure, combinations of structures, and transformations of structures. Their sculptures are often made from many pieces of identical parts, and each enjoys having groups gather and pitch in to assemble the intricate creations and then celebrate the surprising result. Both use the computer in an essential way in planning and previewing their sculptures. Roelofs uses the program *Rhinoceros* to produce digital (virtual) sculpture as well as to develop sketches of ideas that just a few years ago would have been sketched with pencil. Hart uses some commercial programs, but mostly writes his own software, developing algorithms that have led to new classes of symmetric polyhedra that he calls 'symmetrohedra' [**CH01**], 'zonish' [**H98**] and 'propellorized' [**H00**] (Figure 22). Both are accomplished in working in a variety of media: paper, wood, metal, acrylic, and most recently, rapid prototyping, in which a "3-d printer" makes a model in wax, metal, nylon, or plastic, directly from the artists' computer data file of the sculpture. Hart also assembles polyhedral sculptures from plastic cutlery, compact disks, toothbrushes, and other common objects, playfully naming them with a pun. Roelofs has observed that although their interests and their sculptures have similarities, the two sculptors have a different approach to making spheres from various elements. "In George's spheres the elements are (mostly)in the plane of a polyhedron. You can call this tangent. And in mine the elements (mostly) are radial [travel along paths of great circles]." They plan joint work in 2005: "let's see what happens if we combine those two approaches." [**RS**]

Hart and Roelofs are each prolific, with busy schedules of exhibits, conferences and competitions (both have won prestigious awards), installations of commissioned work, and public events that resemble "barn-raising." Their respective web sites [**H**] and [**R**] give an overview of current work and publications; Hart's web site also has

FIGURE 22. Left: George Hart. (Left) *Roads Untaken*, 1998. Hardwoods (yellowheart, paela, and padauk) with walnut "grout." 17 in diam. An "exploded propellorized truncated icosahedron." (Right) *Bouquet*, 2003. Acrylic plastic (plexiglass), assembled from 30 identical S-shaped pieces. 9 in diam.

FIGURE 23. Right: Rinus Roelofs. *Four interwoven triangles circumscribed by a cuboctahedron*, 1993. Digital sculpture.

an excellent compendium of illustrated information on polyhedra. Each of these sculptors also has connections to one of our previously-mentioned artists. Like Pelletier, Hart is an expert on building polyhedra with *Zometool*, and is coauthor, with Henri Picciotto, of *Zome Geometry* [**HP01**]. In 1993, Roelofs, independently of Odom, and unaware of Holden's work, discovered the interlocking of four triangles as in Odom's sculpture, while exploring the possibilities of connecting midpoints of edges of a cube. He also discovered what Coxeter had observed [**C94**, pp. 27–28], that the four intertwined triangles would be circumscribed by a cuboctahedron (Figure 23).

Hart has been working on a treatise, "Polyhedra and Art" for more than ten years, and hopes to complete it when he has a chunk of time. The first page begins with a quote:

> *The chief reason for studying regular polyhedra is still the same as in the time of the Pythagoreans, namely, that their symmetrical shapes appeal to one's artistic sense. —H.S.M. Coxeter*

All of these artists testify to the truth of that observation. In a letter to Coxeter (17-11-92), John Robinson wrote, "To be an Artist and know Mathematicians is so rewarding." Donald might respond that the dual sentiment is just as true.

Acknowledgements. I am grateful to the artists, their families, and friends who answered many questions and provided information and pictures. I also want to thank Susan Coxeter Thomas, Siobhan Roberts, Donald Crowe, Douglas Dunham, Asia Weiss, Walter Whiteley, Ronald Brown, Joseph Malkevitch, Alison Conway, and The M.C. Escher Company for their help and cooperation.

References

[B70]   Baer, Stephen C., Zome primer, in *Zomeworks*, Albuquerque, 1970.

[B92]   Baer, Stephen C., The discovery of space frames with fivefold symmetry, in *Fivefold Symmetry*, World Scientific, Singapore, 1992, pp. 205–219.

[B38]   Ball, W.W. Rouse and Coxeter, H.S.M., *Mathematical Recreations & Essays*, 11th edition, Macmillan, London, 1939; 12th edition, University of Toronto Press, 1974; 13th edition, Dover, New York, 1987.

[BC]   Bomford, A.G. and Coxeter, H.S.M., Correspondence, courtesy of the family.

[B82]   Bool, F.H., Kist, J.B., Locher, J.L., and Wierda, F., *M.C. Escher, His Life and Complete Graphic Work*, Harry Abrams, New York, 1982.

[Bo92]   Booth, David, The new zome primer, in *Fivefold Symmetry*, World Scientific, Singapore, 1992, pp. 221–233.

[Br94]   Brown, Ronald, Sculptures by John Robinson at the University of Wales, Bangor, *Math. Intell.*, **16**, no 3 (1994) 62–64.

[BFG96] Burgiel, H., Franzblau, D.S., and Gutschera, K.R., The mystery of the linked triangles, *Math. Magazine*, **69**, no 2 (1996) 94–102.

[CH01]   Kaplan, Craig S. and Hart, G., Symmetrohedra: Polyhedra from symmetric placement of regular polygons, in *Bridges: Mathematical Connections in Art, Music, and Science, Conference Proceedings*, Southwestern College, Winfield, Kansas, 2001, pp. 21–29.

[CDFP38] Coxeter, H.S.M., Du Val, P., Flather, H.T., and Petrie, J.F., *The Fifty-Nine Icosahedra*, No. 6, University of Toronto Studies (Mathematical Series), University of Toronto Press, 1938. Springer-Verlag, New York, 1982.

[CM57]   Coxeter, H.S.M. and Moser, W.O., *Generators and Relations for Discrete Groups*, Springer-Verlag, New York, 1957; 2nd ed. 1965.

[C49]   Coxeter, H.S.M., *Regular Polytopes*, Methuen, London, 1948, Pitman, New York, 1949. Second ed., Macmillan, London and New York, 1963. Reprinted (with corrections) by Dover Books, New York, 1973.

[CHM54] Coxeter, H.S.M., Longuet-Higgins, M.S., and Miller, J.C.P., Uniform polyhedra, *Philos. Trans. Roy. Soc. London Ser. A* **246** (1954) 401–450.

[C57]   Coxeter, H.S.M., Crystal Symmetry and its generalizations (Presidential Address), *Trans. Roy. Soc. Canada Sect. III* (3) **51** (1957), 1–13.

[C61]   Coxeter, H.S.M., *Introduction to Geometry*, Wiley, New York, 1961. 2nd edition, 1969.

[C68]   Coxeter, H.S.M., The mathematical implications of Escher's prints, *Catalog of Retrospective Exhibition M.C. Escher*, 8 June – 21 July 1968, Haags Gemeentemuseum, The Hague, The Netherlands, pp. 87–89.

[C71]   Coxeter, H.S.M., Virus macromolecules and geodesic domes, in *A Spectrum of Mathematics: Essays presented to H.G. Forder*, (Ed. J.C. Butcher), Auckland and Oxford University Presses, Auckland, 1971, pp. 98–107.

[C71a]   Coxeter, H.S.M., The mathematical implications of Escher's prints, in *The World of M. C. Escher*, (Ed. J. L. Locher), Harry Abrams, 1971, 49–52.

[C74]   Coxeter, H.S.M., *Regular Complex Polytopes*, Cambridge University Press, 1974, 2nd ed. 1991.

[C79]   Coxeter, H.S.M., The non-Euclidean symmetry of Escher's picture 'Circle Limit III', *Leonardo* **12** (1979), 19–25, 32.

[C81]   Coxeter, H.S.M., Angels and devils, in *The Mathematical Gardner* (Ed. David A. Klarner), Prindle, Weber & Schmidt, Boston, 1981, pp. 197–209 & Plate IV. Reprinted as *Mathematical Recreations: A Collection in Honor of Martin Gardner*, Dover, Mineloa, NY, 1998.

[C85]   Coxeter, H.S.M., A special book review: M.C. Escher: His life and complete graphic work, *Math. Intell.*, **7** no 1 (1985), 59–69.

[C86]   Coxeter, H.S.M., Coloured symmetry, in *M. C. Escher: Art and Science*, (Eds. H. S. M. Coxeter, M. Emmer, R. Penrose, M. Teuber), North-Holland, Amsterdam, 1986, pp. 15–33.

[C87]   Coxeter, H.S.M., Alicia Boole Stott, in *Women of Mathematics: A Biobibliographic Sourcebook*, (Eds. Louise Grinstein and Paul Campbell), Greenwood Press, 1987, pp. 220–224.

[C88]   Coxeter, H.S.M. Escher's lizards, *Structural Topology*, **15** (1988), 23–30.

[C94]   Coxeter, H.S.M., Symmetrical combinations of three or four hollow triangles, *Math. Intell.*, **16**, no 3 (1994) 25–30.

[C95]   Coxeter, H.S.M., Symmetrical combinations of triangles: postscript, *Math. Intell.*, **17**, no 1 (1995) 4.

[C96]   Coxeter, H.S.M., The trigonometry of Escher's woodcut Circle Limit III, *Math. Intell.*, **18** no 4 (1996), 42–46. Edited and reprinted in *HyperSpace* **6** no 2 (1997), 53–57. Edited and reprinted in *M.C. Escher's Legacy: A Centennial Celebration*, (Eds. D. Schattschneider and M. Emmer), Springer-Verlag, Heidelberg, 2003, pp. 297–304. .

[C97]   Coxeter, H.S.M., Numerical distances among the spheres in a loxodromic sequence, *Math. Intell.*, **19**, no 4 (1997) 41–47.

[C00]   Coxeter, H.S.M., Five spheres in mutual contact, *J. Geom. and Graphics*, **4**, no 2 (2000) 109–114.

[C05]   Coxeter, H.S.M., An absolute property of four mutually tangent circles, in *Non-Euclidean Geometries, János Bolyai Memorial Volume* (Eds. A. Prékopa and E. Molnár), Kluwer Academic Pub., 2005.

[CR51]   Cundy, H. Martyn, and Rollett, A.P., *Mathematical Models*, Oxford University Press, Oxford, 1951; 2nd ed. 1961.

[dB54]   de Bruijn, N.G., Preface, in *Catalog for the Exhibition M.C. Escher*, cat. 118, Stedelijk Museum Amsterdam, 1954.

[DC35]   Donchian, P.S. and Coxeter, H.S.M., An n-dimensional extension of Pythagoras' theorem, *Math. Gazette*, **19** (1935) 206.

[D86]   Dunham, Douglas, Double torus tessellated with 24 fish, based on Escher's Circle Limit III, in *M. C. Escher: Art and Science* (Eds. H.S.M. Coxeter, M. Emmer, R. Penrose, and M. Teuber), North-Holland, 1986, Color plate p. 393. Also in *M.C. Escher's Legacy: A Centennial Celebration* (Eds. D. Schattschneider and M. Emmer), Springer-Verlag, Heidelberg, 2003, Plate 18.

[D03]   Dunham, Douglas, Families of Escher patterns, in *M.C. Escher's Legacy: A Centennial Celebration* (Eds. D. Schattschneider and M. Emmer), Springer-Verlag, Heidelberg, 2003, pp. 286–296 and CD Rom.

[D04]   Dunham, Douglas, Tony Bomford's hyperbolic hooked rugs, in *Bridges: Mathematical Connections in Art, Music, and Science* (Eds. Reza Sarhangi and Carlo Séquin), Conference Proceedings, Winfield, Kansas, 2004, pp. 309–314.

[E76]   Ernst, Bruno, *The Magic Mirror of M.C. Escher*, Random House, New York, 1976. Republished by Taschen America, 1995.

[Es57]   Escher, M.C., *Regular Division of the Plane VI*, woodcut, June 1957, in Bool et al., p. 315.

[Es60]   Escher, M.C., *Grafiek en Tekeningen M.C. Escher*, Koniklijke Uitgeverij van de Erven J.J. Tijl N.V., Zwolle, The Netherlands, 1960. American Edition: *The Graphic Work of M.C. Escher*, (translated from the Dutch by John E. Brigham), Meredith Books, New York, 1960.

[EC]   Escher, M.C. and Coxeter, H.S.M., Correspondence, M.C. Escher Archives, National Gallery of Art, Washington, D.C.

[F97]   Fields Institute web site for the Feb. 9, 1997 (Coxeter's 90th birthday) dedication of the sculpture *Intuition*, by John Robinson. www.fields.utoronto.ca/programs/scientific/96-97/coxeter/

[F02]   Fields Institute web site for the Feb. 15, 2002 celebration of the 95th birthday of H.S.M. Coxeter. www.fields.utoronto.ca/programs/scientific/01-02/coxeter95/

[F02s]   *Sculpture Dedication Ceremony (February 15, 2002)*. Fields Institute web site for presentations by H.S.M. (Donald) Coxeter, John H. Conway, and Marc Pelletier. www.fields.utoronto.ca/audio/01-02/#sculpture.

[G61]   Gardner, Martin, Concerning the diversions in a new book on geometry, *Sci. Amer.* **204** (April 1961), pp. 164–175. Reprinted and updated as *H.S.M. Coxeter, New Mathematical Diversions from Scientific American*, Simon & Schuster, New York, 1966, pp. 196–209. Reprinted in the collection of Martin Gardner's Mathematical Games, Math. Assn. Amer., Washington DC, 2005.

[H79]   Hancock, Geoff, The many sides of Donald Coxeter, *The Graduate, University of Toronto Alumni*, **VII**, no 1 (1979) 10–12.

[H96]   Hargittai, István, Life-long Symmetry: A Conversation with H.S.M. Coxeter, *Math. Intell.*, **18** no 4 (1996), 35–41.

[HL00]   Hargittai, I. and Laurent, T.C., Eds. *Symmetry 2000*, Parts I and II, Wenner-Gren International Series, vol. 80, Portland Press, London, 2002. Contains 52 articles, including four by Coxeter, Longuet-Higgins, Wenninger, and Schattschneider.

[HP01]   Hart, George W., and Picciotto, Henri, *Zome Geometry, Hands-on Learning with Zome Models*, Key Curriculum Press, 2001.

[H]   Hart, George, Web site, www.georgehart.com.

[H98]   Hart, George, Zonish polyhedra, in *Proceedings of Mathematics and Design '98* (Ed. Barrallo Javier), University of the Basque Country, San Sebastian, Spain, 1998. Online version at www.georgehart.com/zonish/zonish.html.

[H00]   Hart, George, Sculpture based on propellorized polyhedra, in *Proceedings of MOSAIC 2000*, Seattle, WA, August, 2000, pp. 61–70. Online version at www.georgehart.com/propello/propello.html.

[H35]   *Hartford Courant*, January 20, 1935.

[H71]   Holden, Alan, *Shapes, Space, and Symmetry*, Columbia University Press, New York, 1971.

[H80]   Holden, Alan, Regular polylinks, *Structural Topology*, **4** (1980) 41–45.

[H04]   Longuet-Higgins, Michael S., Encounters with polytopes, *Symmetry: Culture and Science*, **13** nos 1-2 (2004) 17–31.

[MA03]   *Meeting Alhambra. ISMA – Bridges 2003*. Conference Proceedings, (Eds. Barrallo, J., Friedman, N., Sarhangi, R., Séquin, C., Martìnez, J., and Maldonado, J.), University of Granada, Granada, Spain, 2003. The conference was dedicated to the memory of sculptors Eduardo Chillida and George Rickey, and to mathematician H.S.M. Coxeter, all of whom died in 2003.

[O83]   Odom, George, Problem E 3007. *Amer. Math. Monthly*, **90** (1983), 482.

[OC]   Odom, George and Coxeter, H.S.M., Correspondence, courtesy of Magnus Wenninger.

[OW]   Odom, George and Wenninger, Magnus, Correspondence, courtesy of Magnus Wenninger.

[P]   Peterson, Ivars, Polyhedron man, *Science News*, **160**, no. 25/26 (2001) 396-398.

[R03]   Roberts, Siobhan, Figure head, *Toronto Life*, **37**, no. 1 (2003) 82–88.

[RW04]   Roberts, Siobhan, and Weiss, Asia Ivić, Donald in wonderland: The many-faceted life of H.S.M. Coxeter, *Math. Intell.*, **26**, no 3 (2004) 17–25.

[Ro92]   Robinson, John, *Symbolic Sculpture*, Edition Limitée, Carouge-Geneva, Geneva, Switzerland, 1992. See also the web site: www.popmath.org.uk/sculpture/sculpture.html.

[RoF]   Robinson, John, *Firmament*, www.cpm.informatics.bangor.ac.uk/sculpture/pages/5firm.html.

[RoI]   Robinson, John, *Intuition*, www.cpm.informatics.bangor.ac.uk/sculpture/pages/2intuiti.html.

[RoC]   Robinson, John and Coxeter, H.S.M., Correspondence, courtesy of John Robinson and Susan Coxeter Thomas.

[R]   Roelofs, Rinus, Web site, www.rinusroelofs.nl.

[R04]   Rowe, David E., Coxeter on people and polytopes, *Math. Intell.*, **26**, no.3 (2004) 26–30.

[RS]   Roelofs, Rinus, and Schattschneider, Doris, Private communication.

[S90]   Schattschneider, Doris, *M.C. Escher: Visions of Symmetry*, W.H. Freeman, New York, 1990. New edition, Harry Abrams, New York, 2004.

[S05]   Schattschneider, Doris, Coxeter and the artists: Two-way inspiration, Part 2, in *Renaissance Banff Bridges: Mathematical Connections in Art Music, and Science & Coxeter Day, Conference Proceedings* (Eds. Reza Sarhangi and Robert V. Moody), Winfield, Kansas, 2005.

[S36]   Soddy, Frederick, The kiss precise, *Nature*, **137** (1936) 1021.

[vdC86]   van de Craats, Jan, The golden ratio from an equilateral triangle and its circumcircle, *Amer. Math. Monthly*, **93** (1986), 572.

[WC]   Wenninger, Magnus and Coxeter, H.S.M., Correspondence, courtesy of M. Wenninger.

[W66]   Wenninger, Magnus, *Polyhedron Models for the Classroom*, National Council of Teachers of Mathematics, 1966, 2nd ed. 1975.

[W71]   Wenninger, Magnus, *Polyhedron Models*, Cambridge University Press, London and
          New York, 1971.
[W79]   Wenninger, Magnus, *Spherical Models*, Cambridge University Press, London and New
          York, 1979.
[W83]   Wenninger, Magnus, *Dual Models*, Cambridge University Press, London and New York,
          1983.
[Z]   *Zometool* web site. www.zometool.com/about/quotes.html. Coxeter sent this endorse-
          ment to Marc Pelletier in a letter.

DEPARTMENT OF MATHEMATICS AND COMPUTER SCIENCE, MORAVIAN COLLEGE, PPHAC,
1200 MAIN ST., BETHLEHEM, PA 18018-6650 USA
    E-mail address: schattdo@moravian.edu

The Visual Mind: Art, Mathematics and Cinema

Michele Emmer

To Valeria

ABSTRACT. The encounter of Coxeter with the graphic artist M. C. Escher, the making of my film on Escher with the cooperation of Coxeter, these are the topics of this paper. How the "visual mind" of Coxeter helped to make visible the interior "visual ideas" of Escher in a "visual" film.

Introduction: Coxeter met Escher

" 'A mathematician, like a painter or a poet, is a maker of structures, of models—and like poems or paintings, the models must be beautiful; beauty is the fundamental test—there is no place in the world for ugly mathematics.' These are words of the English mathematician Hardy [**29**], but they also apply to many artists that have been interested in aspects of mathematics such as, for example, Platonic solids. Nobody knows who was the first one noticing that the number of regular polygons: - triangles, squares, pentagons, hexagons and so on - goes on to infinity, but the really interesting discovery is that the number of regular solids is finite. This fact fascinated Plato, who related them to the structure of the world and the elements of physical space." [**13**], [**16**], [**31**]

These were the first words spoken by Donald Coxeter at the University of Rome 'La Sapienza', in introducing one of the films of the series *Art and Mathematics*, dedicated to the Dutch graphic artist Maurits Cornelis Escher.

We had previously discussed the subject of the first part of the takes, but he surprised me with this phrase when we started shooting the film. It was the first time I had heard of Hardy and his book *A Mathematician's Apology*. That sentence was to become one of my favourites. Coxeter had perfectly understood the spirit of the project *Art and Mathematics* [**17**], [**20**] , [**19**] and had prepared himself for the filming. It was January 1979. I too had prepared myself. I had studied 'Coxeter' in the previous months. At that time, Escher - who had died in 1972 - was already well known to the scientific community.

Coxeter described his encounter with Escher: [**6**] "In September 1954, an exhibition of his work was sponsored by the International Congress of Mathematicians, meeting that year in Amsterdam. Through the previous 17 years Escher had been

2000 *Mathematics Subject Classification*. Primary 01-xx, 00-xx.

FIGURE 1. Model of hyperbolic geometry

making designs in which a drawing of some animal (such as a fish or a reptile or a
bird) is repeated as on wallpaper, with two remarkable innovations: the basic unit
(usually a single animal, or one half of a symmetrical animal or two different animals juxtaposed) is repeated not only by translations but also by other isometries
(or congruence transformations): rotations, reflections, or glide-reflections; and the
replicas ingeniously fit together so that there are no interstices. In the language of
mathematics (a subject in which Escher resolutely claimed to be absolutely inno-
cent of training or knowledge) the basic unit is a *fundamental region* for a *symmetry
group*."

The encounter between Coxeter and Escher at the Amsterdam conference was
to be very significant for both. For Coxeter it meant meeting one of the most
ingenious graphic minds of the XXth century, while for Escher it meant getting in
contact with ideas of geometry that he was to make full use of.

Coxeter had requested Escher's permission to use two of his graphics in one of
the articles he was publishing. Escher was very happy to oblige. He wrote to Coxeter in December 1958 [6]: "Did I ever thank you for sending me *A Symposium on
Symmetry*? I was so pleased with this booklet and proud of the two reproductions
of my plane patterns. Though the text of your article on *Crystal Symmetry and its
generalization* is much too learned for a simple, self-made plane pattern-man like
me, some of the text illustrations and especially Figure 7, page 11, gave me quite
a shock. (Fig. 1)

"For a long time I have been interested in patterns with 'motives' getting
smaller and smaller till they reach the limit of infinite smallness. The question is
relatively simple if the limit is a point in the centre of the pattern. Also a line-limit
is not new to me, but I was never able to make a pattern in which each 'blot' is
getting smaller gradually from a center towards the outside circle-limit, as your
Figure 7 shows ... If you could give me a simple explanation how to construct the
following circles, whose centers approach gradually from outside till they reach the
limit, I should be immensely pleased and very thankful to you!

"Nevertheless, I used your model for a large woodcut (of which I executed only
a sector of 120 degrees in wood, which I printed three times). I am sending you a
copy of it."

This was the picture *Circle Limit I*.

It was inevitable that Escher should encounter mathematicians. He met not only Coxeter but also Roger Penrose at that conference in Amsterdam. A few years later, he wrote on his artistic work in his first book *The Graphic Work of M. C. Escher* [25]. The fundamental event in his life, according to Escher, was in 1938 when the Escher family left Italy after a long stay:

"In Switzerland, Belgium and Holland ... I found the outward appearances of landscape and architecture less striking than those which are particularly to be seen in the southern part of Italy. Thus I felt compelled to withdraw from the more or less direct and realistic illustration of my surroundings. No doubt this circumstance was in a high degree responsible for bringing my inner visions into being."

All Escher's work illustrated in this book, except for the first seven, were done with the idea of communicating a detail of these interior visions.

"The ideas that are basic to them often bear witness to my amazement and wonder at the laws of nature which operate in the world around us. He who wonders discovers that this is in itself a wonder. By keenly confronting the enigmas that surround us, and by considering and analyzing the observations that I had made, I ended up in the domain of mathematics. Although I am absolutely without training or knowledge in the exact sciences, I often seem to have more in common with mathematicians than with my fellow artists."

Evidently, the encounter with Penrose and Coxeter had affected Escher deeply.

Escher divides the work he had done until 1961 into several chapters in *The Graphic Work*. The series *Circle Limit* is shown in a chapter that Escher calls *Infinity of number* [25]. He describes the series as follows:

"So far four examples have been shown with points as limits of infinite smallness. A diminution in the size of the figures progressing in the opposite direction, i.e., from within outwards, leads to a more satisfying result. The limit is no longer a point, but a line which borders the whole complex and gives it a logical boundary. In this way one creates, as it were, a universe, a geometrical enclosure. If the progressive reduction in size radiates in all directions at an equal rate, then the limit becomes a circle."

This was *Circle Limit I*, sent by Escher to Coxeter. Escher was not satisfied [6]:

"This woodcut being a first attempt, displays all sorts of shortcomings. Not only the shape of the fish, still hardly developed from rectilinear abstraction into rudimentary animals, but also their arrangement and relative position leave much to be desired. There is no continuity, no 'traffic flow', no unity of colour in each row."

Thus wrote Escher to Coxeter, in almost identical terms as in *The Graphic Work of Escher*. On the other hand, he was very satisfied with *Circle Limit III*.

Escher had taken into account Coxeter's suggestions [6]:

"Here, the failings of the previous work are as far as possible remedied. White curved lines cut across each other and divide one another into sections, each of which equals the length of a fish. They mark the routes along which series of fish move forward from the infinitely small through the greatest size to infinitely small."

But this is not the end of the story. A few years later, in 1978, Coxeter wrote an article [6] about *The Non-Euclidean Symmetry of Escher's Picture 'Circle Limit III'*. He later returned to the article for an Italian version in 1989 [9]. Still later, in 1999 he took up the topic at a lecture in Banff on *Aspects of Symmetry* [32].

"The topic of my paper," began Coxeter, "is one that has intrigued me and preoccupied me for nearly five decades. It's about what I can call the 'intuitive geometry' of my friend M. C. Escher."

There was no doubt in my mind that for a project of making films on the relationships between art and mathematics, I had to make a film on Escher with Coxeter's help. Films should be as 'visible' as possible. To tell the story of Escher and Coxeter's relationship regarding Poincaré's non-euclidean geometry through image and animation was a magnificent example of cinema. Or it could be. Escher himself contemplated cinema.

Escher and cinema

The title of the first important book containing the collected works of Escher, *The World of M. C. Escher*, 1971 ([26]), is very appropriate. Escher created his own world of meticulously constructed images, a magic world which, though fantastic and imaginative, was also realistic, coherent, and observed with an eye apparently lacking in emotion. Escher was an artist of minute details, tiny elements which create instability and disturb the apparent calm of the whole.

Regarding this, the mathematician C. P. Bruter wrote that the eye creates local images which, little by little, are projected onto a common receptacle and probably reproduced several millions of times. The construction of these individual maps, defined by the trajectories outlined by the viewer's eye movements, enables one to construct an image of the surrounding space that is stable and well ordered. Escher's drawings only appear to be strange, since each individual image faithfully represents reality [3].

This is the first reason for using a movie camera to examine Escher's work: cinematic technique lets us concentrate our attention on what the filmmaker wants us to see. The film forces us to look at Escher's world as if we were part of it. We are not distracted by anything else. The second reason is that Escher's drawing technique is very precise and detailed. Cinematic technique allows us both to isolate these minute details and also to magnify them many times in order to appreciate the precision of the artist's method. Another important element in using film is the fact that a movie forces those who look at it to see things passing quickly on the screen in a definite order. Many of Escher's drawings tell a story that develops and must be observed in the sequence suggested by Escher himself. The movie camera permits a very precise and accurate analysis of Escher's graphics.

Escher himself suggested the use of cinematic technique. In his book *The Regular Division of the Plane* he wrote:

"In this book it is the images and not the words that come first ... For me it remains an open question whether the play of white and black figures as shown in the six woodcuts of this book pertains to the realm of mathematics or that of art ... ([27] and [28])

the first woodcut ... shows clearly that a succession of gradually changing figures can result in the creation of a story in pictures. In a similar way the artists of the Middle Ages depicted the lives of Saints in a series of static tableaux ... The observer was expected to view each stage in sequence. The series of static representations acquired a dynamic character by reason of the space of time needed to follow the whole story. Cinematic projection provides a contrast with this. Images appear, one after the other, on a still screen and the eye of the observer remains fixed and unmoving. Both in the medieval pictorial story and in the developing pattern of a regular division of the plane the images are side by side and the time factor is shifted to the movement the observer's eye makes in following the sequence from picture to picture" [**27**] and [**28**].

A very significant example of 'a succession of gradually changing figures' which sums up the various aspects noted by Escher, is given by his *Metamorphose* prints. In particular, his *Metamorphose II* seemed to me a perfect cinematic sequence. Escher discusses them in the same book in a section entitled *Metamorphose*, where he describes one of the sequences of images as follows:

"First the black insect silhouettes join; at the moment when they touch, their white background has become the shape of a fish. Then figures and background change places and white fish can be seen swimming against a black background ... A succession of figures with a number of metamorphoses acquires a dynamic character. Above I pointed out the difference between a series of cinematographic images projected on a screen and the series of figures in the regular division of plane. Although in the latter the figures are shown all at once, side by side, in both cases the time factor plays a role." ([**27**] and [**28**])

The films of the series *Mathematics and Art*

The film on M. C. Escher was part of the project *Mathematics and Art* which started in 1976. That year, I was at the University of Trento, in the North of Italy. I was working in the area called the Calculus of Variations, in particular Minimal Surfaces and Capillarity problems. Also in 1976, Jean Taylor proved a famous result that settled a conjecture posed on the basis of experiments by the Belgian physicist Joseph Plateau over a hundred years earlier: the types of singularities, of edges, that soap films generate when they meet. Plateau had experimentally observed that the angles generated by soap films are only of two kinds [**30**].

Jean Taylor using the theory of Integral Currents introduced by Federer, and then by Allard and Almgren, was able to prove that the result was true. In 1976, the journal *Scientific American* [**1**] asked Jean Taylor and Fred Almgren to write a paper on the recent results on the topic of minimal surfaces and soap bubbles. A professional photographer was asked to realise the pictures for the paper. In the same year Jean Taylor and Fred Almgren were invited to the University of Trento as visiting professors. When they came to Trento, the issue of the *Scientific American* was just published. The pictures of the article and the cover were very beautiful and interesting. I do not remember why, but looking at the pictures I had the idea of making a movie on soap film in order to show more closely and with slow-motion technique their shapes and geometries.

I must say that for me thinking to make a film was quite natural. My father is a famous Italian filmmaker. Marcello Mastroianni made his first film with him, *Domenica d'agosto*, in 1949. Both Almgren and Jean Taylor were very interested in my project. In any case, my idea was not to make a 'small' scientific film, a sort of scientific spot just to show some little experiments with soap bubbles and soap films. I was attracted by the phenomena of soap films because they are visually interesting and I was thinking that the technique of filming would increase their interest and fascination.

Almost in the same period I discovered the 'topological' sculptures of one of the most important artists of the last century: Max Bill. The topological sculptures of Bill were for me a real discovery. The impression of *Endless Ribbon* [**2**], that enormous Möbius Band in granite, was indeed a revelation. Its shape, its real tridimensionality, lets it come alive in space. A mathematical form—alive! This was the idea that was missing for my project: mathematics, mathematicians in all the historical periods and in all the civilisations have created shapes, forms, relationships.

The project was becoming clearer: to make films in which to compare on the same theme the mathematical and artistic point of view, asking for the opinion of mathematicians and artists. Not just filming a long discussion among artists and scientists on the vague theme of the connections between art and science, but a real confrontation on the visual ideas of the artists and the mathematicians. To make "visible the invisible", as the artist David Brisson says in the film *Dimensions* made in 1984 with Thomas Banchoff. So the general scheme of the project was almost clear: to make films on the visual relationships of the forms created by artists and mathematicians. The themes of the first two films were: soap bubbles and topology, in particular the Möbius Band [**14**], [**12**].

From the beginning of the project there was the idea of focusing on the cultural aspect of mathematics, the influence and the connections of mathematics and culture, of course starting from the point of view that mathematics has always played a relevant role in culture, being an important part of it—all these using the most important visual tool: filming. As these were the general lines of the project it was quite natural to include in it the organisation of exhibitions (there were many in the following years), congresses and seminars, the publishing of books (with many illustrations!) [**18**].

The themes of the first two films were clear enough; then it was necessary to make a choice of mathematicians and artists to involve in the project. Of course, the first thing was to obtain their collaboration.

During these same years I had already discovered the art of the Dutch graphic artist Maurits Cornelis Escher. From the first time that I saw his engravings, my purpose was to make a film only on him and his art. My idea was to use the technique of animation in order to make his work really tridimensional. This was something that Escher himself suggested; he personally was involved in a short film with several animations of his work before his death in 1972.

I thought about making a film in two parts, each one lasting 27 minutes. The film became a 50-minute video in the mid-nineties.

As I pointed out in my essay *Movies on M. C. Escher and their Mathematical Appeal* [**12**], by filming the prints and drawings we arrive at something new:

"The images flow quickly by; they are in movement, not just statically side by side. We have taken Escher's suggestions and extended them to a world of Escher, or better, a world according to Escher, which moves and changes in three-dimensional space. Cinema is illusion, and Escher's technique suggests we not only analyze but reinvent, starting from his ideas. When you make an animation, you need to construct the drawings for the animation frame by frame. Of course you cannot use Escher's orginal drawings and woodcuts and cut them to pieces! The collections of drawings made to produce the animations for the movie posed interesting problems for drawings in which geometry, design, and cinema are combined."

So these were my general ideas for the preparatory stage of the project of the film.

The contact with H. S. M. Coxeter

I started reading Coxeter's and Penrose's books and scientific work before I wrote to them to ask for their collaboration, as I wanted a clearer idea of their work. In particular, I read *Introduction to Geometry* and *Regular Polytopes*. The first page of the former reads [4] (1961):

"For the last thirty or forty years, most Americans have somehow lost interest in geometry. The present book constitutes an attempt to revitalize this sadly neglected subject."

Coxeter was interested in geometry, in what can be called 'classical geometry'. He was interested in visualization, in intuition, but, as he wrote in *Regular Polytopes* (April 1947) [5]:

"Only one or two people have ever attained the ability to visualize hypersolids as simply and naturally as we ordinary mortals visualize solids; but a certain facility in that direction may be acquired by contemplating the analogy between one and two dimensions, then two and three, and so on, three and four. This intuitive approach is very fruitful in suggesting what results could be expected. However, there is some danger of our being led astray unless we check our results with the aid of one of the other two procedures, the axiomatic and the algebraic."

So, intuition and visualization are required, but not only these. We want rigor, to be precise and correct. I was preparing the film *Dimensions* [15] on the fourth dimension at the same time as the film on Escher. So I looked into Coxeter's book *Introduction to Geometry*, Chapter 22 *Four Dimensional geometry* [4]:

"The idea of four-dimensional space has long been surrounded by an attractive aura of mistery. The axiomatic approach dispels the mistery without reducing the fascination." And talking of Ludwig Schläfli, who died in 1895: "The French and English abstracts of his work, which were published in 1855 and 1858, attracted no attention. This may have been because their dry-sounding titles tended to hide the geometrical treasures that they contain, or perhaps it was just because they were ahead of their time, like the art of van Gogh." Escher's first great exhibition at the International Congress of Mathematicians in Amsterdam in 1954 took place concomitantly with a van Gogh exhibition!

The more I read Coxeter's books and articles, the more I was convinced that my film could not be realized without him. Furthermore, in the same period, I had gotten in contact with Frank Malina, the cinematic artist who had, among other things, founded the journal *Leonardo*. This journal was to become the most important journal on the relationship between art, technology, and science. Today it is directed by Frank's son, Roger Malina, an astrophysicist. It is published by MIT Press. Thanks to this I had the chance to read Coxeter's article on the *Circle Limit*, published in *Leonardo*. I had received a draft of the article. In it, Coxeter wrote ([6]):

"Escher's work, based on his intuition, without any computation, is perfect, even though his poetic description of it (*Loodrecht uit de limiet*, perpendicular from the boundary) was only approximate."

Coxeter had written in *Introduction to Geometry* [4]:

"The unifying thread that runs through the whole work is the idea of a group of transformations or, in a single word, symmetry."

I sent my letter to the Department of Mathematics of the University of Toronto on May 18th, 1978. I did not know that at that time Coxeter was visiting professor at the University of Bologna, in the center of Italy. On June 5th, I received Coxeter's reply:

"Dear Dr. Emmer, many thanks for your letter of May 18 which was waiting for me on my return from five weeks in Bologna. If only someone had told you I was there, we might have got together in Italy. By coincidence, just three days before you wrote, I was giving a lecture on the mathematical aspects of Escher's work at Siena [report published in *Leonardo*] by special request of the mathematicians there. Since that time I have been writing up the lecture, concentrating particularly on the four *Circle Limit* pictures for which an old drawing of mine had provided inspiration.

Your idea of making a special film on Escher is very appealing. I believe one such film was made while he was still alive. Someone said it included some animated versions of his repeating patterns. I have always regretted missing it when it was shown on TV.

Yes, I would indeed be interested in helping you in your own film. I expect to be in Toronto this summer. So let us keep in touch.

Sincerely yours,

H. S. M. Coxeter."

In addition, Coxeter mentioned an Italian artist, Lucio Saffaro, who had spent a lifetime painting polyhedra. I had met Saffaro in Bologna. Thanks to Coxeter, I got in touch with him and together we produced two films as well as several books and exhibitions. While the Toronto conference on Coxeter's Legacy was being held in May 2004, a major exhibition on Saffaro took place in Bologna [**33**]. Saffaro had died in 1998. Among other things, Saffaro was the first person to notice the presence of a star-pointed dodecahedron on the floor of St. Mark's Basilica in Venice. This is attributed to Paolo Uccello of the mid-15th century, in other words, 150 years before the official discovery of this star-like solid by Kepler. In 1986, on Saffaro's recommendation, this figure by Paolo Uccello became the emblem of the Venice International Art Biennale dedicated to *Art and Science*.

In a subsequent letter, dated August 16, 1978, Coxeter sent me a preprint of his article to appear in *Leonardo* in which he mentioned *Circle Limit IV* where the crystallographic group is p4g. He suggested several sculptures that "could be gradually turned about their various axes of rotation, illustrating interesting point-groups." Moreover, he referred to the Möbius strips, the spirals and loxodromes as well as the Platonic solids in Escher's work. He ended the letter by saying: "These, it seems to me, are the works that most clearly require mathematical treatment."

As can be seen, just a week or two after getting my letter, Donald began to plan how to deal with the part of the film where he was involved. His visual mind was already at work.

In addition to the film on Escher, I was thinking of making a film on the Möbius strip and another one on Platonic solids. The one on the Möbius strip features several types that Escher devised, in particular the one with the ants which was used for the title frames on the final version of the film on Escher. The film on Platonic solids also includes appearances by Coxeter and Lucio Saffaro.

The filming

The project for the film went ahead. As a general rule, I don't write detailed scripts for my films but just an outline, though quite extensive, so that I can easily change plans depending on the success or otherwise of the images that are filmed. This gives considerably more freedom for invention and creativity. With Coxeter, we decided to do three different parts for three different films: *Platonic Solids*, *M. C. Escher: Symmetry and Space*, and *M. C. Escher: Geometries and Impossible Worlds*.

Shooting began in January 1979 at Rome University 'La Sapienza'. As my study was too small, we used a friend's office in the Genetics department.

On January 9, Coxeter wrote to me again, as always starting the letter with "Dear Dr. Emmer," but for the first time writing by hand instead of using a typewriter. He confirmed that he had received the prepaid ticket and would leave Toronto for Rome on January 28.

"I will bring the Escher correspondence, my own articles about him, the short article on tilings by Grünbaum and Shephard in the *Mathematics Magazine* and my review of it, the typescript of their extended work, and the *Scientific American* article on non-periodic tilings. I can also bring Escher's 'tin box' icosahedron, but I do not have his other solid models, 'the sphere with fishes' and the spherical version of 'Heaven and Hell.' If you could borrow those two, I would like to talk about them."

I asked Coxeter not to bring the icosahedron model which was filmed a few months later at Bruno Ernst's house in Utrecht. Unfortunately it was not possible to get the other two models that Coxeter referred to for filming in Rome. I filmed them at the Gemeentemuseum in The Hague, but only a couple of years later.

Coxeter stayed in Rome for a few days. Shooting went very well and we became friends. When I invited him to dinner, he met my wife Valeria and they, too, got along very well.

As always happens, the amount of film shot was much longer than the actual parts chosen by the director for the final version. Since I no longer have the original reels of film, the only parts remaining are those I chose during the editing process. The film was shot in 16 millimetre format with live sound. The first take was Coxeter's piece for the film on Platonic solids. Part of the text of his talk is given at the beginning of this paper.

Coxeter appears at the beginning of the film and I decided to use his few words as a sort of introduction. Later Coxeter told me he was disappointed to appear so briefly!

Obviously Coxeter played a larger part in the film on Escher. I had decided to split the film into two parts, making two separate films. The first part, *M. C. Escher: Symmetry and Space*, dealt with symmetry and the solids in Escher's work, so I included appearances by crystallographer Caroline MacGillavry and Bruno Ernst. In the second film *M. C. Escher: Geometries and Impossible Worlds*, both Coxeter and Penrose took part. The first piece by Coxeter concerns filling space with octahedrons and tetrahedrons— properties that Escher used in one of his prints.

"The tetrahedron and the octahedron have dihedral angles that are supplementary in such a way that you can put a tetrahedron on the face of an octahedron and you obtain that their triangular faces are in the same plane. If you put another tetrahedron on the opposite face, then the whole figure becomes a rhombohedron that can fill space. All space can really be filled up by tetrahedrons and octahedrons put together in this way. This property can be observed in a drawing by Escher with the title of *Flat Worms*."

Then comes the part devoted to the *Circle Limit* series. We begin with the idea of infinity starting with the classification that Escher himself had given for this series. He classified it in the chapter on *Infinity of Number*:

"If all component parts are equal in size, it is impossible to represent more than a fragment of a regular plane-filling. If one wishes to illustrate an infinite number, then one must have recourse to a gradual reduction in the size of figures, until one reaches—at any rate theoretically—the limit of infinite smallness" [**25**].

The fascination of infinity and of movement. The scene opens with a spiral moving on a sphere, and Coxeter quoting from Shakespeare:

" 'Oh my God, I can be bounded in a nut shell and become the king of infinite space.' These are words by Hamlet, and Escher too, like Hamlet, was fascinated by the idea of infinity, to find infinity in a nut shell.

And the figure that Escher saw, a drawing by myself, was developing exactly this function; and thinking of it he drew *Circle Limit I*. This was a first attempt and he was not too satisfied; so he asked me for more details on the hyperbolic geometry, in which the points infinitely far away in the hyperbolic plane are represented in a circle, and everything in the interior of the circumference is the hyperbolic world. The effect is that all figures are represented truly, distances are distorted. The closer to the circumference, the bigger the distortion. The points on the circumference must be thought of as infinitely far away. The French mathematician Poincaré described a very clever model for this kind of geometry, and in that way Escher was able to see that all these black and white triangles are really the same size when they move further away even though they don't seem to be; it is analogous to the situation on a sphere where, if this is a picture of a sphere, you see that the triangles in the center of the sphere seem to be larger than those near to the circumference. But that is only a perspective foreshortening, these on the sphere are all the same size. Similarly, in the hyperbolic plane these triangles are to be thought of as being all of the same size. And so it is when Escher made the green fishes going this way and the yellow fishes going that way and so on, and this he found to be a very pleasant pattern—as indeed it is. And the symmetry of it is very interesting, because, as a matter of fact, there are points at various positions which can be distinguished in the picture, for instance the points where two yellow fishes and two green fishes come together are all arranged in a very symmetrical fashion. The points in the middle where two yellow and two green fins come together are not essentially different from other points such as this one, where two green and two blue come together, and this one, where two blue and two yellow come together and so on, and all together they form an octagon. Each side of that octagon forms with the center of the whole figure a triangle which from a hyperbolic standpoint is equilateral."

The third part is again devoted to *the idea of infinity*:

"One way to bring infinity to finite terms is to use a transformation called inversion in a circle: a point outside the circle is transformed into a point inside the circle, so that it is at the same diameter, but the distance from the center is the reciprocal of the distance to the original point.

In this way as the point moves outside the circle, the inverse point moves inside. And even if the point moves infinitely far away, its image is still inside the circle. As the point moves further and further away, the image point gets closer and closer to the center and so if one considers the important curve called the equiangular spiral, which is a curve that goes around and around, going further from the center, or going in the opposite direction, gets closer and closer to the center; if we invert that curve, and the circle is in this position, it will become a new kind of spiral with its pole at the inverse of the original pole and of course going the opposite way around. But, the part that goes to infinity makes a second pole at the center and so you have this curious effect of a curve that has two poles. And that is the curve which Escher used in his work."

Obviously, the text alone of Coxeter's talk does not give a proper idea of the film, nor of his visual presence. You can't recount a film - you have to see it. I would like to give an example of what I am saying in Escher's words, on the subject of his woodcut *Metamorphose II* [**25**].

In 1964 Escher visited his son George in Canada. He had been invited by several organizations in the United States, including the Massachusetts Institute of Technology and Bell Laboratories, to give presentations on his work. Shortly after arriving in Canada, Escher had to be admitted to Saint Michael's hospital in Toronto for an emergency operation. All speaking commitments had to be canceled, and Escher would never again have a chance to give his carefully prepared lectures. In his usual meticulous manner, he had written out the complete English text of his lectures and these texts have been preserved. They were published in 1989 in *Escher on Escher: Exploring the Infinite* [**15**]. The chapter entitled *The lectures that were never given* contains this lecture. (We should note that the lecture was given, but not by Escher. The notes and lecture slides were sent to Arthur Loeb at Harvard, who gave the lecture at Ledgemont Laboratory in Lexington, Massachusetts, in Escher's absence.) The final part of this lecture was dedicated to *Metamorphose II.*

"I propose to round off this talk by showing you a woodcut strip with a length of thirteen feet. It's much too long to display in one or even in two slides, so I had it photographed in six parts, which I can present in three successive pairs and which you are invited to look at as if it were one uninterrupted piece of paper. It's a picture story consisting of many successive stages of transformation. The word 'Metamorphose' itself serves as a point of departure. Placed horizontally and vertically in the plane, with the letters O and M as points of intersection, the words are gradually transformed into a mosaic of black and white squares, which, in turn, develop into reptiles. If a comparison with music is allowed, one might say that, up to this point, the melody was written in two-quarter measure.

Now the rhythm changes: bluish elements are added to the white and black, and it turns into a three-quarter measure. By and by each figure simplifies into a regular hexagon. At this point an association of ideas occurs: hexagons are reminiscent of the cells of a honeycomb, and no sooner has this thought occurred than a bee larva begins to stir in every cell. In a flash every adult larva has developed into a mature bee, and soon these insects fly out into space.

The life span of my bees is short, for their black silhouettes soon merge to serve another function, namely, to provide a background for white fishes. These also, in turn, merge into each other, and the interspacings take on the shape of black birds. Then, in the distance, against a white background, appear little red-bird silhouettes. Constantly gaining in size, their contours soon touch those of their black fellow birds. What then remains of the white also takes a bird shape, so that three bird motifs, each with its own specific form and color, now entirely fill the surface in a rhythmic pattern.

Again simplification follows: each bird is transformed into a rhomb, and this gives rise to a second association of ideas: a hexagon made up of three rhombs gives a plastic effect, appearing perspectively as a cube. From cube to house is but one step, and from the houses a town is built up. It's a typical little town of southern Italy on the Mediterranean, with, as commonly seen on the Amalfi coast, a Saracen tower standing in the water and linked to shore by a bridge. [It is the town of Atrani.]

Now emerges the third association of ideas: town and sea are left behind, and interest is now centered on the tower: the rook and the other pieces on a chessboard. Meanwhile, the strip of paper on which 'Metamorphose' is portrayed has grown to some twelve feet in length. It's time to finish the story, and this opportunity is offered by the chessboard, by the white and black squares, which at the start emerged from the letters and which now return to that same word 'Metamorphose'.

Thank you very much for your attention."

So ends Escher's lecture.

When I was making my video I was not aware of this lecture text, but my idea was exactly that described by Escher, perhaps more so: not only a story, but what in cinema is called a storyboard, a precise description of a sequence to be filmed, usually given by drawings and words. So I used Escher's original woodcut as a storyboard for making the animation of *Metamorphose*. What Escher had to describe in words, because it was impossible with slides to show continuously the complete ribbon of the print to the audience, I could describe in the movie without any words, using just the animation to follow the continuously changing forms in the woodcut, accompanied by music, another suggestion by Escher himself! To paraphrase Escher, we can say: In this film it is the images and not the words that come first.

Of course, all this cannot be communicated through words only. It would not be cinema. Much of what Coxeter says in the film is 'backed up' by animations and drawings to make his words as visually interesting as possible. About half the film, lasting fifty minutes, consists of animations, special effects, and drawings.

After the filming

Once back in Toronto, Coxeter wrote his first impressions. In this letter, dated February 4, 1979, he uses "Dear Michele" for the first time.

"It was a great pleasure to visit you and to see something of your activities. It was an interesting experience to be in a movie. I hope you will cut out the parts of the film where I hesitated too long or spoke indistinctly. I also enjoyed meeting your father [Luciano Emmer, the Italian film maker] and seeing some of your work. Sincerely Donald."

In 1979 I received many other letters from him, in particular concerning the possibility of an Italian translation of Coxeter's paper *Groups generated by one rotation and one reflection, Angels and Devils*. Coxeter started to write the paper while he was in Rome. It was dedicated to Martin Gardner on his 65th birthday and published in *The Mathematical Gardner*, edited by David Klarner [7].

In September 1980 I started to publish a book on art and mathematics, to make not only the series of films on the subject. I asked Coxeter's opinion and he answered that he was interested to cooperate. He asked for information on Luca Pacioli and Piero della Francesca. He wrote November 23, 1980 "I had forgotten that Piero occupies a chapter in Coolidge's book *The Great Amateurs*. I actually have that book on my own shelf!"

At the beginning of 1981 I started thinking of an international congress dedicated to Escher and to organize a large exhibition of his work in Rome, where he lived for many years. I asked the opinion of Coxeter, and his first answer was (February 26, 1981): "An International Congress on the various aspects of Escher seems an attractive but very ambitious project. I feel honoured by your wishing me to give advice, but I have many other commitments. I will let you know if I have any further ideas. Sincerely Donald."

This first reaction was not very encouraging. During the summer of 1981, I met Coxeter in the USA and asked him whether he would be interested in a conference on my films at the University of Toronto. He replied on June 17, 1981:

"I am particularly keen to see the living Radiolaria in *Soap Bubbles*, hoping, of course, to find that Haeckel was right when he asserted that some of them look like Platonic solids!"

In fact, Haeckel was partly right [**22**]. The conference, which included the film on Escher, was held at the Mathematics Department of the University of Toronto on November 24, 1981. For the first time, Coxeter saw the film in which he had taken part two years earlier in Rome.

In 1983, the project for the congress and exhibition on Escher began to take shape. On December 20, 1982, Coxeter agreed to join the Scientific Committee for the Escher congress to be held in March 1985. In October 1984, Donald sent out some of the invitation letters, on behalf of the Scientific Committee, to the speakers for the congress to be held the following year.

He knew that the exhibition would be held at the Dutch Institute in Rome at the same time as the congress, and was worried about obtaining permission to reproduce some of Escher's work. He enclosed a letter, dated July 1984, from the editor of the *Mathematical Intelligencer*, John Ewing, saying that he had been asked to pay a fee of $500 for permission to reproduce a single work by Escher. "Frankly, the Intelligencer cannot afford that much money," he said.

The problem of the Proceedings for the Rome congress and of the exhibition catalogue was resolved thanks to the help of the Dutch government. Coxeter wrote to the Springer-Verlag to find out if they would be interested in publishing the proceedings. Their reply was negative and the volume was eventually published (and reprinted several times) by North-Holland Elsevier [**11**].

In the summer of 1984, Coxeter decided what to talk about at the Rome congress the following year. On August 24, 1984, he sent me a letter from the Mathematics Department of Princeton University where he was visiting professor that year:

"For my principal talk at the Escher Conference I propose to concentrate on Escher's use of coloured symmetry, pointing out that his tacit definition of this is more natural than Belov's. You once wrote that you want also a different essay to print in the catalogue of the exhibition. For this purpose I think I may write a few pages about Escher's use of (and fondness for) animals, with references to the various pictures involving lizards, birds, fish, snakes, ants, etc., as well as his own white cat.

Would this be suitable?"

The article for the catalogue was ready in October 1984. The Congress was held at the University of Rome at the end of March 1985. The exhibition at the Dutch Institute lasted for two months and was a great success. The print-run for the catalogue, including Coxeter's article, was 6,000 copies which sold out in a few weeks. It was edited by the Dutch Institute in Rome, and included the following articles (some in Italian, some in English): 'Introduzione' (M. Emmer), 'Escher e l'Italia' (J. Offerhaus, then director of the Dutch Institute in Rome), 'Roman Memories' (G. Escher), 'La fantasia dell'enigma e l'enigma della fantasia' (M. Emmer), 'M. C. Escher, the Man and his Work' (C. H. MacGillavry), 'Escher's Fondness for Animals' (H. S. M. Coxeter). [**24**]

Many years later, in 1998, I organized another conference on Escher, again at Rome University and at the European Centre in Ravello, combined with two exhibitions of Escher's work, one at the university's Contemporary Art Centre and the other in Ravello, a town that Escher loved. For health reasons, Coxeter could not come to this conference. However he sent two articles. The conference proceedings were eventually published in 2003 by Springer-Verlag [**23**]. One article was based on the original published in the 1985 catalogue, 'Escher's Fondness for Animals', with an addition by Escher's son, George [**8**].

At this time, Donald was going through a difficult period because his wife was ill. In a letter dated March 17, 1999, he wrote:

"I too regret that I did not manage to come to your meeting and see Valeria once more. (Valeria died on October 8th, 1998). Yes, I remember Senigallia as a lovely place. It is good that Tommaso is doing so well when one recalls how close to the end he came as a teen-ager. I am glad too that Matteo is an architect in Venice. Perhaps he can help to save that beautiful city!

While tidying my study I found the enclosed essay 'Escher's Fondness for Animals'. Perhaps you can include it in the volume, dedicated to Valeria. With every good wish, Donald."

So his paper is dedicated to Valeria in the Proceedings. The second paper is a new investigation on Escher's *Circle Limit* series starting from the article of 1978 in *Leonardo*: *The Trigonometry of Escher's Woodcut Circle Limit III* [**10**]. In the preface he wrote:

"On one of the occasions when Escher was visiting his son George in Nova Scotia, he gave an illustrated lecture in the Art Gallery of Ontario and spent a few nights with us in our Toronto house. He gave us four original prints, including one of *Circle Limit III*, which inspired this article, as well as an earlier one. In contrast with the other three Circle Limits (I, II, and IV), this employs four colours in addition to black and white, and features arcs which are not orthogonal to the peripheral circle. In an earlier article [**6**], I used hyperbolic trigonometry for my analysis, but several years later I took up the challenge of using Euclidean trigonometry instead. My former student J. Chris Fisher kindly helped by reducing my expressions for the measurements to calculated numbers that could be compared with the actual print on my staircases wall. At first one of the six measurements seemed to be wrong by a few millimeters. Rather than blame Escher, I asked Chris to check his computer again. When he admitted that the mistake was his, I realized that Escher's intuition was completely justified. I still find it almost incredible that he, with no knowledge of algebra or trigonometry, obtained accurately the centres and radii of the three different circles to which the three different axes belong."

I met Donald for the last time in Stockholm for the congress *Symmetry 2000*.

I am grateful to have had the chance to make the film on Escher with him, and I hope my efforts have contributed to keeping alive the memory of a person with such an exceptional visual mind.

References

[1] Almgren, F. and Taylor, J., *The Geometry of Soap Films and Soap Bubbles*, Scientific American, July 1976, pp. 82–93.

[2] Bill, M., *The Mathematical Way of Thinking in the Visual Art of Our Time*, in *The Visual Mind*, Emmer, M. ed., Cambridge, MA, 1993, pp. 5-9.

[3] Bruter, C., *Topologie et Perception*, Maloine-Doin, Paris, 1976.

[4] Coxeter, H. S. M., *Introduction to Geometry*, J. Wiley and Sons, New York, 1961, pp. VII, 396, IX.

[5] Coxeter, H. S. M., *Regular Polytopes*, Dover Publ., New York, 1973, p. 119.

[6] Coxeter, H. S. M., *The Non-Euclidean Symmetry of Escher's Picture 'Circle Limit III'*, Leonardo **12** (1979), pp. 19–25.

[7] Coxeter, H. S. M., *Groups generated by one rotation and one reflection, Angels and Devils*, in *The Mathematical Gardner*, Klarner, D. ed., Prindle, Weber & Schmidt, Boston MA, 1980, pp. 48–53.

[8] Coxeter, H. S. M. *Escher's Fondness for Animals* in *M.C. Escher*, Emmer, M., van Vlanderen, C., eds., Catalogue of the exhibition *M.C. Escher*, Roma, Ist. Olandese, 1985, pp. 28–32.

[9] Coxeter, H. S. M. *La simmetria non-euclidea in 'Circle Limit III' di M. C. Escher*, in Emmer, M. ed., *L'occhio di Horus: itinerari nell'immaginario matematico*, Roma, Ist. Enc. Italiana, 1989, pp. 98–104.

[10] Coxeter, H. S. M. in *M. C. Escher's Legacy*, Emmer, M., Schattschneider, D., eds., Springer-Verlag, Berlin, 2003, pp. 297–304.

[11] Coxeter, H. S. M., Emmer, M., Penrose, R., Teuber, M. eds. *M.C.Escher: Art and Science*, Amsterdam, North-Holland, 1986.

[12] Emmer, M. *Moebius Band*, film and video, series *Art and Mathematics*, 27 minutes, 1984.

[13] Emmer, M. *Platonic Solids*, film and video, series *Art and Mathematics*, 27 minutes, 1984.

[14] Emmer, M. *Soap Bubbles*, film and video, series *Art and Mathematics*, 27 minutes, 1984.

[15] Emmer, M. *Dimensions*, film and video, series *Art and Mathematics*, 27 minutes, 1986.

[16] Emmer, M. *Art and Mathematics: The Platonic Solids*, in *The Visual Mind*, Emmer, M. ed., MIT press, Cambridge, MA, 1993, pp. 215–220.

[17] Emmer, M. *Mathematics and Art: the Film Series*, Mathematics and Visualization series, Bruter, P.C., ed., *Mathematics and Art*, Berlin, Springer-Verlag, 2002, pp. 119–133.

[18] Emmer, M. *The Project Art and Mathematics* see [**17**]. This paper contains a complete list of all the books, films, exhibitions included in the project and in the more recent one, started in 1997 in Venezia, *Mathematics and Culture*.

[19] Emmer, M. *Films: A Communicating Tool for Mathematics*, Mathematics and Visualization Series, vol. 3, Hege, C. and Polthier, K., eds., *Mathematics and Visualization*, Berlin, Springer-Verlag, 2003, pp. 393–405.

[20] Emmer, M. *The 'Mathematics and Culture' Project*, Trends and Challenges in Mathematics Education, Wang, J. and Xu, B., eds., East China Normal Univ. Press, Shanghai, 2004, pp. 85–103.

[21] Emmer, M. *Movies on M.C. Escher and Their Mathematical Appeal*, in [**8**], pp. 249–262.

[22] Emmer, M. *Bolle di sapone: un viaggio tra arte, fantasia e scienza*, Firenze, La Nuova Italia, 1991.

[23] Emmer, M., Schattschneider, D., eds. *M.C. Escher's Legacy*, Springer-Verlag, Berlin, 2003.

[24] Emmer, M., van Vlanderen, C., eds. *M.C. Escher*, Catalogue of the exhibition *M.C. Escher*, Roma, Ist. Olandese, 1985.

[25] Escher, M. C., *The Graphic Work of M.C. Escher*, MacDonald, London, 1961; pp.7, 11, 48.

[26] Escher, M. C., *The world of M. C. Escher*, H. N. Abrams, New York, 1971.

[27] Escher, M. C., *M. C. Escher: His Life and Complete Graphic Work*, Bool, F. H., Kist, J. R., Locher, J. L., Wierda, F., eds., H. N. Abrams, New York, 1982; pp. 155, 158, 170.

[28] Escher, M. C., *M. C. Escher on Escher - Exploring the Infinite*, New York, H. N. Abrams, 1986; pp. 92, 98, 120.

[29] Hardy, G. H., *A Mathematician's Apology*, Cambridge University Press, Cambridge, 1940, reprinted 1988, pp. 84, 85. The original text is: "A mathematician, like a painter or a poet, is a maker of patterns ... The mathematician's patterns, like the painter's or the poet's, must be beautiful ... Beauty is the first test: there is no permanent place in the world for ugly mathematics."

[30] Plateau, J., *Statique expérimentale et théorique des liquides soumis aux seules forces moléculaires*, Gauthier-Villars, Paris, 1873.

[31] Plato, *Timaeus*, The Great Book of the Western World, vol. 7, Plato, Encyclopedia Britannica, London, 1952, pp. 442–477.

[32] Roberts, S. and Weiss, A. I., *Donald in Wonderland: the Many-Faceted Life of H. S. M. Coxeter*, The Mathematical Intelligencer **26**, no. 3 (2004), pp. 17–25.

[33] Saffaro, L., *Le forme del pensiero*, catalogue of the exhibition, Accame, G. M. , ed., Edizioni Aspasia, Bologna, 2004.

DIPARTIMENTO DI MATEMATICA, UNIVERSITÀ DI ROMA "LA SAPIENZA", P.LE A. MORO 2, 00185 ROME, ITALY

E-mail address: `emmer@mat.uniroma1.it`

Publications of H. S. M. Coxeter

1. H. S. M. Coxeter.
   The pure Archimedean polytopes in six and seven dimensions.
   Math. Proc. Cambridge Philos. Soc., 24:1–9, 1927.

2. H. S. M. Coxeter.
   The densities of the regular polytopes.
   Math. Proc. Cambridge Philos. Soc., 27:201–211, 1930.

3. H. S. M. Coxeter.
   The polytopes with regular-prismatic vertex figures, Part I.
   Philos. Trans. Roy. Soc. London Ser. A, 229:329–425, 1930.

4. H. S. M. Coxeter.
   The densities of the regular polytopes, Part II.
   Math. Proc. Cambridge Philos. Soc., 28:509–521, 1931.

5. H. S. M. Coxeter.
   Groups whose fundamental regions are simplexes.
   J. London Math. Soc., 6:132–136, 1931.

6. H. S. M. Coxeter.
   The densities of the regular polytopes, Part III.
   Math. Proc. Cambridge Philos. Soc., 29:1–22, 1932.

7. H. S. M. Coxeter.
   The polytopes with regular-prismatic vertex figures, Part II.
   Proc. London Math. Soc., 34(2):126–189, 1932.

8. H. S. M. Coxeter.
   Finite groups generated by reflections, and their subgroups generated by reflections.
   Math. Proc. Cambridge Philos. Soc., 30:466–482, 1933.

9. H. S. M. Coxeter.
   Regular compound polytopes in more than four dimensions.
   J. Math. and Phys., 12:334–345, 1933.

10. H. S. M. Coxeter and J. A. Todd.
    On points with arbitrarily assigned mutual distances.
    Math. Proc. Cambridge Philos. Soc., 30:1–3, 1933.

11. H. S. M. Coxeter.
    On simple isomorphisms between abstract groups.
    J. London Math. Soc., 9:211–212, 1934.

12. H. S. M. Coxeter.
    Abstract groups of the form $V_1^k = V_j^3 = (V_i V_j)^2 = 1$.
    J. London Math. Soc., 9:213–219, 1934.

13. H. S. M. Coxeter.
Discrete groups generated by reflections.
Ann. of Math. (2), 35(3):588–621, 1934.

14. H. S. M. Coxeter and P. S. Donchian.
An n-dimensional extension of Pythagoras' theorem.
Math. Gaz., 19:206, 1935.

15. H. S. M. Coxeter.
The complete enumeration of finite groups $R_i^2 = (R_i R_j)^{k_{ij}} = 1$.
J. London Math. Soc., 10:21–25, 1935.

16. H. S. M. Coxeter.
The functions of Schläfli and Lobatschefsky.
Quart. J. Math. Oxford, 6:13–29, 1935.

17. H. S. M. Coxeter.
Wythoff's construction for uniform polytopes.
Proc. London Math. Soc., 38:327–339, 1935.

18. H. S. M. Coxeter.
An abstract definition for the alternating group in terms of two generators.
J. London Math. Soc., 11:150–156, 1936.

19. H. S. M. Coxeter.
The abstract groups $R^m = S^m = (R^j S^j)^{p_j} = 1$,
$S^m = T^2 = (S^j T)^{2p_j} = 1$ and $S^m = T^2 = (S^{-j} T S^j T)^{p_j} = 1$.
Proc. London Math. Soc., 41(2):278–301, 1936.

20. H. S. M. Coxeter.
The groups determined by the relations $S^l = T^m = (S^{-1} T^{-1} S T)^p = 1$.
Duke Math. J., 2:61–73, 1936.

21. H. S. M. Coxeter.
On Schläfli's generalization of Napier's pentagramma mirificum.
Bull. Calcutta Math. Soc., 28:123–144, 1936.

22. H. S. M. Coxeter.
The representation of conformal space on a quadric.
Ann. of Math. (2), 37(2):416–426, 1936.

23. H. S. M. Coxeter and J. A. Todd.
Abstract definitions for the symmetry groups of the regular polytopes, in terms of two generators, Part I: the complete groups.
Math. Proc. Cambridge Philos. Soc., 32:194–200, 1936.

24. H. S. M. Coxeter and J. A. Todd.
A practical method of enumerating cosets of a finite abstract group.
Proc. Edinburgh Math. Soc., 5(2):26–34, 1936.

25. H. S. M. Coxeter.
Abstract definitions for the symmetry groups of the regular polytopes, in terms of two generators, Part II: the rotation groups.
Math. Proc. Cambridge Philos. Soc., 33:315–324, 1937.

26. H. S. M. Coxeter.
Regular skew polyhedra in three and four dimensions, and their topological analogues.
Proc. London Math. Soc., 43(2):33–62, 1937.

27. H. S. M. Coxeter, P. Du Val, H. T. Flather, and J. F. Petrie.
The Fifty-nine Icosahedra.
University of Toronto Studies (Math. Series, No. 6). 1938.
Springer-Verlag, Berlin, New York, reprint 1982.

28. H. S. M. Coxeter.
An easy method for constructing polyhedral group-pictures.
Amer. Math. Monthly, 45:522–525, 1938.

29. H. S. M. Coxeter.
The abstract groups $G^{m,n,p}$.
Trans. Amer. Math. Soc., 45(1):73–150, 1939.

30. H. S. M. Coxeter.
The regular sponges, or skew polyhedra.
Scripta Math., 6:240–244, 1939.

31. Richard Brauer and H. S. M. Coxeter.
A generalization of theorems of Schönhardt and Mehmke on polytopes.
Trans. Roy. Soc. Canada. Sect. III. (3), 34:29–34, 1940.

32. H. S. M. Coxeter.
The binary polyhedral groups, and other generalizations
of the quaternion group.
Duke Math. J., 7:367–379, 1940.

33. H. S. M. Coxeter.
A method for proving certain abstract groups to be infinite.
Bull. Amer. Math. Soc., 46:246–251, 1940.

34. H. S. M. Coxeter.
The polytope 2_{21}, whose twenty-seven vertices correspond
to the lines on the general cubic surface.
Amer. J. Math., 62:457–486, 1940.

35. H. S. M. Coxeter.
Regular and semi-regular polytopes. I.
Math. Z., 46:380–407, 1940.

36. H. S. M. Coxeter.
Non-Euclidean Geometry.
Mathematical Expositions, no. 2,
University of Toronto Press, Toronto, Ont., 1942.
MAA Spectrum, Mathematical Association of America,
Washington, D.C., sixth edition 1998.

37. H. S. M. Coxeter.
A geometrical background for de Sitter's world.
Amer. Math. Monthly, 50:217–228, 1943.

38. H. S. M. Coxeter.
The map-coloring of unorientable surfaces.
Duke Math. J., 10:293–304, 1943.

39. H. S. M. Coxeter.
Integral Cayley numbers.
Duke Math. J., 13:561–578, 1946.

40. H. S. M. Coxeter.
  The nine regular solids.
  In *Proc. First Canadian Math. Congress, Montreal, 1945*, pages 252–264.
  University of Toronto Press, Toronto, 1946.

41. H. S. M. Coxeter.
  Quaternions and reflections.
  Amer. Math. Monthly, 53:136–146, 1946.

42. H. S. M. Coxeter.
  The product of three reflections.
  Quart. Appl. Math., 5:217–222, 1947.

43. H. S. M. Coxeter.
  A problem of collinear points.
  Amer. Math. Monthly, 55:26–28, 1948.

44. H. S. M. Coxeter.
  Regular Polytopes.
  Methuen & Co. Ltd., London, 1948.
  Dover Publications Inc., New York, third edition 1973.

45. H. S. M. Coxeter.
  Configurations and maps.
  Rep. Math. Colloquium (2), 8:18–38, 1949.

46. H. S. M. Coxeter.
  Projective geometry.
  Math. Mag., 23:79–97, 1949.

47. H. S. M. Coxeter.
  The Real Projective Plane.
  McGraw-Hill Book Company, Inc., New York, N. Y., 1949.
  Springer-Verlag, New York, third edition 1993.

48. H. S. M. Coxeter.
  Self-dual configurations and regular graphs.
  Bull. Amer. Math. Soc., 56:413–455, 1950.

49. H. S. M. Coxeter and G. J. Whitrow.
  World-structure and non-Euclidean honeycombs.
  Proc. Roy. Soc. London. Ser. A., 201:417–437, 1950.

50. H. S. M. Coxeter.
  Extreme forms.
  Canadian J. Math., 3:391–441, 1951.

51. H. S. M. Coxeter.
  The product of the generators of a finite group generated by reflections.
  Duke Math. J., 18:765–782, 1951.

52. H. S. M. Coxeter.
  Interlocked rings of spheres.
  Scripta Math., 18:113–121, 1952.

53. H. S. M. Coxeter.
  Rouse Ball's unpublished notes on three fours.
  Scripta Math., 18:85–86, 1952.

54. H. S. M. Coxeter.
The golden section, phyllotaxis, and Wythoff's game.
Scripta Math., 19:135–143, 1953.

55. H. S. M. Coxeter and J. A. Todd.
An extreme duodenary form.
Canadian J. Math., 5:384–392, 1953.

56. H. S. M. Coxeter, M. S. Longuet-Higgins, and J. C. P. Miller.
Uniform polyhedra.
Philos. Trans. Roy. Soc. London. Ser. A., 246:401–450 (6 plates), 1954.

57. H. S. M. Coxeter.
Arrangements of equal spheres in non-Euclidean spaces.
Acta Math. Acad. Sci. Hungar., 5:263–274, 1954.

58. H. S. M. Coxeter.
An extension of Pascal's theorem.
Amer. Math. Monthly, 61:723, 1954.

59. H. S. M. Coxeter.
Regular honeycombs in elliptic space.
Proc. London Math. Soc. (3), 4:471–501, 1954.

60. H. S. M. Coxeter.
Six uniform polyhedra.
Scripta Math., 20:227, 1954.

61. H. S. M. Coxeter.
The affine plane.
Scripta Math., 21:5–14, 1955.

62. H. S. M. Coxeter.
On Laves' graph of girth ten.
Canad. J. Math., 7:18–23, 1955.

63. H. S. M. Coxeter.
The collineation groups of the finite affine and projective planes
with four lines through each point.
Abh. Math. Sem. Univ. Hamburg, 20:165–177, 1956.

64. H. S. M. Coxeter.
Hyperbolic triangles.
Scripta Math., 22:5–13, 1956.

65. H. S. M. Coxeter.
Regular honeycombs in hyperbolic space.
In *Proceedings of the International Congress of Mathematicians, 1954,
Amsterdam, vol. III*, pages 155–169.
Erven P. Noordhoff N.V., Groningen, 1956.

66. H. S. M. Coxeter and W. O. J. Moser.
Generators and relations for discrete groups.
Springer-Verlag, Berlin, 1957.
Fourth edition 1980.

67. H. S. M. Coxeter.
Crystal symmetry and its generalizations (Presidential Address).
Trans. Roy. Soc. Canada Sect. III, 51(3):1–13, 1957.

68. H. S. M. Coxeter.
Groups generated by unitary reflections of period two.
Canad. J. Math., 9:243–272, 1957.

69. H. S. M. Coxeter.
Map-coloring problems.
Scripta Math., 23:11–25, 1957.

70. H. S. M. Coxeter.
The chords of the non-ruled quadric in *PG*(3, 3).
Canad. J. Math., 10:484–488, 1958.

71. H. S. M. Coxeter.
Close-packing and froth.
Illinois J. Math., 2:746–758, 1958.

72. H. S. M. Coxeter.
On subgroups of the modular group.
J. Math. Pures Appl. (9), 37:317–319, 1958.

73. H. S. M. Coxeter.
Twelve points in PG(5, 3) with 95040 self-transformations.
Proc. Roy. Soc. London. Ser. A, 247:279–293, 1958.

74. H. S. M. Coxeter, L. Few, and C. A. Rogers.
Covering space with equal spheres.
Mathematika, 6:147–157, 1959.

75. H. S. M. Coxeter.
Factor groups of the braid group.
In *Proc. 4th Canad. Math. Congress, Banff (1957)*, pages 95–122.
Univ. Toronto Press, Toronto, Ont., 1959.

76. H. S. M. Coxeter.
The four-color map problem, 1840–1890.
Math. Teacher, 52:283–289, 1959.

77. H. S. M. Coxeter.
Polytopes over GF(2) and their relevance for the cubic surface group.
Canad. J. Math., 11:646–650, 1959.

78. H. S. M. Coxeter.
Symmetrical definitions for the binary polyhedral groups.
In *Proc. Sympos. Pure Math., Vol. 1*, pages 64–87.
American Mathematical Society, Providence, R.I., 1959.

79. H. S. M. Coxeter.
Introduction to geometry.
John Wiley & Sons Inc., New York, 1961.
Second edition 1969, reprint 1989.

80. H. S. M. Coxeter.
On Wigner's problem of reflected light signals in de Sitter space.
Proc. Roy. Soc. Ser. A, 261:435–442, 1961.

81. H. S. M. Coxeter.
Similarities and conformal transformations.
Ann. Mat. Pura Appl. (4), 53:165–172, 1961.

82. H. S. M. Coxeter.
The abstract group $G^{3,7,16}$.
Proc. Edinburgh Math. Soc. (2), 13:47–61 and 189, 1962.

83. H. S. M. Coxeter.
The classification of zonohedra by means of projective diagrams.
J. Math. Pures Appl. (9), 41:137–156, 1962.

84. H. S. M. Coxeter.
The problem of packing a number of equal nonoverlapping circles
on a sphere.
Trans. New York Acad. Sci., 24(II):320–331, 1962.

85. H. S. M. Coxeter.
Projective line geometry.
Math. Notae, 1:197–216, 1962.

86. H. S. M. Coxeter.
The symmetry groups of the regular complex polygons.
Arch. Math., 13:86–97, 1962.

87. H. S. M. Coxeter.
The total length of the edges of a non-Euclidean polyhedron.
In *Studies in mathematical analysis and related topics*, pages 62–69.
Stanford Univ. Press, Stanford, Calif, 1962.

88. B. L. Chilton and H. S. M. Coxeter.
Polar zonohedra.
Amer. Math. Monthly, 70:946–951, 1963.

89. H. S. M. Coxeter and L. Fejes Tóth.
The total length of the edges of a non-Euclidean polyhedron with
triangular faces.
Quart. J. Math. Oxford Ser. (2), 14:273–284, 1963.

90. H. S. M. Coxeter.
An upper bound for the number of equal nonoverlapping spheres
that can touch another of the same size.
In *Proc. Sympos. Pure Math., Vol. VII*, pages 53–71.
Amer. Math. Soc., Providence, R.I., 1963.

91. H. S. M. Coxeter.
Projective geometry.
Blaisdell Publishing Co. Ginn and Co. New York-London-Toronto, 1964.
University of Toronto Press, Toronto, Ont., second edition 1974.
Springer-Verlag, New York, N.Y., reprint 1994.

92. H. S. M. Coxeter.
Regular compound tessellations of the hyperbolic plane.
Proc. Roy. Soc. Ser. A, 278:147–167, 1964.

93. H. S. M. Coxeter.
Geometry.
In *Lectures on Modern Mathematics, Vol. III*, pages 58–94.
Wiley, New York, 1965.

94. H. S. M. Coxeter.
Inversive distance.
Ann. Mat. Pura Appl. (4), 71:73–83, 1966.

95. H. S. M. Coxeter.
The inversive plane and hyperbolic space.
Abh. Math. Sem. Univ. Hamburg, 29:217–242, 1966.

96. H. S. M. Coxeter.
Reflected light signals.
In *Perspectives in Geometry (Essays in Honor of V. Hlavatý)*, pages
58–70. Indiana Univ. Press, Bloomington, Ind., 1966.

97. H. S. M. Coxeter and S. L. Greitzer.
Geometry Revisited.
Number 19 in New Math Library. Random House, New York, N.Y., and
Mathematical Association of America, Washington, D.C., 1967.

98. H. S. M. Coxeter.
Finite groups generated by unitary reflections.
Abh. Math. Sem. Univ. Hamburg, 31:125–135, 1967.

99. H. S. M. Coxeter.
The Lorentz group and the group of homographies.
In *Proc. Internat. Conf. Theory of Groups (Canberra, 1965)*, pages 73–77.
Gordon and Breach, New York, N.Y., 1967.

100. H. S. M. Coxeter.
Transformation groups from the geometric viewpoint.
Proc. CUPM Geometry Conference, pages 1–72, 1967.

101. H. S. M. Coxeter.
Loxodromic sequences of tangent spheres.
Aequationes Math., 1:104–121, 1968.

102. H. S. M. Coxeter.
Mid-circles and loxodromes.
Math. Gaz., 52:1–8, 1968.

103. H. S. M. Coxeter.
The problem of Apollonius.
Amer. Math. Monthly, 75:5–15, 1968.

104. H. S. M. Coxeter.
Twelve geometric essays.
Southern Illinois University Press, Carbondale, Ill., 1968.
Dover Publications Inc., Mineola, N.Y., reprint
with title "The Beauty of Geometry", 1999.

105. H. S. M. Coxeter.
Helices and concho-spirals.
Nobel Symp., 11:29–34, 1969.
(Eds. A. Engström and B. Strandberg), Almqvist and Wiskell, Stockholm.

106. H. S. M. Coxeter.
Affinely regular polygons.
Abh. Math. Sem. Univ. Hamburg, 34:38–58, 1969/1970.

107. H. S. M. Coxeter.
Products of shears in an affine Pappian plane.
Rend. Mat. (6), 3:161–166, 1970.

108. H. S. M. Coxeter.
Twisted honeycombs.
American Mathematical Society, Providence, R.I., 1970.
Conference Board of the Mathematical Sciences Regional Conference
Series in Mathematics, No. 4.

109. H. S. M. Coxeter.
The finite inversive plane with four points on each circle.
In *Studies in Pure Mathematics (Presented to Richard Rado)*,
pages 39–51. Academic Press, London, 1971.

110. H. S. M. Coxeter.
Frieze patterns.
Acta Arith., 18:297–310, 1971.

111. H. S. M. Coxeter.
Inversive geometry.
Educ. Stud. Math., 3:310–321, 1971.

112. H. S. M. Coxeter.
The mathematical implications of Escher's prints.
The World of M. C. Escher (Ed. J. L. Locher), Abrams, New York,
pages 49–52, 1971.

113. H. S. M. Coxeter.
The mathematics of map coloring.
Leonardo, 4:273–277, 1971.

114. H. S. M. Coxeter.
Virus macromolecules and geodesic domes.
In *A spectrum of mathematics (Essays presented to H. G. Forder)*,
pages 98–107. Auckland Univ. Press, Auckland, 1971.

115. J. H. Conway, H. S. M. Coxeter, and G. C. Shephard.
The centre of a finitely generated group.
Tensor (N.S.), 25:405–418; erratum, ibid. (N.S.) **26** (1972), 477, 1972.
Commemoration volumes for Prof. Dr. Akitsugu Kawaguchi's seventieth
birthday, Vol. II.

116. H. S. M. Coxeter.
The role of intermediate convergents in Tait's explanation for phyllotaxis.
J. Algebra, 20:167–175, 1972.

117. J. H. Conway and H. S. M. Coxeter.
Triangulated polygons and frieze patterns.
Math. Gaz., 57(400):87–94, 1973.

118. J. H. Conway and H. S. M. Coxeter.
Triangulated polygons and frieze patterns.
Math. Gaz., 57(401):175–183, 1973.

119. H. S. M. Coxeter.
Cayley diagrams and regular complex polygons.
In *A survey of combinatorial theory (dedicated to R. C. Bose on the occasion of his seventieth birthday, Proc. Internat. Sympos., Colorado State Univ., Fort Collins, Colo., 1971)*, pages 85–93.
North-Holland, Amsterdam, 1973.

120. H. S. M. Coxeter.
The Dirac matrix group and other generalizations of the quaternion group.
Comm. Pure Appl. Math., 26:693–698, 1973.
Collection of articles dedicated to Wilhelm Magnus.

121. H. S. M. Coxeter.
The equianharmonic surface and the Hessian polyhedron.
Ann. Mat. Pura Appl. (4), 98:77–92, 1974.

122. H. S. M. Coxeter.
Kepler and mathematics.
Vistas in Astronomy (Ed. Arthur Beer, London), 19, 1974.

123. H. S. M. Coxeter.
Polyhedral numbers.
In *For Dirk Struik*, Boston Stud. Philos. Sci., XV, pages 25–35.
Reidel, Dordrecht, 1974.

124. H. S. M. Coxeter.
Regular complex polytopes.
Cambridge University Press, London, 1974.
Second edition, 1991.

125. W. W. Rouse Ball and H. S. M. Coxeter.
Mathematical recreations and essays.
University of Toronto Press, Toronto, Ont., twelfth edition, 1974.
Dover Publications Inc., New York, thirteenth edition 1987.

126. H. S. M. Coxeter.
Desargues configurations and their collineation groups.
Math. Proc. Cambridge Philos. Soc., 78(2):227–246, 1975.

127. H. S. M. Coxeter.
The space-time continuum.
Historia Math., 2:289–298, 1975.

128. H. S. M. Coxeter.
The Pappus configuration and its groups.
Nederl. Akad. Wetensch. Verslag Afd. Natuurk., 85(4):44–46, 1976.

129. C. M. Campbell, H. S. M. Coxeter, and E. F. Robertson.
Some families of finite groups having two generators and two relations.
Proc. Roy. Soc. London Ser. A, 357(1691):423–438, 1977.

130. H. S. M. Coxeter.
The Erlangen program.
Math. Intelligencer, 0:22, 1977.

131. H. S. M. Coxeter.
Gauss as a geometer.
Historia Math., 4(4):379–396, 1977.

132. H. S. M. Coxeter.
The Pappus configuration and its groups.
Pi Mu Epsilon J., 6(6):331–336, 1977.

133. H. S. M. Coxeter.
The Pappus configuration and the self-inscribed octagon. I.
Nederl. Akad. Wetensch. Proc. Ser. A **80**=*Indag. Math.*, 39(4):256–269,
1977.

134. H. S. M. Coxeter.
The Pappus configuration and the self-inscribed octagon. II.
Nederl. Akad. Wetensch. Proc. Ser. A **80**=*Indag. Math.*, 39(4):270–284,
1977.

135. H. S. M. Coxeter.
The Pappus configuration and the self-inscribed octagon. III.
Nederl. Akad. Wetensch. Proc. Ser. A **80**=*Indag. Math.*, 39(4):285–300,
1977.

136. H. S. M. Coxeter and G. C. Shephard.
Regular 3-complexes with toroidal cells.
J. Combinatorial Theory Ser. B, 22(2):131–138, 1977.

137. H. S. M. Coxeter.
The amplitude of a Petrie polygon.
C. R. Math. Rep. Acad. Sci. Canada, 1(1):9–12, 1978.

138. H. S. M. Coxeter.
Parallel lines.
Canad. Math. Bull., 21(4):385–397, 1978.

139. H. S. M. Coxeter.
Polytopes in the Netherlands.
Nieuw Arch. Wisk. (3), 26(1):116–141, 1978.

140. Stillman Drake, H. S. M. Coxeter, Charles V. Jones, Henry S. Tropp,
Christoph J. Scriba, B. Sinclair, Michael S. Mahoney, and Dirk J. Struik.
A memorial tribute to Kenneth O. May.
Historia Math., 5(1):3–12, 1978.

141. Pieter Huybers and H. S. M. Coxeter.
A new approach to the chiral Archimedean solids.
C. R. Math. Rep. Acad. Sci. Canada, 1(5):269–274, 1978.

142. H. S. M. Coxeter and Roberto W. Frucht.
A new trivalent symmetrical graph with 110 vertices.
In *Second International Conference on Combinatorial Mathematics (New
York, 1978)*, volume 319 of *Ann. New York Acad. Sci.*, pages 141–152.
New York Acad. Sci., New York, N.Y., 1979.

143. H. S. M. Coxeter.
The non-Euclidean symmetry of Escher's picture "Circle Limit III".
Leonardo, 12:19–25, 1979.

144. H. S. M. Coxeter.
On R. M. Foster's regular maps with large faces.
In *Relations between combinatorics and other parts of mathematics (Proc. Sympos. Pure Math., Ohio State Univ., Columbus, Ohio, 1978)*, Proc. Sympos. Pure Math., XXXIV, pages 117–128.
Amer. Math. Soc., Providence, R.I., 1979.

145. H. S. M. Coxeter.
Angles and arcs in the hyperbolic plane.
Math. Chronicle, 9:17–33, 1980.
H. G. Forder 90th birthday volume.

146. H. S. M. Coxeter.
The edges and faces of a 4-dimensional polytope.
In *Proceedings of the Eleventh Southeastern Conference on Combinatorics, Graph Theory and Computing (Florida Atlantic Univ., Boca Raton, Fla., 1980), Vol. I*, volume 28, pages 309–334, 1980.

147. H. S. M. Coxeter.
Higher-dimensional analogues of the tetrahedrite crystal twin.
Match, 9:67–72, 1980.

148. H. S. M. Coxeter, R. Frucht, and D. L. Powers.
Zero-symmetric Graphs: trivalent graphical regular representations of groups.
Academic Press, New York, 1981.

149. H. S. M. Coxeter.
Angels and devils.
In *The Mathematical Gardner* (Ed. David Klarner), pages 197–209.
Wadsworth, Belmont, California, 1981.

150. H. S. M. Coxeter.
The derivation of Schoenberg's star-polytopes
from Schoute's simplex nets.
In *The geometric vein*, pages 149–164. Springer, New York, 1981.

151. H. S. M. Coxeter.
A systematic notation for the Coxeter graph.
C. R. Math. Rep. Acad. Sci. Canada, 3(6):329–332, 1981.

152. H. S. M. Coxeter.
Rational spherical triangles.
Math. Gaz., 66(436):145–147, 1982.

153. H. S. M. Coxeter.
Ten toroids and fifty-seven hemidodecahedra.
Geom. Dedicata, 13(1):87–99, 1982.

154. H. S. M. Coxeter and W. L. Edge.
The simple groups PSL(2, 7) and PSL(2, 11).
C. R. Math. Rep. Acad. Sci. Canada, 5(5):201–206, 1983.

155. H. S. M. Coxeter.
The affine aspect of Yaglom's Galilean Feuerbach.
Nieuw Arch. Wisk. (4), 1(2):212–223, 1983.

156. H. S. M. Coxeter.
My graph.
Proc. London Math. Soc. (3), 46(1):117–136, 1983.

157. H. S. M. Coxeter.
The twenty-seven lines on the cubic surface.
In *Convexity and its applications*, pages 111–119. Birkhäuser, Basel, 1983.

158. H. S. M. Coxeter.
The group of genus two.
In *Proceedings of the conference on combinatorial and incidence geometry: principles and applications (La Mendola, 1982)*, volume 7 of *Rend. Sem. Mat. Brescia*, pages 219–248, Milan, 1984. Vita e Pensiero.

159. H. S. M. Coxeter.
Surprising relationships among unitary reflection groups.
Proc. Edinburgh Math. Soc. (2), 27(2):185–194, 1984.

160. H. S. M. Coxeter.
A symmetrical arrangement of eleven hemi-icosahedra.
In *Convexity and graph theory (Jerusalem, 1981)*, volume 87 of *North-Holland Math. Stud.*, pages 103–114. North-Holland, Amsterdam, 1984.

161. H. S. M. Coxeter and Asia Ivić Weiss.
Twisted honeycombs $\{3,5,3\}_t$ and their groups.
Geom. Dedicata, 17(2):169–179, 1984.

162. H. S. M. Coxeter.
A special book review of *M. C. Escher: His life and complete graphic work* (Ed. J. L. Lochner) Abrams, New York, N.Y., 1982.
Math. Intelligencer, 7:59–69, 1985.

163. H. S. M. Coxeter.
The Lehmus inequality.
Aequationes Math., 28(1-2):4–12, 1985.

164. H. S. M. Coxeter.
Polytopes, kaleidoscopes, Pythagoras and the future.
C. R. Math. Rep. Acad. Sci. Canada, 7(2):107–114, 1985.

165. H. S. M. Coxeter.
Regular and semiregular polytopes. II.
Math. Z., 188(4):559–591, 1985.

166. H. S. M. Coxeter.
The seventeen black and white frieze types.
C. R. Math. Rep. Acad. Sci. Canada, 7(5):327–331, 1985.

167. H. S. M. Coxeter.
The simplicial helix and the equation $\tan n\theta = n \tan \theta$.
Canad. Math. Bull., 28(4):385–393, 1985.

168. H. S. M. Coxeter.
Coloured symmetry.
In *M. C. Escher: art and science (Rome, 1985)*, pages 15–33.
North-Holland, Amsterdam, 1986.

169. H. S. M. Coxeter.
The generalized Petersen graph $G(24, 5)$.
Comput. Math. Appl. Part B, 12(3-4):579–583, 1986.

170. H. S. M. Coxeter.
Alicia Boole Stott (1860–1940).
In *Women of mathematics*, pages 220–224.
Greenwood, Westport, CT, 1987.

171. H. S. M. Coxeter.
On Miller's generalized dihedral group.
C. R. Math. Rep. Acad. Sci. Canada, 9(5):265–269, 1987.

172. H. S. M. Coxeter.
A packing of 840 balls of radius $9°0'19''$ on the 3-sphere.
In *Intuitive geometry (Siófok, 1985)*, volume 48 of *Colloq. Math. Soc. János Bolyai*, pages 127–137. North-Holland, Amsterdam, 1987.

173. H. S. M. Coxeter.
A simple introduction to colored symmetry.
Internat. J. Quantum Chem., 31(3):455–461, 1987.

174. H. S. M. Coxeter.
Escher's lizards.
Structural Topology, (15):23–30, 1988.
Dual English-French text.

175. H. S. M. Coxeter.
Regular and semi-regular polytopes. III.
Math. Z., 200(1):3–45, 1988.

176. H. S. M. Coxeter.
Regular and semiregular polyhedra.
In *Shaping space (Northampton, Mass., 1984)*, Design Sci. Collect., pages 67–79. Birkhäuser Boston, Boston, MA, 1988.

177. H. S. M. Coxeter, Larry Carmack, and Gerald Schrag.
Research problems.
Discrete Math., 73(3):311–312, 1989.

178. H. S. M. Coxeter.
Review of *Sphere packings, Lattices and Groups* by J. H. Conway and N. J. A. Sloane.
Amer. Math. Monthly, 96:538–544, 1989.

179. H. S. M. Coxeter.
Star polytopes and the Schläfli function $f(\alpha, \beta, \gamma)$.
Elem. Math., 44(2):25–36, 1989.

180. H. S. M. Coxeter.
Trisecting an orthoscheme.
Comput. Math. Appl., 17(1-3):59–71, 1989.

181. H. S. M. Coxeter.
Affine regularity.
C. R. Math. Rep. Acad. Sci. Canada, 13(5):181–186, 1991.

182. H. S. M. Coxeter.
The evolution of Coxeter-Dynkin diagrams.
Nieuw Arch. Wisk. (4), 9(3):233–248, 1991.

183. H. S. M. Coxeter.
Orthogonal trees.
Bull. Inst. Combin. Appl., 3:83–91, 1991.

184. H. S. M. Coxeter and G. C. Shephard.
Some regular maps and their polyhedral realizations.
In *Applied geometry and discrete mathematics*, volume 4 of *DIMACS Ser. Discrete Math. Theoret. Comput. Sci.*, pages 157–174.
Amer. Math. Soc., Providence, RI, 1991.

185. H. S. M. Coxeter.
Affine regularity.
Abh. Math. Sem. Univ. Hamburg, 62:249–253, 1992.

186. H. S. M. Coxeter.
Errata: "The evolution of Coxeter-Dynkin diagrams".
Nieuw Arch. Wisk. (4), 10(1-2):iii, 1992.

187. H. S. M. Coxeter.
Cubic curves related to a quadrangle.
C. R. Math. Rep. Acad. Sci. Canada, 15(6):237–242, 1993.

188. H. S. M. Coxeter.
Cyclotomic integers, nondiscrete tessellations, and quasicrystals.
Indag. Math. (N.S.), 4(1):27–38, 1993.

189. H. S. M. Coxeter and G. C. Shephard.
Portraits of a family of complex polytopes.
In *The visual mind*, Leonardo Book Ser., pages 19–26.
MIT Press, Cambridge, MA, 1993.

190. H. S. M. Coxeter and Jan van de Craats.
Philon lines in non-Euclidean planes.
J. Geom., 48(1-2):26–55, 1993.

191. H. S. M. Coxeter.
The evolution of Coxeter-Dynkin diagrams.
In *Polytopes: abstract, convex and computational (Scarborough, Ont., 1993)*, volume 440 of *NATO Adv. Sci. Inst. Ser. C Math. Phys. Sci.*, pages 21–42. Kluwer Acad. Publ., Dordrecht, 1994.

192. H. S. M. Coxeter.
Symmetrical combinations of three or four hollow triangles.
Math. Intelligencer, 16(3):25–30, 1994, and 17(1):4, 1995.

193. H. S. M. Coxeter.
Kaleidoscopes.
Canadian Mathematical Society Series of Monographs and Advanced Texts. John Wiley & Sons Inc., New York, 1995.
Selected writings of H. S. M. Coxeter, Edited by F. Arthur Sherk, Peter McMullen, Anthony C. Thompson and Asia Ivić Weiss,
A Wiley-Interscience Publication.

194. H. S. M. Coxeter.
Some applications of trilinear coordinates.
Linear Algebra Appl., 226/228:375–388, 1995.

195. H. S. M. Coxeter.
Close-packing and froth.
Forma, 11(3):271–285, 1996.
Reprint of *Illinois J. Math.*, 2:746–758, 1958.

196. H. S. M. Coxeter.
The trigonometry of Escher's woodcut "Circle Limit III".
Math. Intelligencer, 18(4):42–46, 1996.

197. H. S. M. Coxeter, J. Chris Fisher, and J. B. Wilker.
Coordinates for the regular complex polygons.
J. London Math. Soc. (2), 55(3):527–548, 1997.

198. H. S. M. Coxeter.
Erratum: "The trigonometry of Escher's woodcut 'Circle Limit III'"
[*Math. Intelligencer*, 18(4):42–46, 1996].
Math. Intelligencer, 19(1):79, 1997.

199. H. S. M. Coxeter.
Numerical distances among the spheres in a loxodromic sequence.
Math. Intelligencer, 19(4):41–47, 1997.

200. H. S. M. Coxeter.
Reciprocating the regular polytopes.
J. London Math. Soc. (2), 55(3):549–557, 1997.

201. H. S. M. Coxeter.
The trigonometry of hyperbolic tessellations.
Canad. Math. Bull., 40(2):158–168, 1997.

202. H. S. M. Coxeter and Branko Grünbaum.
Face-transitive polyhedra with rectangular faces.
C. R. Math. Acad. Sci. Soc. R. Can., 20(1):16–21, 1998.

203. H. S. M. Coxeter.
Numerical distances among the circles in a loxodromic sequence.
Nieuw Arch. Wisk. (4), 16(1-2):1–9, 1998.

204. H. S. M. Coxeter.
Seven cubes and ten 24-cells.
Discrete Comput. Geom., 19(2):151–157, 1998.

205. H. S. M. Coxeter.
Whence does an ellipse look like a circle?
C. R. Math. Acad. Sci. Soc. R. Can., 20(4):124–127, 1998.

206. H. S. M. Coxeter and A. V. Kharchenko.
Frieze patterns for regular star polytopes and statistical honeycombs.
Period. Math. Hungar., 39(1-3):51–63, 1999.
Discrete geometry and rigidity (Budapest, 1999).

207. H. S. M. Coxeter.
Five spheres in mutual contact.
J. Geom. Graph., 4(2):109–114, 2000.

208. H. S. M. Coxeter and Branko Grünbaum.
Face-transitive polyhedra with rectangular faces and
icosahedral symmetry.
Discrete Comput. Geom., 25(2):163–172, 2001.

Index